SLUDGE—
HEALTH RISKS OF
LAND APPLICATION

SLUDGE— HEALTH RISKS OF LAND APPLICATION

Edited by
Gabriel Bitton
Bobby L. Damron
George T. Edds
James M. Davidson

ANN ARBOR SCIENCE
PUBLISHERS INC / THE BUTTERWORTH GROUP

Copyright © 1980 by Ann Arbor Science Publishers, Inc.
230 Collingwood, P.O. Box 1425, Ann Arbor, Michigan 48106

Library of Congress Catalog Card Number 80-68822
ISBN 0-250-40374-9

Manufactured in the United States of America
All Rights Reserved

PREFACE

Urbanization and the resulting population centers are exerting an ever-increasing demand on technology for procedures to manage municipal wastes in an environmentally "safe" and acceptable manner. Society's awareness of our need to reuse, when possible, resources contained in wastes has led to an intense interest in the application of municipal sewage sludge to agricultural land. This procedure has been shown to be beneficial to crop growth because of the essential plant nutrients contained in sludges. However, out of concern that toxic chemicals and/or pathogenic organisms present in municipal wastes might be harmful to humans consuming food produced on sludge-amended land, the procedure has been brought into question. These proceedings represent an assessment of the state-of-the-art regarding the health risks associated with the use of sewage sludge in a soil-plant-animal food production system.

At the present time, approximately 25% of the sludge produced in the United States is applied to land—not all of which is used to produce edible crops or meats. Energy and economical constraints may increase this percentage substantially in the future. It is interesting to note, however, that even if all the sludge projected for 1985 were applied to agricultural cropland at a rate suitable to meet the nitrogen requirements of crops, only 2% of the present cropland would be required. But, where the population is concentrated (e.g., the Northeast), the quantity of available cropland may be limited. For example, it is estimated that 65% of the cropland in Connecticut would be needed to meet the 1985 projected sludge production in that state. States such as Maryland and Rhode Island would not have sufficient cropland to handle their 1985 municipal sludge production estimates. Thus, only in some areas of the United States will the amount of cropland available for sludge application be a limiting factor. The above discussion assumes that all the sludge would be applied to cropland and does not consider the vast acreage associated with range and forests.

The survival and transmission of pathogenic organisms in aquatic and terrestrial environments is well recognized. Because of this, the application of municipal sludge to cropland poses a potential health risk to the public. This concern is intensified in regions where municipal and domestic water supplies are drawn from shallow aquifers. Of equal concern are those crops grown on sludge-treated land in which the marketed portion of the crop is in direct or indirect contact with the sludge. Conventional sewage treatment procedures remove most of the microbial pathogens associated with municipal sewage, but these procedures are not infallible nor 100% effective. More effective methods for pathogen inactivation are available; however, they are both capital and energy intensive. Thus, well managed land-spreading systems which would not contaminate surface and/or ground water supplies are a preferable alternative.

There is equal concern that toxic chemicals (metals and refractory organic compounds) generally present in sludges may be toxic to crops and/or the quantity of toxic substances found naturally in crops may increase sufficiently to have deleterious effects on animals and humans. The long-term impact of repeated applications of municipal sludge to land dedicated to food production could be substantially reduced by growing corn or other crops harvested for their edible seeds or fruits rather than forages or leafy vegetables. Thus, the impact of toxic chemicals in sludges applied to cropland can be reduced significantly by proper crop selection. In general, leafy vegetable tissues accumulate larger quantities of toxic metals than do grains. The benefits of growing nonedible (fiber) or processed crops such as sugar beets and sugarcane are obvious. The impact of toxic chemicals in sludges can also be reduced by using proper management procedures. For example, if the soil is allowed to become acid, the solubility of a number of the toxic metal compounds is increased, and an unnecessary accumulation of these substances in plants may result. Pretreatment of industrial wastes prior to their introduction into municipal waste treatment plants will reduce much of the hazard associated with toxic chemicals in sludges applied to cropland.

These proceedings represent an assessment of the health risks associated with animal feeding and/or land application of municipal sludge. The authors were assigned specific topics and asked to provide the most recent information available for evaluating the hazards posed by land application of municipal sludges. Papers have been edited for typographical errors and format, but not for content. The abstracts at the end of the proceedings represent current research activities and are from talks presented at the

Workshop, "Evaluation of Health Risks Associated with Animal Feeding and/or Land Application of Municipal Sludge," held in Tampa, Florida, April 29-May 1, 1980.

> G. Bitton
> B. L. Damron
> G. T. Edds
> J. M. Davidson

ACKNOWLEDGMENTS

The editors are indebted to the technical and financial support of the U.S. Environmental Protection Agency, Health Effects Research Laboratory, Cincinnati, Ohio. We are especially appreciative of Mr. Herb Pahren, who has offered constructive suggestions on many municipal waste disposal studies throughout the United States, as well as the project at the University of Florida. He was also instrumental in planning the workshop entitled "Evaluation of Health Risks Associated with Animal Feeding and/or Land Application of Municipal Sludge" held in Tampa, Florida, April 29-May 1, 1980.

The Center for Environmental and Natural Resources Programs of the Institute of Food and Agricultural Sciences (IFAS) at the University of Florida supported portions of this workshop, as well as research on the use and health risks associated with the application of municipal sewage sludge to agricultural land. Dr. J. F. Gerber, Director of IFAS Grants, was particularly helpful in identifying the workshop topic as an area of major concern to Florida producers and municipalities. We also thank the City of Pensacola for its support of portions of the research on the use of sludge on agricultural land at the Agricultural Research Center, Jay, Florida.

A special thank you is extended to the session moderators: N. P. Thompson, R. L. Shirley, C. A. Sorber, J. J. Street, J. Zoltek, Jr., S. Farrah, R. L. Ward, and R. Chaney, who ably and adroitly moderated and guided their sessions. The authors and speakers for this workshop are commended for their willingness to participate, to meet some stringent manuscript requirements, and for their promptness in manuscript submission. A special debt of gratitude is extended to our colleagues who spent long hours preparing for and executing the program associated with this workshop.

The editors are grateful to Mrs. Donna Todd, who corrected and typed much of the material contained in these proceedings.

G. Bitton
B. L. Damron
G. T. Edds
J. M. Davidson

DEDICATED TO OUR COLLEAGUES AND THE
GRADUATE STUDENTS AND STAFF
WHO WORKED ON THIS PROJECT

BITTON   DAMRON

GABRIEL BITTON is an Associate Professor in the Department of
Environmental Engineering Sciences at the University of
Florida. His research activities include adsorption of microorganisms to surfaces, detection of viruses in environmental
samples, survival and transport of pathogens through soils
following application of wastewater effluents and residuals.
Dr. Bitton teaches graduate level courses in environmental
microbiology and toxicology. He received his PhD in Microbiology from the Hebrew University, Israel, his MS from Laval
University, Canada, his Agricultural Engineering degree from
Ecole Nationale Supérieure Agronomique de Toulouse, France,
and his BS from the University of Toulouse, France.
Dr. Bitton has served on various national and international
committees dealing with environmental virology. He has
authored more than 50 scientific papers in his research area.
He is the editor (with K. C. Marshall) of Adsorption of
Microorganisms to Surfaces, Wiley, N.Y. (1980). He is also
the author of the Wiley forthcoming textbook, Introduction
to Environmental Virology.

BOBBY L. DAMRON is an Associate Professor in the Department
of Poultry Science at the University of Florida. A large
portion of his research has been devoted to the study of
mineral interrelationships and requirements for chicks,
broilers and laying hens. He has also participated in experi-
mentation concerning the amino acid and energy requirements
of laying hens and the evaluation of various feed additives
and nutrient sources for use in poultry diets. Dr. Damron
teaches both undergraduate and graduate courses in poultry
nutrition at the University of Florida. He received the BS,
MS degrees from the University of Florida in 1963 and 1964,
respectively, with a PhD in Animal Science also being awarded
by that institution in 1968. Dr. Damron is a member of the
Poultry Science Association, the Society of Experimental
Biology and Medicine and the Florida Academy of Sciences and
has served as a reviewer for the Poultry Science Journal.
He has authored or coauthored 78 refereed scientific papers
and 56 abstracts along with a number of popular articles and
monographs.

EDDS   DAVIDSON

GEORGE T. EDDS, Veterinarian and Toxicologist, Department of Preventive Medicine, College of Veterinary Medicine, University of Florida, Gainesville, has served as co-director of the EPA sponsored project "Sewage Sludge - Viral and Pathogenic Agents in Soil-Plant-Animal Systems." Samples of soil, forage plants and grains produced on sludge-amended soils, and cattle, swine and poultry tissues consuming such plant material, sewage sludges or equivalent levels of minerals have been examined for hazardous pathogens or residues. Growth performance and health effects in animals, as well as examination of tissues for residues affecting reproductive performance in mice have been observed during the 3-year trial. Dr. Edds has also been involved in drug safety and efficacy studies and, more recently, the toxic effects of mycotoxin residues in grain or food products. He received his DVM degree from Texas A&M University and his PhD in Toxicology from the University of Minnesota. He has published about 90 scientific papers, has lectured on Toxicology in scientific meetings in the U.S. and several foreign countries. He is currently Chairman, Environmental Residues Committee, of the U.S. Animal Health Association.

JAMES M. DAVIDSON is Professor of Soil Physics and Assistant Dean for Research in the Institute of Food and Agricultural Sciences at the University of Florida. His research activities include measurement and simulation of pesticide adsorption and movement in water-saturated and -unsaturated soils, and development and evaluation of mathematical relationships for describing the fate of various nitrogen species associated with commercial fertilizers and/or wastes applied to soils. He received a PhD in Soil Physics in 1965 from the University of California at Davis, and his MS and BS degrees from Oregon State University. Dr. Davidson has served as Chairman of the Environmental Quality Division of the American Society of Agronomy, and currently serves as Associate Editor for the Soil Science Society of America Journal. He has also served on various national and international environmental quality and water and solute transport committees. Dr. Davidson has authored more than 80 scientific papers in his research field. He co-edited, with Dr. Michael Overcash, <u>Environmental Impact of Nonpoint Source Pollution</u>, published in 1980 by Ann Arbor Science.

TABLE OF CONTENTS

Introductions:

Overview of the Problem, H.R. Pahren 1

Regulatory Aspects of Sludge Application
to Land, J.F. Robens 7

SECTION I. HEALTH RISKS ASSOCIATED WITH SPECIFIC
AGENTS IN MUNICIPAL SLUDGE

1. Health Risks Associated with Microbial Agents
   in Municipal Sludge, B.P. Sagik, S.M. Duboise,
   and C.A. Sorber 15

2. Agents of Health Significance: Parasites,
   M.D. Little 47

3. Agents of Health Significance: Toxic Metals,
   R. Chaney 59

4. Potential Health Hazards of Toxic Organic Residues
   in Sludge, J.C. Dacre 85

SECTION II. HEALTH RISKS ASSOCIATED WITH USING
MUNICIPAL SLUDGE FOR PRODUCTION OF
FOOD CHAIN SUBSTANCES

5. Toxic Metals in Agricultural Crops, L.E. Sommers . 105

6. Agricultural Crops: Pathogens, D.O. Cliver . . . 141

7. Effects of Toxic Chemicals Present in Sewage
   Sludge on Animal Health, E.W. Kienholz 153

8. Risks to Animal Health from Pathogens in Municipal
   Sludge, J.G. Yeager 173

9. Disease Transmission by Wild Animals from Sludge-Amended Land, A.K. *Prestwood* 201

    SECTION III. OCCUPATIONAL HAZARDS ASSOCIATED WITH MUNICIPAL SLUDGE

10. Occupational Hazards Associated with Sludge Handling, C.S. *Clark*, H.S. *Bjornson*, J.W. *Holland*, T.L. *Huge*, V.A. *Majeti* and P.S. *Gartside*. . . . 215

11. Health Aspects of Composting: Primary and Secondary Pathogens, W.D. *Burge*, and P.D. *Millner* 245

    SECTION IV. MAJOR FINDINGS FROM THREE SLUDGE PROJECTS

12. Observations on the Health of Some Animals Exposed to Anaerobically Digested Sludge Originating in the Metropolitan Sanitary District of Greater Chicago System, P.R. *Fitzgerald* 267

13. Uptake of Trace Metals and Persistent Organics Into Bovine Tissues from Sewage Sludge-Denver Project, J.C. *Baxter*, D.E. *Johnson*, and E.W. *Kienholz* 285

14. Health Effects of Sewage Sludge for Plant Production or Direct Feeding to Cattle, Swine, Poultry or Animal Tissue to Mice, G.T. *Edds*, O. *Osuna*, C.F. *Simpson*. 311

    SECTION V. ABSTRACTS

A. University of Florida Project 329

B. Abstracts of Papers Presented at the Symposium. . 341

    INDEX 361

OVERVIEW OF THE PROBLEM

Herbert R. Pahren.   Health Effects Research Laboratory, US EPA, Cincinnati, Ohio.

INTRODUCTION

Human wastes have been applied to land since the beginning of mankind. Until relatively recent times, these wastes were applied in the form of raw sewage, either from individual households or communities. Persons utilizing such materials on agricultural lands were aware of the nutrient value of the sewage and periodically added the waste products to grow better crops. With advances in the understanding of the causes of disease, it became apparent that when food crops were grown on soils to which untreated human wastes were applied, increased disease rates were associated with consumption of such foods. Thus, the use of sewage on land acquired a stigma that is still with us today.

The first simple sewage treatment plant in the United States, with the resultant production of sludge, was built only 100 years ago. Fifty years ago only 18 million persons were served by a sewage treatment plant [1]. Sludge was applied to land even in the early days of sewage treatment. The City of Baltimore, for example, gave away digested sludge to farmers in three forms: liquid, air dried, and heat dried, as early as 1914 [2,3]. The problem of municipal sludge disposal, however, is a recent one, and is related to the passage of legislation requiring increasing levels of treatment with resultant increased production of sludge before wastewater is discharged to receiving streams.

Basically, sludge disposal can be handled in four ways. It can be burned, discharged to the ocean, buried, or applied to land. A fifth way, use as an animal feed, is a novel additional approach which is at the research stage and which is to be discussed at this symposium. No method is without problems. Incineration not only contributes to air pollution but is energy dependent and destroys a potential resource. Ocean disposal is being phased out. Burial results in the loss of nutrients, and, without proper management, could result in groundwater problems. Land application is being considered by many as an alternative which not only solves the sludge disposal problem, but also reclaims nutrients. While this fact has been known for years, the low cost of synthetic commercial fertilizers and soil amendment materials inhibited the use of sewage sludge on a large scale. Recent economic and energy developments are changing the balance. Increasingly higher costs of fertilizers have resulted in the farming community beginning to consider acceptance of treated sewage sludge in areas where this has not been done previously. The fertilizer value of the nitrogen, phosphorus, and potassium in the sludge produced on a national basis amounts to several hundred million dollars per year. It is recognized that not all of the sludge can or should be utilized for land application. However, in addition to saving the natural resources that would be used to produce the commercial fertilizer saved, the monetary benefits of the sludge nutrients are substantial.

Economics and energy also play a role in the processing and ultimate disposal of the sludge. Sludge handling and disposal are the most problematic and costly unit processes in wastewater treatment. These costs could be half the total annual operation and maintenance budget. Costs are quite site specific, but there generally is a large range between the least and most expensive options available to a municipality. For example, the unit cost for sludge disposal in one community for sludge incineration was 72% higher than for land application [4].

CONGRESSIONAL STIMULUS

Land application of both wastewater and municipal sludge received encouragement from the Congress in the Federal Water Pollution Control Act Amendments of 1972, and in the Senate report which served as the basis for the Act. When the Congress found a reluctance to use land treatment or other recycling techniques to handle municipal wastes, it passed the Clean Water Act of 1977 which, among other things, amended the construction grants part of the law to leave no doubt of the direction it wants the US EPA and the municipalities to take. A quotation from the Act is worth repeating.

> The Administrator shall not make grants from funds authorized for any fiscal year beginning after September 30, 1978, to any State, municipality, or intermunicipal or interstate agency for the erection, building, acquisition, alteration, remodeling, improvement, or extension of treatment works unless the grant applicant has satisfactorily demonstrated to

the Administrator that innovative and alternative wastewater treatment processes and techniques which provide for the reclaiming and reuse of water, otherwise eliminate the discharge of pollutants, and utilize recycling techniques, land treatment, new or improved methods of waste treatment management for municipal and industrial waste (discharged into municipal systems) and the confined disposal of pollutants, so that pollutants will not migrate to cause water or other environmental pollution, have been fully studied and evaluated by the applicant taking into account section 201(d) of this Act and taking into account and allowing to the extent practicable the more efficient use of energy and resources.

Recognizing that the utilization of sludge for various purposes may cause problems if done improperly, Section 405 of the Clean Water Act requires that regulations be issued which provide guidelines for the disposal and utilization of sludge.

POTENTIAL PROBLEM AREAS

From a public health standpoint, the problem of the potential for human or animal disease as a result of pathogenic organisms being transmitted from improperly managed municipal sludge is probably the most important. Acute diseases resulting from the land application or animal feeding of municipal sludge can be completely avoided, but at a cost. A number of treatment processes can significantly reduce pathogens. It is expected that sewage sludge applied to the land surface or incorporated into the soil will all receive such pretreatment in order to provide protection of public health. If necessary to provide further public health protection due to the specific use of the sludge, additional processes are available that will result in a very high degree of inactivation of even the most hardy organisms.

Natural environmental conditions will also reduce the density of microorganisms. A number of factors, such as sunlight, soil moisture, and temperature, affect the persistence of the organisms and the ability to survive. Bacteria, protozoa, and viruses generally are inactivated in a few days to a few months but helminth ova, under conditions of high moisture and shade, could survive several years. When a sludge management system is considered, a proper balance must be struck for reduction of the microorganisms between natural and treatment processes, and it would depend primarily on the specific use of the sludge.

Another area of interest and potential concern is with the content of metals in the sludge and the possibility that some metals may be passed through the food chain if food crops are grown on sludge amended soils. Cadmium has been identified as the metal of most concern because it can readily be taken up by plants under acid soil conditions and there is a relatively low safety factor between amounts being ingested normally and the amount whereby symptoms of disease appear with long-term exposure. Most municipal sludges have very low concentrations of cadmium and would not be expected to present a problem, especially with good management practices, such as pH control, limiting the amount of sludge applied, and proper selection of the crop grown. There are situations, however, where metals in the sludge are sufficiently high that land application is not desirable and other means of sludge disposal should be selected.

There are persons who believe that the use of the animal pathway in the food chain increases the safety factor for people. In this case, plants grown on sludge amended soils would be fed to animals which in turn would be used for human food. Metals do not seek the edible muscle tissues and less would enter the food chain.

Trace organics find their way into sludge from industrial or other wastes discharged into the municipal sewerage system. Relatively little research has been conducted to date on the uptake of organics by plants in sludge amended soils, passage through the food chain, or the effect in animals which consume sludge as part of their diet. Hopefully, papers at this symposium will provide insight on this subject.

Organic chemicals are routinely applied directly to plants and soils for the control of plant pests. Residuals of these chemicals may pass through the food chain and reach the tissues of persons consuming the crops. Trace amounts of some of these same chemicals could appear in municipal sludge and reach agricultural lands. The problem is not so much one of uptake by the plant but adherence to the plant. Large organic molecules do not tend to pass the semipermeable membrane of plant roots but if sludge is applied to grazing lands, the animal could ingest the sludge adhering to the grass [5]. Many organics have a tendency to accumulate in lipid-rich tissues and fluids but not in muscle tissue.

Whether or not municipal sludge contains sufficient organics to render it unsuitable for land application or animal feeding depends primarily on the amount of industrial wastewater containing synthetic organic chemicals being received by the sewage treatment plant. Since many of these compounds are quite persistent to degradation, and their toxicological properties could result in adverse effects to animals or man, uncontrolled release to the environment would not be desirable. Sludge primarily of domestic origin would be expected to have little or none of these organics and would most likely be a good candidate for recycle. An evaluation must be made, however, before a decision is made on the most appropriate method of sludge disposal.

The goal of the National sludge management program is to utilize available resources while at the same time reducing public health risks to an acceptable level. Achievement of that goal will require a better understanding of the potential health problems, methods of treating sludge, and the costs of sludge management so that the health problems can be designed out of a sludge management system. Since we do not live in a risk-free society, methods for evaluating risk and determining what is an acceptable risk must also be better understood. The problems will not go away. We must be able to apply our research knowledge and manage the sludge residuals so as to maintain the quality of life and the environment.

REFERENCES

1. Babbitt, H. E. Sewerage and Sewage Treatment, Seventh Ed. (New York: John Wiley & Sons, Inc., 1953), p. 4.
2. Schaetzle, T. C. "Nine Years Operation of the Baltimore Sewage Works," Eng. News Record. 87:50 (July 14-21, 1921).
3. Gregory, J. H. "Removal and Disposal of Air-dried Sludge at Baltimore," Eng. News Record. 93:13 (July 3, 1924).
4. Baxter, J. C., W. J. Martin, B. R. Sabey, W. E. Hart, D. B. Cohen and C. F. Calkins. "Comprehensive Summary of Sludge Disposal Recycling History," Municipal Environmental Research Laboratory, US EPA Report-600/2-77-054 (1977).
5. Pahren, H. R., J. B. Lucas, J. A. Ryan and G. K. Dotson. "Health Risks Associated with Land Application of Municipal Sludge," Jour. Water Poll. Control Fed. 51:2588-2601 (1979).

Regulatory Aspects of Sludge
Application to Land

Jane F. Robens, DVM,
Food and Drug Administration

Thank you for inviting me to participate in this symposium on the health risks of utilizing municipal sludge in animal diets or on land on which animal food is raised.

Dr. Pahren has given us a historical perspective and an overview of the problem of disposal of the tons of sludge that are generated each day from our municipalities. Disposal of sludge in landfills, incineration, ocean dumping and application to productive land all are associated with certain problems of potential environmental hazard and compliance with local, state and federal regulations.
I will try to explain some of the regulatory health concerns of the Food and Drug Administration in disposing of sewage sludges.

The Food and Drug Administration has worked with the Environmental Protection Agency to recommend safe methods by which sludges can be applied to food and feed crops to assure the continuing safety and wholesomeness of the food supply. We believe that under certain controlled conditions of use, disposal of sludge on land may prove to be acceptable. We are aware that a certain amount of sludge has been applied to lands for growing crops for years without known harm to consumers. However, the acreages involved have been relatively small. Now that many municipalities including the large industrial metropolitan centers are interested in such disposal, the possibility of introducing unsafe amounts of residues has been increased geometrically.

Since this Symposium is concerned both with application of sewage sludge to land and the use of sludge in animal diets, let me explain the regulatory position of the FDA in regard to each. The Food, Drug and Cosmetic Act gives the FDA the primary regulatory authority and the responsibility for assuring the safety of food for both man and other animals; and of course other authorities in relation to drugs, devices, diagnostics and cosmetics with which we are not concerned here today.

Thus, when sludge is used as a direct component of animal feed it becomes a feed (food) additive within the meaning of Section 409 of the FD&C Act. This is because it is not generally recognized as safe for this purpose nor was it commonly used in animal feed prior to the passage of the Food Additive amendment of 1958.

If use of sludge in animal feeds were proposed in a food additive petition one of the first difficulties would be characterizing the material since the FDA must of necessity know what constitutes a given food or feed additive. The composition of sludge varies widely depending first on the municipal area from which it is produced, and secondly it may change within a given area with time from such variables as domestic sources, seasonal variation, addition or closing of particular industries, and accidents or even storms which may carry new material into the sewers. Input of contaminants can result from continuous seepage or inflow from industries or because of spills, that is, a drum of something being dumped into the sewer. In addition, sewage plant operation is not a precise science but it is greatly affected by temperatures and nature of the inflow. Spills of phenol or other antibacterials, for instance, may affect the ability of the plant to alter or destroy some chemicals which will then pass unchanged or in only slightly altered form into the sludge. Industrial pretreatment programs may substantially reduce the metal content of sludge, however this cannot be depended upon to produce a uniform sludge.

There is another provision of the food additive regulations which states that a feed additive ingredient must accomplish its intended effect. I have difficulty in envisioning what the intended effect of sludge would be since my impression from reading abstracts of the work performed here in Florida and to be presented later in this Symposium is that at least anaerobically digested sludge is not a very useful feed ingredient. Generally 30 to 60% of the total nitrogen in these sludges is present in the ammoniacal form and the remainder is present in the organic form. I realize that when radiation was used to replace digestion, the resulting sludge in a few trials had greater potential feeding value.

Finally we come to the question of direct toxicity. While we may have difficulty in measuring a direct detrimental effect on the animal (other than that resulting from dilution of the diet with a poorly utilized material), the FDA must also be concerned with the change in the composition of foods for humans produced by the animal, that is, residues of heavy metals or chlorinated organics in meat, milk or eggs. In regard to organic chemical compounds, a number of them, including some pesticides, are refractory and will not be destroyed in the sewage treatment process, or will be altered to products that still have potentially toxic effects. Compounds found most commonly in sludge are the polychlorinated biphenyls (PCB's) and the chlorinated pesticides. PCB's have been found in dried sludges at levels up to 352 ppm.

At this time cadmium and lead are the metals in sludge that present the greatest hazards to the safety of the food supply. Our Bureau of Foods has data indicating that lead levels have reached around 5000 ppm in dried sludge with 50% of the sludges containing lead at a level below 600 ppm. Cadmium concentrations of up to about 1000 ppm in the dried sludges have been reported with 50% of the sludges containing less than 20 ppm cadmium. These and other metals present in sludge often result in increased tissue concentrations in animals either eating the sludge directly or consuming feed grown on sludge-treated land. Cadmium is the element of greatest concern since any increase in daily intake may put the daily levels above the total tolerable daily intake for humans of 57-71 micrograms per day proposed by FAO-WHO. The tolerable daily intake however, is below that known to cause adult kidney damage, and dietary cadmium concentrations would have to be significantly increased for several years for this effect to occur.

In the studies where metropolitan Denver sewage sludge was fed to feedlot steers, increases in 9 of 10 inorganic elements were observed in various tissues. For instance, arsenic was increased in liver, kidney and muscle, cadmium was increased in liver and kidney, mercury was increased in liver, kidney and muscle and lead was increased in liver, kidney, bone and blood. We would not expect that withdrawal periods could be reliably used to decrease these residues. There are also unresolved questions about transmission of viable microorganisms through sludge such as ascaris ova, Salmonella species, <u>Mycobacterium</u> <u>tuberculosis</u> and human viruses. Thus my candid, but of course unofficial opinion is that interstate use of sewage sludge as an ingredient of animal feed faces a very tough road.

The second proposed disposal of sludge that is, its application to land used for raising crops for direct human or animal consumption, also directly impinges on the quality of our food supply. Animal feed may be contaminated by either direct application of sludges to growing plants or by uptake of sludge constituents by the plants from the soil. There are wide differences in uptake, translocation and accumulation of potentially toxic elements by plants. Plant factors which bring about these variations include differences in selectivity among plant species and varieties and plant parts while soil factors include pH, content of clay and sesquioxides, cation-exchange capacity, redox potential, and texture.

FDA does not directly regulate materials spread on the land such as fertilizers. This for the most part is the concern of the Environmental Protection Agency in their enforcement of the Water Pollution Control Act and Resources Conservation Recovery Act. However, the USDA is also interested because of the potential long term detrimental effects on agricultural land, that is, the possibility of having to withdraw large acreages from food production because of the presence of hazardous levels of contaminants in the soil; and the Consumer Product Safety Commission is concerned because sludges may potentially find their way to the home gardener.

FDA is concerned with the effects of application of sludge to land because of the potential of this practice to directly or indirectly affect the character of the food supply, that is, to result in toxic concentrations of chlorinated organics and/or heavy metals. Since we do not have direct regulatory authority we work with the EPA, by providing scientific expertise in their development and promulgation of regulations to control the practice.

Cadmium is of great concern in application of sludges to land just as it is in direct feeding of sludge. The chemistry of cadmium in soil is not well understood but its liability in soil is reduced by organic matter, clay, hydrous iron oxides, high pH and reducing conditions. Other heavy metals which can accumulate in plants and which may pose a hazard to plants, animals and/or humans under certain conditions include copper, molybdenum, nickel and zinc. The EPA has issued regulations which delineate specific conditions for disposal of sludge on land used for food chain crops to prevent cadmium and PCB's from entering the food supply in toxic amounts.

The evaluation of the toxicity of sludge is important and I want to make a few comments regarding the material sent me prior to this Symposium. From this material I learned that the University of Florida has conducted rather extensive relay feeding studies with sludge in addition to direct feeding studies in cattle and swine. A relay feeding study is one in which edible tissues (to include milk or eggs) harvested from animals fed the potentially toxic feed ingredient are fed to laboratory animals and their effects, such as the lesions, survival, reproduction or other parameters in the laboratory animals, are measured.

What I want to emphasize are the problems facing FDA when they are presented with the results of relay feeding studies. Such studies have been submitted regarding complex substances other than sludges whose composition was not well characterized. Since a thorough chemical analysis of the tissue to locate and quantitate the residues of concern is incredibly expensive, the sponsor with or without FDA concurrence has sometimes chosen to conduct relay feeding studies. I have reviewed a few of these studies and I seriously question whether the results can be extrapolated to evaluate the safety for humans of the original substance fed to the food producing animal. One of the most important tenets of any toxicity study is that we know the identity of the test substance. We do not know the identity in the case of relay toxicity studies and this limitation is not abrogated by the fact that this is the precise reason for performing the study. Secondly, a highly abnormal diet is fed to the laboratory animals in these relay toxicity studies and it is generally one in which we have no background information with which to compare specific but limited results. Thus, I question the value of such studies in establishing the safety of feeding even a sludge of known composition.

In summary FDA believes there are certain defined ways in which sewage sludge can be safely applied to agricultural lands, but direct use of this material in animal diets faces a great many very serious questions such as definition of the feed ingredient, and its uniformity, toxicity and potential for producing hazardous residues.

SECTION I

HEALTH RISKS ASSOCIATED WITH SPECIFIC
AGENTS IN MUNICIPAL SLUDGE

CHAPTER 1

HEALTH RISKS ASSOCIATED WITH MICROBIAL AGENTS IN
MUNICIPAL SLUDGE*

Bernard P. Sagik. College of Sciences and Mathematics,
The University of Texas at San Antonio.[+]

S. Monroe Duboise. Center for Applied Research and
Technology, The University of Texas at San Antonio.

Charles A. Sorber. Center for Applied Research and
Technology, The University of Texas at San Antonio.

INTRODUCTION

Human excreta usually accounts for a minute portion of wastewater volume, but is almost entirely responsible for the risks of infectious disease associated with sewage. Although most persons consume less than one gallon of liquid and solid foods per day, average daily indoor water use in the U.S. is 59 gallons per person (51). Most of this water ultimately is mixed with excreta in the sewers. Further dilution occurs in many sewers as a result of infiltration or the addition of storm water.

With incentives provided by declining water quality, increasing population, and the increasingly rapid recycling of water, the need for abandoning the common practice of diluting pollutants into surface waters has been recognized and addressed in legislation. The Federal Water Pollution Control Act as amended in 1972 (P.L. 92-500) called for considerable water quality improvements by 1983 and elimination of pollutant discharge by 1985. Recent amendments to this legislation have given impetus to the application of wastewater effluents and sludges to land.

In addition to serving as a disposal method, land application of wastewater is often viewed as an added treatment process in which the soil acts as a "living filter" (53, 63, 71). Much enthusiasm for the land application of wastewater effluents and sludges has derived from the idea that such practices involve not only disposal but also resource recycling, i.e., plant nutrients and water are returned to the land. Unfortunately, however, land application as a simple extension of current waste treatment practices cannot be viewed as a thoroughly sound resource recovery method as effluents and sludges often contain other materials which do not belong in soil any more than they belong in surface waters. Industrial chemicals, pesticides, heavy metals, and other toxic or non-biodegradable materials may pose serious difficulties for long term land disposal operations. Another limitation of land treatment practices as a well-founded resource recycling method is the presence of pathogens.

If ground water contamination by human pathogens occurs readily, serious problems may already exist in some areas. Land disposal of human excreta has a long history and is a continuing and widespread practice. In addition to pathogens in treated municipal wastewater applied to land, significant contamination of soil occurs from other sources. Cesspools or septic tanks serve approximately 10 percent of the urban and 66 percent of the rural year-round housing units in the U.S. (99). Another one percent of urban and 15 percent of rural housing units are served by privies or similar facilities.

APPLICATION TO LAND: HISTORY, CURRENT PRACTICE, AND CONCERNS

Human beings have been aware of waste disposal problems for millenia. The long history of poor sanitary practices lends credibility to the speculation of Freud (31) that "the incitement to cleanliness originates in an urge to get rid of excreta, which have become disagreeable to the sense perceptions" and that hygienic considerations serve as "ex post facto justification." The development in the West of a basic understanding of the relation of environmental conditions to health has often been traced to the writings of Hippocrates in the fifth century B.C. (104). The water-borne etiology of diseases such as cholera and typhoid fever was not, however, clearly recognized until the nineteenth century. In 1873, Austin Flint (reprinted in 104) discussed the transmission of typhoid fever in drinking water and concluded as follows:

> The discovery of the causation of typhoid fever through the medium naturally has led to the inquiry whether other diseases may not be traced to drinking water which either contains viruses of contagion, or is polluted by divers kinds of morbific matter. The facts to which it has been the object of this paper to call attention have opened up a new field for investigation in etiology, and further researches in this direction may shed much light on the causation of numerous diseases. Already in the opinion of many, there is ground for assuming that epidemic cholera is diffused

> by means of a contagium, derived from the
> alimentary canal, with which drinking water
> is liable to become infected. This opinion
> is based on analogical reasoning rather than
> on logical proof.

Thus, throughout most of human history the disposal of excreta and other wastes has been accomplished without adequate knowledge of the possible impact upon health. Nevertheless, concern with the purity of water and with cleanliness in other respects has many historical and anthropological precedents.

The social regulation of waste disposal practices has a long history. The biblical command (Deuteronomy 23:12-14, King James Version) was given:

> Thou shalt have a place also without the
> camp, whither thou shalt go forth abroad:
> And thou shalt have a paddle upon thy
> weapon: and it shall be, when thou wilt
> ease thyself abroad, thou shalt dig therewith, and shalt turn back and cover that
> which cometh from thee: For the Lord thy
> God walketh in the midst of thy camp, to
> deliver thee, and to give up thine enemies
> before thee; therefore shall thy camp be
> holy: that he see no unclean thing in
> thee, and turn away from thee.

The desire to remove excrement from human camps has, however, also been a critical factor in the development of other methods of disposal. As communities have become larger and more permanent, people have faced the difficulties of greater amounts of wastes being produced in more limited space and the increasing inconvenience and even impossibility of going forth abroad routinely. Various cultures have pursued widely divergent strategies for dealing with such problems. Distribution of wastes onto land, into water, or into water followed by disposal on land are each basic strategies of considerable antiquity (73, 106).

The Chinese and other cultural groups including some Europeans traditionally returned undiluted excreta to their land as fertilizer (32, 73, 106). Among those who are believed to have used land disposal methods are the ancient Egyptians who are also known to have been particularly concerned with sanitation and the protection of water supplies (73, 106). Water-carriage systems for sewage date back at least to the Minoans and Romans.

The value of excreta as fertilizer was recognized clearly in ancient times. The Chinese farmer, who has generally had few farm animals to produce manure, has placed "upon the soil every ounce of manure, vegetable, animal, and human, that he can procure" (36). The use of night-soil remains important in China. The National Program for Agricultural Development 1956-1967 for the People's Republic of China states that "the fullest use should be made of night-soil from the cities and countryside and garbage and miscellaneous refuse which can be utilized as fertilizer." The use of excreta as fertilizer is also an ancient practice in the West. Pliny, in his Natural History, traces the practices back as far as Homeric times and then discusses the various opinions concerning the relative value of different types of manure including human. Pliny also describes Roman composting practices as follows:

> They recommend making dung-heaps in the open air in a hole in the ground made so as to collect moisture, and covering the heaps with straw to prevent their drying up in the sun, after driving a hard-oak state into the ground, which keeps snakes from breeding in the dung.

According to Fussell (1965), such classical composting practices persisted during the Middle Ages in Europe. Fussell states

> A more unusual application was known as "beer", i.e., sewage from the towns and other centres of population. The use of this material, which became famous much later, is first mentioned in the Chronicle of the Abbey of Affligem at the end of the eleventh century.

The development of sewers, especially those that also carried storm waters, made the large scale discharge of wastes into surface waters more expedient than ever before. Sewerage did not put an end to the land application of wastes, however. Winslow (1952) points out that "in the nineteenth century the conservation of nitrogen was considered almost as important an object of sewage treatment as the avoidance of nuisance." Thus, land disposal of sewage was still considered appropriate by many including Chadwick who, according to Reynolds (1946), put forth the maxim: "The rainfall to the river, the sewage to the soil." The tradition of fertilization with excreta and the nuisance and disease problems associated with contaminated waters contributed to the findings of a Royal Commission report in 1858 which recommended sewage farming. According to Winslow (1952) sewage farming developed rapidly throughout England as a result of the report. Sewage farming was also adopted by Paris and Berlin in the late 1800's and continued into the twentieth century.

The incentives for land application of sewage were not nearly as great in the United States where population was less dense and resources were more abundant. Most sewage farming in the U.S., until recently, was done in arid regions where, as Winslow (1952) states, "the water value rather than the nitrogen value of the sewage" has been important.

The development of modern biological methods for the treatment of wastewater led to the production of greatly improved effluents which encouraged the expediency of discharge into waterways by lessening the accompanying nuisance problems and the risks of epidemics due to bacterial pathogens. In addition, chemical fertilizer production led to the abandonment of "conservation of nitrogen" as a goal of waste management. Aside from economic reasons for the decline of land disposal practices early in the twentieth century, Egeland (1973) states that "an equally compelling reason for the change in favor of water disposal was that centuries of experience had shown that land disposal practices very often resulted in a marked deterioration of the soil." The relevance of Egeland's assertion to well-managed land disposal systems currently in operation has not yet been fully assessed. It has become clear, however, that current treatment practices, which generally involve discharge into surface waters, are to be modified.

As a result, the surface application of wastewater and sludges to land has once again become a popular disposal strategy. One of the most important benefits cited is the improvement of wastewater quality during percolation through soils (28, 63, 65, 71, 94). While many advocates of land application of wastewater may prefer the identification of land-based alternatives as "treatment" rather than "disposal" processes, both purposes are, no doubt, served to varying extents depending upon management and other circumstances. The fact that soils may remove substances from percolating solutions has been recognized clearly since at least the middle of the nineteenth century. J. Thomas Way (1850), in one of the earliest publications describing ion exchange phenomena, introduced his own investigations by mentioning the work of Huxtable who

> ... stated that he had made an experiment in the filtration of liquid manure in his tanks through a bed of an ordinary loamy soil; and that after its passage through the filter bed, the urine was found to be deprived of colour and smell -- in fact, that it went in manure and came out water.

Way also called attention to much earlier work including a description by Lord Bacon of the use of soil filtration for desalination on the Barbary coast. Bacon is quoted as follows:

> Digge a hole on the sea-shore somewhat above high-water mark, and as deep as low-water mark, which when the tide cometh will be filled with water fresh and potable.

In addition to the benefits of percolation of wastewater through soils as a treatment process, several other advantages of land application are generally noted. The value of effluents and particularly of sludges as fertilizers or soil conditioners has been reported by a number of authors (19, 23, 52, 55, 69, 84). Some even have euphemistically labelled digested sludge as "liquid fertilizer" (62). Effluents and sludges have been used with some success in experimental attempts to restore acid spoil banks associated with the strip mining of coal (59, 83). Vegetation responses to wastewater irrigation have not been uniformly beneficial, however. Sopper and Kardos (1973) noted that red pines irrigated for several years at a rate of two inches per week became vulnerable to high winds. In fact, during a wet snowfall accompanied by high winds, every tree on a one acre plot was blown over. Some adverse effects on vegetation or the fitness of crops for consumption by humans or livestock may also result from toxic materials, such as heavy metals, which are present in many wastewaters and sludges (60, 85).

Land application of either effluents or sludges may introduce both deleterious chemicals and pathogenic microorganisms into soil systems. The chemical and microbiological quality of sludges and effluents from treatment plants is dependent both upon the nature of incoming raw wastewaters and the effectiveness of the treatment train employed.

The concentration of biological contaminants in domestic sewage is influenced by several complex factors, including the age and health of the contributing population, as well as the season of year. The presence and survival of potentially pathogenic organisms in wastewater were reviewed recently (1). Whereas certain pathogenic organisms such as _Salmonella typhosa_ have a relatively brief survival time in wastewater, other pathogens including Mycobacteria, _Ascaris_ ova, and certain enteric viruses, are highly resistant to many environmental stresses.

Published results summarized by Sproul (1978) indicate that primary treatment alone does not remove the pathogen load in domestic wastewater. On the other hand, the sludge biomass generated by conventional secondary treatment may be expected to contain a large portion of that microbial population which has been removed from incoming wastewater.

Generally, disinfection (specifically chlorination) of effluents before discharge provides the last step in the treatment scheme. However, certain microorganisms, including Mycobacteria, amebic cysts, certain enteric viruses, and the agent of waterborne hepatitis, are more chlorine-resistant than are indicator coliform organisms. Nevertheless, chlorination of effluents can provide an additional 90 to 99+% removal of certain potential pathogens, as illustrated by field studies reported by Sorber et al. (1974).

If domestic sewage is subjected to adequate secondary

treatment followed by disinfection, the resultant effluent may be expected to contain on a per-unit-basis low levels of potential human pathogens. The continual application of effluents to soils may allow the retention and accumulation of such organisms within a given soil profile. The dimensions of the problems inherent in land disposal of effluents have been suggested by Foster and Engelbrecht (1973) as they attempted to relate treatment effectiveness to application rate for land disposal systems. Thus, Foster and Engelbrecht calculated that with an application rate of 5 cm/week and assuming 50% removal by primary settling, 85 to 95% removal by secondary treatment, and 99.5% kill by disinfection, there would be 3.9×10^3 Salmonellae, 1.2×10^2 Mycobacteria, and 1.6×10^4 viruses applied per acre per day in wastewater.

Residual sludges also can transport large quantities of potentially pathogenic organisms onto a land disposal site. Assuming 1×10^3 enteric viruses per liter mixed liquor suspended solids (MLSS) and a solids level of 0.2 to 0.4%, one can estimate the potential transport of human enteric viruses to land disposal sites. Even with a hypothetical 99% reduction in virus level in the anaerobic digestion process, the use of 10 dry t/acre/yr (22.4 t/ha) as soil additive implies the addition of more than 2×10^7 plaque-forming units per acre (PFU/acre) [or about 1×10^3 PFU/ft^3, assuming injection or plowing to a 6-in. (15 cm) depth]. The ultimate public health importance of these and other organisms in the soil environment would depend upon their survival rate, their potential for movement to surface or groundwaters, and the uses to which the site is to be put.

In a recent review, Gerba et al. (1975) listed the factors affecting the survival of enteric bacteria in soil as moisture content, moisture-holding capacity, temperature, pH, sunlight, organic matter, and antagonism from soil microflora. Early studies by Beard (1938, 1940) demonstrated that <u>Salmonella typhosa</u> could be recovered from loam and peat soils for periods up to 85 days, while survival of this organism in drying sand was only 4 to 7 days. Mycobacteria, because of their high content of waxy substances, can survive even dry conditions for long periods of time. Greenberg and Kupka (1957) in a review of available literature cited survival times ranging from 150 days to 15 months for Mycobacteria in soil.

Survival of viruses in soils are influenced by many of the same parameters described above, although at this time little direct evidence supports viral inactivation by antagonistic microorganisms. As expected, lower temperatures favor longer survival times. An optimal soil moisture content favors poliovirus survival in soil, while desiccation results in a more rapid loss of virus recoverability. Bagdasaryan (1964), working with a wide variety of human enteroviruses, including polioviruses, Coxsackie viruses, and echoviruses, reported survival times ranging from 110 to 170 days at a soil pH of 7.5 and a soil temperature of 3 to 10°C.

Bacterial movement through soils has been demonstrated at several field sites. Reporting from the available literature reports, Gerba et al. (1975) noted coliform movements in a variety of soils for distances ranging from 3 to 1500 feet (0.9 to 456 m). Release and movement of microorganisms would be expected, since physical adsorption of particulates is a reversible phenomenon and, in part, ion-dependent. Duboise (1977) has monitored the movement of a genetically distinguishable coliform organism through soil cores during cyclic applications of secondary effluent followed by distilled water. The release and subsequent movement of this organism was consistent with decreasing conductivity of the core effluents. Changes in the ionic nature of percolate waters would be expected to have the same effect in field situations.

The phenomenon of adsorption as a mechanism for the retention of viruses in soil systems was demonstrated by Drewry and Eliassen (1968). The results they obtained using bacteriophage systems showed that virus adsorption followed typical Freundlich isotherms. In general, virus adsorption by soils increased with increasing ion-exchange capacity, clay content, organic carbon, and glycerol-retention capacity. The movement of poliovirus I (Chat) through 20-cm-length nonsterile cores taken from a sandy forest soil was monitored using simulated cycles of effluent application and rainfall (22). Results show a burst of released virus detected in the core effluent as the specific conductance of the percolating water began to decrease. Additionally, Duboise et al. (1976b) found that the capacity of surviving virions to migrate through the soil columns during an 84-day period (during which time the natural soil moisture was maintained) was unchanged. Similar movement of poliovirus in 150-cm columns packed with calcareous sand were reported by Lance et al. (1976). While most of the virus inoculum applied to the column surface in secondary effluent was adsorbed in the top 5 cm of soil, subsequent application of deionized water resulted in virus desorption and movement to a depth of 160 cm. In this study, drying for one day between viral application and flooding with deionized water reportedly prevented desorption (or enhanced viral inactivation).

FIELD STUDIES AND REPORTS

One of the largest land disposal sites for the recycling of anaerobically digested sludges is located in Fulton County, Illinois, and is operated by the Metropolitan Sanitary District of Greater Chicago. Biological testing at the site included fecal coliform and virus monitoring at selected surface water sites (107). The greatest increases in fecal coliform levels due to sludge disposal could be seen in the minimal to maximal counts in field runoff water. While geometric mean values differed little, the maximal fecal coliform counts per 100-ml volumes before and after the application of digested sludge were 2.3×10^4 and 1.2×10^5, respectively. Other bacterial and viral data reported for surface streams and reservoirs showed no dramatic increases at points downstream from the disposal site when compared to upstream values.

Another sludge disposal site is that operated by the East Bay Municipal Utility District in Solano County, California (47). Anaerobically digested sludge is sprayed on both row crop test plots and irrigated and dryland pasture at application rates ranging from 3.3 dry tons/acre (7.4 t/ha) to 32.3 dry tons/acre (72.4 t/ha). In most instances, the applied sludges are then plowed into the upper soil layer. Three parasitic helminths (Ascaris lumbricoides, Strongyloides stercoralis, and Hymanolepsis nana) were found represented in the soil/sludge samples from the row crop plots. Helminth ova densities ranged from 1 to 50 per gram of soil/sludge. Significant number of total and fecal coliform, fecal Streptococci, Salmonellae, and Shigellae survived for as long as 7 months. Streptococcus faecalis, Clostridium tetani, Clostridium perfringens, and butyl butyric Clostridia were found in small numbers on both irrigated and dryland pasture seven months after sludge application during the winter season. Clostridium botulinum was isolated at the same time from the dryland pasture.

Additional reports of pathogen isolation from wasted sludge have been published by Wellings et al. (1976) in Florida. One poliovirus type 3 isolate was recovered from 500 ml of sludge after 48 hours on a spray field. Twenty-four isolates identified as echovirus 7 were recovered from 250 g of sludge after a 13-day period on a sludge-drying bed.

In a study using anaerobically digested sludge to recover a forest clearcut area in northwest Washington, Edmonds (1976) monitored the survival and movement of indigenous coliform bacteria. Fecal coliform counts in sludge applied in summer decreased from 1.08×10^5 to 358/g in 204 days and the organisms were undetected after 267 days. Fecal coliforms were isolated from both a spring draining beneath the sludge application site and a groundwater well.

Several recent reports have considered the public health aspects of land disposal of treated effluents. The Flushing Meadows Wastewater Renovation Project near Phoenix, Arizona, has been operational since 1967. Research initiated in 1974 sought to ascertain the fate of fecal coliforms, fecal Streptococci, Salmonellae, and enteric viruses (34). Both sewage effluent applied to the infiltration basins and

renovated well water taken from depths of 6 to 9 m were
screened for these microorganisms. No viruses or Salmonellae
were detected in well samples; fecal coliform and fecal
Streptococci levels were diminished by 99.9%. However, difficulties attributable to the viral concentration methodology
used (filter clogging and precipitate formation during
reconcentration of eluates) were noted.

In contrast to these findings, Wellings et al. (1975)
repeatedly isolated enteric viruses from groundwater. Monitoring a wastewater spray irrigation site at St. Petersburg,
Florida, this group showed virus to have moved through 5 ft
(1.5 m) of sandy soil. Following heavy rains, viruses were
isolated at this site from 10- to 20-ft (3- to 6-m) wells.
Similar isolations were made during a study of a cypress dome
receiving secondary effluents. Water from 10-ft (3 m)-deep
monitoring wells were shown to contain virus. Two of the
three positive isolations reported coincided with a period of
heavy rainfall 28 days after the last application of sewage
effluent.

Using methodologies developed in the CART laboratories at
The University of Texas at San Antonio, Dudley et al. (1980)
have provided quantitative data on the population densities
in sludge of fluorescent Pseudomonas species. Staplylococcus
aureus, Mycobacterium species, Clostridium species, and
Klebsiella species, as well as semiquantitative data on
Shigella and Salmonella species. Typical data from sludges
undergoing land application are given in Tables 1-3.

In addition to bacterial populations, the levels of human
enteric viruses capable of forming plaques on HeLa or Hep-2
cells were sought by Dudley et al. (1980), using the methods
reported by Moore et al. (1979). These results (see Table 4)
are, of course, minimal values as the method does not pretend
to recover such epidemiologically significant agents as human
rotaviruses, Norwalk agent, or the virus of infectious hepatitis.

The application of wastewater residuals to food crops
raises the obvious question of possible ingestion of surviving pathogenic microorganisms. No conclusive data exist to
support the uptake and subsequent translocation of pathogens
from contaminated soil into nontraumatized edible plant
tissues. However, various studies have shown survival times
ranging from hours to months for microbial pathogens applied
to the surface of fruits and vegetables (24, 75, 76, 77).
Bagdasaryan (1964) reported survival of enteroviruses on
artificially contaminated tomatoes and radishes over a period
of two weeks under household storage conditions. Larkin et
al. (1976) added poliovirus I to wasted sludge and secondary
effluent used to spray-irrigate a series of test plots
planted with lettuce and radishes. As expected, greater
viral numbers were observed on the sludge-irrigated plants

MICROBIAL AGENTS   25

Table I. Enumeration of Bacteria in Sludges Undergoing Land Application[a]

	Enumeration (CFU/g)[b]		
	Site 1 (digested sludge)	Site 2 (lagooned sludge)	Site 3 (digested sludge)
Total aerobic count	4.8×10^8	2.2×10^9	1.2×10^8
Total coliforms	4.5×10^6	6.1×10^7	5.1×10^6
Fecal coliforms	4.8×10^5	4.7×10^6	3.1×10^6
Fecal streptococci	2.4×10^5	4.5×10^5	6.9×10^4
Fluorescent pseudomonads	6.4×10^4	7.1×10^4	2.0×10^4
Staphylococcus sp.	$< 7.9 \times 10^4$ (ND)	1.2×10^6	$< 6.7 \times 10^4$ (ND)
Clostridium perfringens (MPN/g)			
Vegetative (room temp)	2.6×10^7	2.2×10^7	3.1×10^7
Sporulated (80°C)	2.6×10^7	4.7×10^6	4.9×10^6
Mycobacterium sp.	4.0×10^7	2.4×10^9	4.1×10^7
Salmonella sp.	$\geq 2.4 - < 24$	> 2.0	≥ 2.0
Shigella sp.	ND	ND	≥ 20
Klebsiella sp.	8.8×10^6	1.2×10^7	4.7×10^5
Total suspended solids (mg/liter)	41,900	50,700	49,100

[a] Sludge from site 1 had undergone digestion in a two-stage high-rate anaerobic digestion system. Sludge from site 2 was lagooned after undergoing anaerobic digestion for 11 days. Sludge from site 3 had undergone standard anaerobic digestion for 20 to 30 days with no mixing or heating.

[b] CFU/g, Colony-forming units per gram of total suspended solids, unless indicated otherwise. ND, None detected, -, Not detected.

[c] Highest numbers were recovered off either Mac-Conkey or XLD agar. Approximately 200 colonies (100 from each agar) were isolated from appropriate dilutions of each sludge and inoculated to API 20E strips.

Table II. Enumeration of Bacteria in Sludges Undergoing Land Application[a]

	Enumeration (CFU/g)[b]		
	Site 1 (digested sludge)	Site 2 (lagooned sludge)	Site 3 (digested sludge)
Enterobacteriaceae[c]			
Citrobacter diversus subsp. levinae	-	-	1.0×10^6
C. freundii	7.1×10^5	5.9×10^6	6.1×10^5
Enterobacter aerogenes	-	2.0×10^6	-
E. agglomerans	2.4×10^5	4.9×10^6	-
E. cloacae	1.6×10^5	1.6×10^7	1.0×10^6
E. sakazakii	-	9.8×10^5	2.0×10^5
Escherichia coli	1.7×10^6	8.8×10^6	2.0×10^5
Hafnia alvei	-	-	8.0×10^6
Klebsiella oxytoca	6.2×10^5	9.8×10^5	6.1×10^5
K. ozaenae	7.9×10^4	9.8×10^5	2.7×10^6
K. pneumoniae	1.6×10^5	2.0×10^6	6.1×10^5
Proteus morganii	-	-	2.0×10^5
Serratia liquefaciens	7.9×10^4	3.9×10^6	-
S. marcescens	-	9.8×10^5	-
S. rubidaea	-	2.0×10^6	-
Yersinia enterocolitica	-	-	2.0×10^5
Y. ruckeri	-	-	1.0×10^6
Total suspended solids (mg/liter)	41,900	50,700	49,100

[a] Sludge from site 1 had undergone digestion in a two-stage high-rate anaerobic digestion system. Sludge from site 2 was lagooned after undergoing anaerobic digestion for 11 days. Sludge from site 3 had undergone standard anaerobic digestion for 20 to 30 days with no mixing or heating.

[b] CFU/g, Colony-forming units per gram of total suspended solids, unless indicated otherwise. ND, None Detected, -, Not detected.

[c] Highest numbers were recovered off either Mac-Conkey or XLD agar. Approximately 200 colonies (100 from each agar) were isolated from appropriate dilutions of each sludge and inoculated to API 20E strips.

TABLE III. ENUMERATION OF BACTERIA IN SLUDGES UNDERGOING LAND APPLICATION[a]

	ENUMERATION (CFU/G)[b]		
	SITE 1 (DIGESTED SLUDGE)	SITE 2 (LAGOONED SLUDGE)	SITE 3 (DIGESTED SLUDGE)
OXIDASE-POSITIVE, GRAM-NEGATIVE ENTERIC BACTERIA[c]			
ACHROMOBACTER SP.	-	-	2.0×10^5
A. XYLOSOXIDANS	-	-	6.1×10^5
ACINETOBACTER CALCOACETICUS VAR. LWOFFI	-	2.0×10^6	-
AEROMONAS HYDROPHILA	2.9×10^6	5.3×10^7	2.0×10^6
ALCALIGENES SP.	7.9×10^4	-	2.0×10^5
BORDETELLA BRONCHISEPTICA	-	-	2.0×10^5
CDC GROUP V E-1	7.9×10^4	-	2.0×10^5
FLAVOBACTERIUM ODORATUM	1.6×10^5	-	-
PSEUDOMONAS AERUGINOSA	7.9×10^4	-	-
P. CEPACIA	5.5×10^5	9.8×10^5	-
P. FLUORESCENS	3.1×10^5	9.8×10^6	-
P. MALTOPHILIA	-	9.8×10^5	-
P. PAUCIMOBILIS	-	9.8×10^5	-
P. PUTIDA	6.2×10^5	1.3×10^7	4.7×10^6
P. PUTREFACIENS	-	9.8×10^5	-
P. STUTZERI	3.1×10^5	9.8×10^5	-
VIBRIO ALGINOLYTICUS	-	9.8×10^5	2.0×10^5
TOTAL SUSPENDED SOLIDS (MG/LITER)	41,900	50,700	49,100

[a] SLUDGE FROM SITE 1 HAD UNDERGONE DIGESTION IN A TWO-STAGE HIGH-RATE ANAEROBIC DIGESTION SYSTEM. SLUDGE FROM SITE 2 WAS LAGOONED AFTER UNDERGOING ANAEROBIC DIGESTION FOR 11 DAYS. SLUDGE FROM SITE 3 HAD UNDERGONE STANDARD ANAEROBIC DIGESTION FOR 20 TO 30 DAYS WITH NO MIXING OR HEATING.

[b] CFU/G, COLONY-FORMING UNITS PER GRAM OF TOTAL SUSPENDED SOLIDS, UNLESS INDICATED OTHERWISE. ND, NONE DETECTED, -, NOT DETECTED.

[c] HIGHEST NUMBERS WERE RECOVERED OFF EITHER MAC-CONKEY OR XLD AGAR. APPROXIMATELY 200 COLONIES (100 FROM EACH AGAR) WERE ISOLATED FROM APPROPRIATE DILUTIONS OF EACH SLUDGE AND INOCULATED TO API 20E STRIPS.

due to retention of particulates.

A cursory reading of the scientific and clinical literature of the last two years has yielded enough data to suggest that a reasoned approach be taken to land disposal and subsequent land use. For purposes of this review, selection has been for reports of infections by several of the species isolated by Dudley et al. (1980) from domestic U.S. sludges being distributed to land.

Earlier, some of the species (and their population densities) isolated from sludge being disposed to land were noted. The levels recovered are not inconsistent with the levels excreted (72). The fraction of warm-blooded animals excreting some common enteric pathogens has also been estimated by the same author. Very few data, however, are available for human minimal infective doses of these organisms except in healthy volunteers and with laboratory strains.

Guentzel (in press, 1980) has written that the Gram-negative bacilli of the genera Escherichia, Klebsiella, Enterobacter, Serratia, and Citrobacter "are found routinely as components of the normal flora of the human and animal

TABLE IV. MAXIMAL VIRUS RECOVERY FROM LIQUID SLUDGES APPLIED TO LAND

SITE	MIDWEST	MOUNTAIN STATES	PACIFIC NORTHWEST	SEMI-TROPICAL STATE	SOUTHWEST	GULF STATE
TREATMENT		TWO STAGE ANAEROBIC HIGH-RATE DIGESTION, MIXING ($\pm$ HEATING)	ANAEROBIC DIGESTION	ANAEROBICALLY DIGESTED, LAGOONED	ANAEROBIC DIGESTION	ANEROBICALLY DIGESTED, 11-DAY DETENTION
COLIPHAGES*	1.1×10^5	5.2×10^4	1.7×10^4	8.5×10^4	2.2×10^3	4.8×10^4
HUMAN ENTERIC VIRUSES*	1.7×10^1	1.1	2.2	1.2	6.7	5.0

*PFU/GM DRY SLUDGE

intestinal tract and may be isolated from a variety of environmental sources .... At one time these microorganisms were dismissed as harmless commensals. Today they are responsible for major health problems throughout the world. The increasing incidence of the coliforms, as well as other Gram-negative organisms, in disease reflects in part a better understanding of their pathogenic potential, but more importantly, the changing ecology of bacterial disease. The widespread .... use of antibiotics had led to the appearance of drug-resistant Gram-negative bacilli which readily acquire multiple resistance through self-transmissable drug resistance plasmids ...." Genetic analysis of transferable drug resistance in enteric bacteria in man and animals shows a common origin for the plasmids (8). "The development of new surgical procedures, health support techniques, and therapeutic regimens has provided new portals of entry and compromised many host defenses. As opportunistic pathogens, the coliforms utilize the opportunity of weakened host defenses to colonize and to elicit a variety of disease states" Guentzel concluded.

Coliform bacilli are responsible for about one-third of all hospital acquired infections. Nearly 4% of all acute-care patients had acquired such nosocomial infections (about 400,000 cases in 1979). Table 5 shows the organisms involved.

Carrington (1977) has noted the persistence of sludge-borne pathogens in the environment. He cites the work of Thunegard (95) who applied Salmonella in a farm slurry to soil which then was sown with barley. He recovered S. typhimurium over a period of 57 days. In another study, the same author recovered identifiable Salmonella serotypes applied in similar fashion for 300 days after application.

Thus, the survival of sludge-borne bacterial pathogens in nature is not unknown. As noted earlier, it is dependent upon temperature, sunlight, moisture, and the nature of the soil (i.e., pH). Similar findings have been reported for poliovirus by Moore (1978). Subrahmanyan (1977) too, noted viral persistence in digested sludges for prolonged periods. Steinmann (1978) reported the persistence of poliovirus in chemical sludge from a tertiary treatment process.

While reports of potentially pathogenic enteric viruses in shellfish growing waters have long been in the literature (64), it should be noted that Vibrio parahaemolyticus also has been isolated from seafood and implicated in acute gastroenteritis (87). Aeromonas primary wound infection of a diver in polluted waters was reported in 1979 (50). Two species were cultured. It is noteworthy that one of these isolates was resistant to tetracycline.

The use of sludges, which have neither been heat-treated, nor irradiated, nor composted, may serve to disseminate microbial pathogens to agricultural lands. A recent survey

TABLE V. Frequency of Selected Pathogens Causing Nosocomial Infections

Pathogen	Primary Bacteremia	Surgical Wound	Lower Respiratory	Urinary Tract	Cutaneous	Other	All Sites
E. coli	14.3	15.7	7.2	31.7	8.5	9.2	19.4
Klebsiella	10.6	5.2	11.1	8.8	4.3	4.6	7.6
Enterobacter	5.2	4.0	6.5	4.1	3.1	2.4	4.3
Serratia	3.0	1.2	2.9	2.1	1.4	1.3	1.9
Proteus-Providencia	3.6	7.1	5.6	10.4	5.5	5.7	7.8
Pseudomonas aeruginosa	5.2	4.1	7.3	8.9	4.7	4.8	6.6
Staphylococcus aureus	12.8	15.5	10.4	1.7	33.7	12.6	10.0
Group D Streptococcus	7.4	10.3	1.2	13.9	6.1	5.9	9.6
Other Pathogens	36.1	25.7	21.0	16.1	19.1	30.2	21.6
No Culture or No Pathogen	1.8	11.2	26.8	2.3	13.6	23.3	11.2
Isolates/10,000 Discharges	16.6	121.9	71.4	174.6	25.7	45.9	456.1

Data obtained from Center for Disease Control: National Nosocomial Infections Study Report, Annual Summary 1976, Issued February 1978.

(92) of the hygienic quality of vegetables grown or imported into the Netherlands make it abundantly clear that once contaminated, table vegetables retain viable E. coli, fecal Streptococci and Salmonellae. Of the 260 samples examined, 11% contained >10^4 E. coli per 100 gm and 14% >10^6 fecal Streptococci per 100 gm. Salmonellae were isolated from 22% of samples evaluated. In one case, S. typhosa was isolated from vegetables imported from the tropics.

Morris and Feeley (1976) reviewed the role of Yersinia enterocolitica in food hygiene. They noted its increased frequency of isolation from man, domestic animals, and some human foods. Infections with Y. enterocolitica were seen as a cause of concern in human and veterinary medicine. The organism has been found in slaughterhouses, market meat, dairy products, shellfish and nonchlorinated water used for drinking purposes (10, 41, 43, 44, 45, 46, 79, 80). Only 23 human infections were reported in 1966. In 1974, over 4000 were reported. Asakawa et al. (1979) found 0.87% carriers among meat handlers and 1.4% carriers among kindergarten children, but none showed clinical symptoms. Isolation rates from swine in this study were 5.7% with about 60% being the 3 and 5 strains which are common in human infection.

Wauters (1979) has demonstrated high levels of Y. enterocolitica on pig's tongues purchased in meat markets. Of 302 samples, 168 yielded Y. enterocolitica of serotype 3, the most common human isolate. The frequency of isolation from the tongue was far greater than from the feces of pigs. As the organism grows to relatively high levels at refrigeration temperatures (see Figure 1), this may be a possible explanation of its association with and spread via milk and processed meat products.

Bovine salmonellosis is an economically important disease, as well as a public health concern in most countries. Smith et al. (1979) fed calves S. typhimurium and elicited the disease with doses ranging from 10^4 to 10^{11} organisms. Survival was inversely related to inoculum size. Fecal cultures were positive for Salmonellae. The animals were all less than nine weeks of age at the time of feeding. Hall and Jones (1977) fed 10-12 month old heifers raw sludge containing up to 10^5 naturally occurring Salmonellae/liter at the rate of either 1 liter/day or 1 liter/week. No animals became ill; none shed virus. Animals fed sterilized sludge to which S. dublin was added to 10^5/liter remained healthy, but several did shed S. dublin and yielded the organism in tissue necropsy. Smith et al. (1979) had noted that susceptibility was inversely age-dependent in their study. Hess and Breer (1975) reported that the seasonally heavy use of sludge increased the isolation of Salmonellae from cattle in Switzerland. Taylor (1979) reported persistent Salmonella excretors in a small dairy herd in Wales. Most Salmonella infections other than with S. dublin are self-curing and do

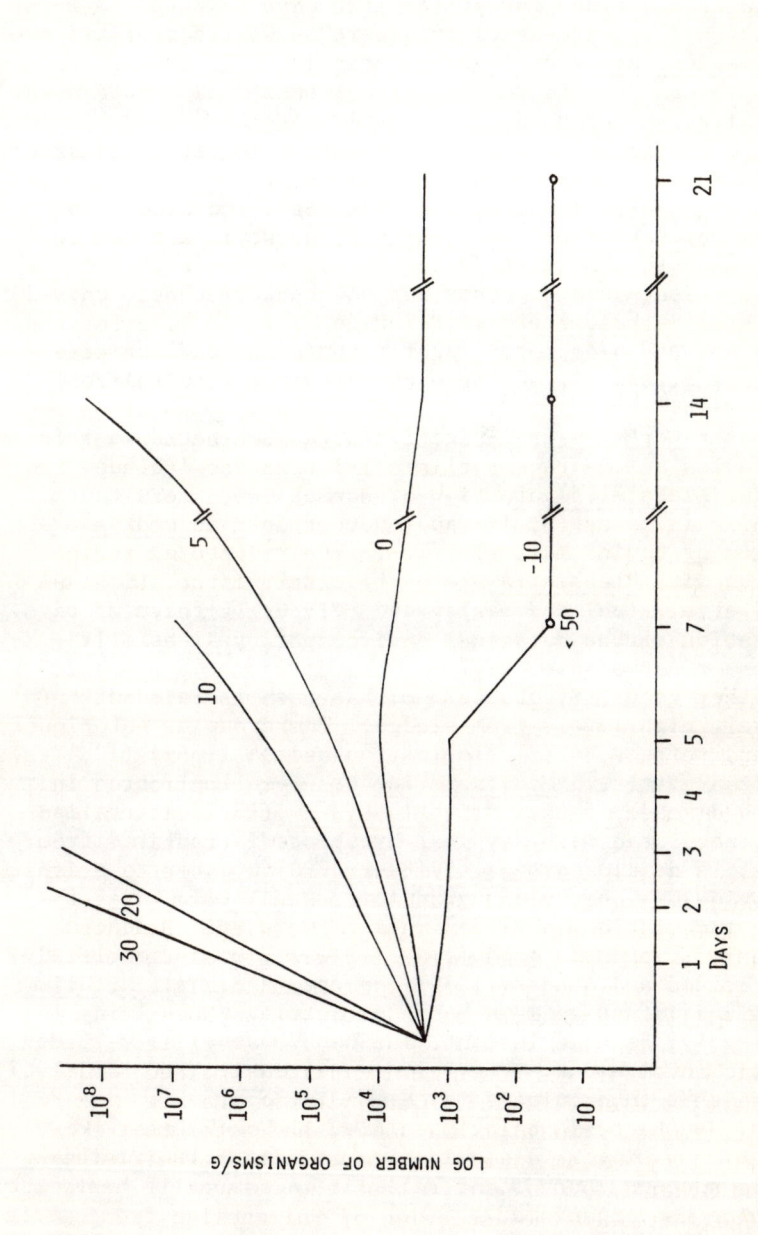

Figure 1. Growth of Y. enterocolitica on sliced ham at various temperatures (°C). Asakawa, Y., S. Akahane, K. Shiozawa and T. Honma. Contr. Microbiol. Immunol. Vol. 5, pp. 115-121 (Karger, Basel, 1979).

not establish carrier states, he noted, in reporting this case.

Parenthetically, while reviewing the large body of literature on Salmonella isolation published in the last two years there was noted survival of Salmonella and coliforms in apple juice and cider for at least 30 days at pH 3.6. This publication (35) cites earlier papers on gastroenteritis due to S. typhimurium in chilled non-sterile apple juice. The organisms were thought to have come from animal manure used to fertilize the orchard.

While not pretending to an exhaustive review we found that Salmonellae have been recovered from outbreaks in birds (38), minced meat (98), scrapings from butcher's shops (25), pork sausage (96), chicken carcasses (97), as well as other less likely human food sources.

The patient-related costs for one cheddar cheese-caused outbreak of S. heidelberg in Colorado averaged $125 for those who did not see a doctor, $222 for those who did but were not hospitalized, and $1,750 for each of the 68 hospitalized patients (14).

Other reports on Gram-negative organisms noted in this brief review of the recent clinical literature included two reporting Klebsiella infection of horses, with resultant barrenness in mares (9, 17) and another on Pseudomonas causation of bovine mastitis (70). The reports on equine infection with Klebsiella are particularly interesting as Knittel et al. (56) and Bagley and Seidler (4) have shown colonization of the botanical environment by Klebsiella isolates of pathogenic origin.

Another genus, Mycobacterium, has been isolated at relatively high levels from sludge. The group is notoriously hardy and refractory to chlorination and to liming (12, 74). In African cattle, a Mycobacterium has been implicated in farcy, a disease of the lymphatic system (13). Australian cattle inoculated with atypical Mycobacteria isolated from local soils developed sensitivity to PPD (greater to avian than bovine). Tuberculous granulomas were produced at the site of inoculation in all animals. Two strains induced mesenteric lymph node granulomas, as well, when inoculated (16). Friend and others (1979) reported a natural M. avium generalized infection in a Labrador Retriever resulting in splenomegaly, hepatic lesions, and nephropathy, as well as lymphatic involvement. They cite an extensive literature to indicate that tuberculosis is common in dogs, with avian and atypical Mycobacteria being possible, although less likely agents. They suggest dissemination in this case could have been via a wound, but do not rule out ingestion of foodstuff containing the organisms.

Steadham (1980) reported the isolation of potentially pathogenic high-catalase strains of Mycobacterium kansasii from 8 of 19 representative water outlets in a small Texas town. M. gordonae was isolated from all samples and M. fortuitum from two. The town was selected for study because of an increased frequency of infections and peculiar skin reactions.

DISCUSSION

Johnson (1979) has written of the nutritional aspects of refeeding cattle manure to ruminants. Smith et al. (1976) and Kienholz et al. (1976) have discussed recycling sewage solids as feedstuffs for livestock. Others (48) have presented the FDA's concerns for microbiological contamination of crops and of livestock. Bastian and Whittington (1976) have stated that from the EPA viewpoint the work most urgently needed in the municipal sewage sludge management field includes
1. resolution of health effect issues
2. continued emphasis on innovative technology leading to beneficial use and resource recovery
3. breakthrough in public acceptance
4. information dissemination to both design engineers/ operators and elected and government officials.

Reasonable evidence has been presented to suggest that sludges are not inocuous from a public health viewpoint. To resolve the health effects issue and achieve a breakthrough in acceptance may not require innovative technology, however. It may suffice, perhaps, that we consider composting, a rather old and low level alternate technology. The risk from sludge application can be reduced by some processing interposed between the municipal treatment plant and the receiving field.

William Lowrance (1976) has been quoted on acceptable risk before. To keep things in perspective, it is worth reading him again.

"We are disturbed by what sometimes appear to be haphazard and irresponsible regulatory actions, and we can't help being suspicious of all the assaults on our freedoms made in the name of safety. We hardly know which cries of "Wolf!" to respond to; but we dare not forget that even in the fairy tale, the wolf really did come.

In 1900 .... "some rivers were so filthy with raw sewage and industrial waste that, as the saying went, 'bait died on the hook.' Industrial towns were black with coal soot, as were people's lungs. Workers labored at their own peril.

"The principal fatal diseases were pneumonia, influenza, and tuberculosis. Infant mortality was high; in 1900, more than thirteen percent of all American children died before their first birthday.

No need to belabor the point; in many ways we are better

off than we used to be. In a sense we now have the luxury to worry about subtle hazards which at one time, even if detected, would have been given only low priority beside the much greater hazards of the day.

"At the same time and partly because of scientific advance, people's values and expectations [have] changed. Discomforting discoveries were forced by the extraordinary growth in our social and physical scale. We have been startled into profound realizations, no less profound for having become commonplaces .... that 'there is no longer any 'away' into which to throw things;' that it is crucial that we stop fouling our earthly nest ....

"Much of the widespread confusion about the nature of safety decisions would be dispelled if the meaning of the term safety were clarified. For a concept so deeply rooted in both technical and popular usage, safety has remained dismayingly ill-defined.

"We will define safety as a judgment of the acceptability of risk, and risk, in turn, as a measure of the probability and severity of harm to human health.
A thing is safe if its risks are judged to be acceptable.

"By its preciseness and connotative power this definition contrasts sharply with simplistic dictionary definitions that have "safe" meaning something like "free from risk." Nothing can be absolutely free of risk. One can't think of anything that isn't, under some circumstances, able to cause harm. Because nothing can be absolutely free of risk, nothing can be said to be absolutely safe. There are degrees of risk, and consequently there are degrees of safety.

"Notice that this definition emphasizes the relativity and judgmental nature of the concept of safety. It also implies that two very different activities are required for determining how safe things are: measuring risk, an objective but probabilistic pursuit; and judging the acceptability of that risk (judging safety), a matter of personal and social value judgment.

".... Safety is obviously a highly relative attribute that can change from time to time and be judged differently in different contexts. Knowledge of risks evolves, and so do our personal and social standards of acceptability .... A power saw that is safe for an adult may not be safe in the hands of a child. Partly because we have discovered adverse health effects we didn't know about earlier, and partly because more sensitive techniques are now available, X-ray doses thought safe forty years ago are now deemed intolerably risky. DDT is essentially banned from most uses in the United States, where we have access to and can afford more costly alternatives, are not so much exposed to tropical diseases, and can afford to be concerned about even a slight risk of cancer -- but in contrast, DDT is the pesticide of choice in many tropical, less wealthy countries where every scrap of food has to be protected from insect predators, and where malaria and other diseases carried by DDT-susceptible insects are more imminent threats to life than the remote possibility of cancer.

".... Since the taking of both personal and societal risks is inherent in human activity, there can be no hope of reducing all risks to zero. Rather, as when steering any course, we must continually adjust our heading so as to enjoy the greatest benefit at the lowest risk and cost."

FOOTNOTE

*Portions of this paper were presented earlier at a Symposium on Utilization of Municipal Sewage Effluent and Sludge on Forest and Disturbed Land conducted by The Pennsylvania State University in 1977 and published in 1979 (William E. Sopper and Sonja N. Kerr, Editors).

+After September 1, 1980, Drexel University, Philadelphia, Pennsylvania.

REFERENCES

1. Akin, E.A., H.P. Pahren, W. Jakubowski, J.B. Lucas. "Health Hazards Associated with Wastewater Effluents and Sludges: Microbiological Considerations," in Proceedings of the Conference on Risk Assessment and Health Effects of Land Application of Municipal Wastewater and Sludges. Center for Applied Research and Technology, The University of Texas at San Antonio. (1978)

2. Asakawa, Y., S. Akahane, K. Shiozawa, and T. Homma. "Investigations of Sources and Route of Yersinia enterocolitica Infection." Contributions to Microbiology and Immunology 5: 115-121. (1979)

3. Bagdasaryan, G.A. "Survival of Viruses of the Enterovirus Group (poliomyelitis, echo, coxsackie) in Soil and on Vegetables." Hyg. Epidemiol. Microbiol. Immunol. 8: 497. (1964)

4. Bagley, S.T. and R.J. Seidler. "Significance of Fecal Coliform. Positive Klebsiella." Applied and Environmental Microbiology 33: 1141-1148. (1977)

5. Bastian, R.K. and W.A. Whittington. "EPA Guidance on Disposal of Municipal Sewage Sludge onto Land," in Proceedings of the Third National Conference on Sludge

Management Disposal and Utilization. Information Transfer, Inc., Rockville, Maryland. pp. 32-34. (1977)

6. Beard, P.J. "The Survival of Typhoid in Nature." *J. Amer. Water Works Assoc.* 30: 124. (1938)

7. Beard, P.J. "Longevity of Eberthella Typhosus in Various Soils." *Amer. J. Public Health* 30: 1077. (1940)

8. Black, R.E., R.J. Jackson, T. Tsai, M. Medvesky, M. Shayegani, J.C. Feeley, K.I.E. MacLeod, and A.M. Wakelee. "Epidemic *Yersinia enterocolitic* Infection due to Contaminated Chocolate Milk." *New England J. of Medicine* 298: 76-79. (1978)

9. Brown, J.E., R.E. Corstvet, L.G. Stratton. "A Study of *Klebsiella pneumoniae* Infection in the Uterus of the Mare." *Am. J. of Veterinary Research* 40: 1523-1530. (1979)

10. Caprioli, T., A.J. Drapeau, and S. Kasatiya. "*Yersinia enterocolitica*: Serotypes and Biotypes Isolated from Humans and the Environment in Quebec, Canada." *J. of Clinical Microbiology* 8: 7-11. (1978)

11. Carrington, E.G. "The Contribution of Sewage Sludges to the Dissemination of Pathogenic Microorganisms in the Environment." Water Res. Centre, Stevenage Laboratory, Hertfordshire. Unpublished. (1978)

12. Carson, L.A., N.J. Petersen, M.S. Favero, and S.M. Aguero. "Growth Characteristics of Atypical Mycobacteria in Water and Their Comparative Resistance to Disinfectants." *Applied and Environmental Microbiology* 36: 839-846. (1978)

13. Chamoiseau, G. "Etiology of Farcy in African Bovines: Nomenclature of the Causal Organisms *Mycobacterium farinogenes* Chamoiseau and *Mycobacterium senegalese* (Chamoiseau) comb. nov." *Int'l J. of Systematic Bacteriology* 29: 407-410. (1979)

14. Cohen, M.L., R.E. Fontaine, R.A. Pollard, S.D. Von Allmen, T.M. Vernon, and E.J. Gangarosa. "An Assessment of Patient-Related Economic Costs in an Outbreak of Salmonellasis." *New England J. of Medicine* 299: 459-460. (1978)

15. Coker, E.G. "The Value of Liquid Digested Sewage Sludge I. The Effect of Liquid Sewage Sludge on Growth and Composition of Grass-Clover Swards in Southeast England." *J. Agric. Sc.* 67: 91-97. (1966)

16. Corner, L.A. and C.W. Pearson. "Response of Cattle to Inoculation with Atypical Mycobacteria Isolated from Soil." Australian Veterinary J. 55: 6-9. (1979)

17. Crouch, J.R.F. "Klebsiella Infections in Mares." The Veterinary Record. 105, No. 14. Report. (1979)

18. Davies, M. and P.R. Stewart. "Transferable Drug Resistance in Man and Animals: Genetic Relationship Between R- Plasmids in Enteric Bacteria from Man and Domestic Pets." Australian Veterinary J. 54: 507-512. (1978)

19. Day, A.D. and R.M. Kirkpatrick. "Effects of Treated Municipal Wastewater on Oat Forage and Grain." J. Environ. Qual. 2: 282-284. (1973)

20. Drewry, W.A., and R. Eliassen. "Virus Movement in Groundwater." J. Water Poll. Control Fed. 40: R257 (1968)

21. Duboise, S.M., B.E. Moore, B.P. Sagik and C.A. Sorber. "The Effects of Temperature and Specific Conductance on Poliovirus Survival and Transport in Soil." Abstract for the National Conference on Environmental Research, Development, and Design. University of Washington, Seattle. (1976a)

22. Duboise, S.M., B.E. Moore, and B.P. Sagik. "Poliovirus Survival and Movement in a Sandy Forest Soil." Appl. Environ. Microbiol. 31: 536. (1976b)

23. Dudley, D.J., M.N. Guentzel, M.J. Ibarra, B.E. Moore, and B.P. Sagik. "Enumeration of Potentially Pathogenic Bacteria from Sewage Sludges." Applied and Environ. Microbiol. 39: 118-126. (1980)

24. Dunlop, S.G. "Survival of Pathogens and Related Disease Hazards," in Proceedings, Municipal Sewage Effluent for Irrigation. C.W. Wilson and F.E. Beckett (Eds.). (Louisiana Tech. Alumni Foundation). (1968)

25. Edel, W.M. Van Schothorst, I.M. Van Leusden, and E.H. Kampelmacher. "Epidemiological Studies on Salmonella in a Certain Area ("Walcheren Project") III. The Presence of Salmonella in Man, Insects, Chopping-Block Scrapings from Butcher's Shops, Effluent of Sewage Treatment Plants and Drains of Butcher's Shops." Zbl. Bakt, Hyg. I. Abt. Orig. A 242: 468-480. (1978)

38 HEALTH RISKS OF SPECIFIC AGENTS

26. Edmonds, R.L. "Survival of Coliform Bacteria in Sewage Sludge Applied to a Forest Clearcut and Potential Movement into Groundwater." Appl. Environ. Microbiol. 32: 537. (1976)

27. Egeland, D.R. "Land Disposal I: A Giant Step Backward." J. Water Pollu. Control Fed. 45: 1465-1475. (1973)

28. Ellis, B.G. "The Soil as a Chemical Filter," in Recycling Treated Municipal Wastewater and Sludge through Forest and Cropland, W.E. Sopper and L.T. Kardos, Eds. (Pennsylvania State University Press, University Park). (1973)

29. Foster, D.H., and R.S. Engelbrecht. "Microbial Hazards in Disposing of Wastewater on Soil," in Recycling Treated Municipal Wastewater and Sludge through Forest and Cropland, W.E. Sopper and L.T. Kardos, Eds. (Pennsylvania State University Press, University Park). (1973)

30. Friend, S.C.E., E.G. Russell, W.J. Hartley, and P. Everist. "Infection of a Dog with Mycobacterium avium Serotype II." Veterinary Pathology 16: 381-384. (1979)

31. Freud, S. "Civilization and Its Discontents." Translated by James Strachey, 1961, W.W. Norton and Company, Inc., New York. (1930)

32. Fussell, G.E. Farming Technique from Prehistoric to Modern Times. Pergamon Press, Oxford. (1965).

33. Gerba, C.P., C. Wallis and J.L. Melnick. "Fate of Wastewater Bacteria and Viruses in Soil." J. Irrigation Div. A.S.C.E. 101: IR3. (1975)

34. Gilbert, R.G., C.P. Gerba, R.C. Rice, H. Bowver, C. Wallis, and J.L. Melnick. "Virus and Bacteria Removal from Wastewater by Land Treatment." Appl. Environ. Microbiol. 32: 333. (1976)

35. Goverd, K.A., F.W. Beech, R.P. Hobbs and R. Shannon. "The Occurrence and Survival of Coliforms and Salmonellas in Apple Juice and Cider." J. of Applied Bacteriology 46: 521-530. (1979)

36. Gras, N.S.B. A History of Agriculture. F.S. Crofts and Company, Publishers, New York. (1925)

37. Greenberg, A.E., and E. Kupka. "Tuberculosis Transmission by Wastewaters—a Review." Sew. Ind. Wastes 29: 524. (1957)

38. Grimes, T.M. "Observations on Salmonella Infections of Birds." Australian Veterinary J. 55: 16-18. (1979)

39. Guentzel, M.N. "Escherichia, Klebsiella, Enterobacter, Serratia and Citrobacter in Community and Hospital - Acquired (Nosocomial) Infections," in Baron, S. Microbiology Principals and Practice, Addison-Wesley, Menlo Park, CA. (1980)

40. Hall, G.A. and P.W. Jones. "A Study of the Susceptibility of Cattle to Oral Infection by Salmonellas Contained in Raw Sewage Sludge." J. Hyg., Camb. 80: 409-414. (1978)

41. Hanna, M.O., J.C. Stewart, Z.L. Carpenter, D.L. Zink, and C. Vanderzant. "Isolation and Characteristics of Yersinia enterocolitica-like Bacteria from Meats." Contributions to Microbiology and Immunology 5: 234-242. (1979)

42. Hess, E. and C. Breer. "Epidemiology of Salmonellae and Fertilizing Grassland with Sewage Sludge." Zentralblatt for Bakteriologie, Parasitenkunde, Infectionskrankheiten und Hygiene (I. Abt. Orig. B) 161: 54-60. (1975)

43. Highsmith, A.K., J.C. Feeley, and G.K. Morris. "Isolation of Yersinia enterocolitica from Water," in Bacterial Indicators/Health Hazards Associated with Water, ASTM STP 635, A.W. Hoadley and B.J. Dutka, Eds., pp. 265-274. (1977)

44. Highsmith, A.K., J.C. Feeley, P. Skaliy, J.G. Wells, and B.T. Wood. "Isolation of Yersinia enterocolitica from Well Water and Growth in Distilled Water." Applied and Environmental Microbiology 34: 745-750. (1977)

45. Hughes, Denise. "Isolation of Yersinia enterocolitica from Milk and a Dairy Farm in Australia." J. of Applied Bacteriology 46: 125-130. (1979)

46. Hurvell, B., V. Glatthard, and E. Thal. "Isolation of Yersinia enterocolitica from Swine at an Abattoir in Sweden." Contributions to Microbiology and Immunology 5: 243-248. (1979)

47. Hyde, H.C. "Utilization of Wastewater Sludge for Agricultural Soil Enrichment." J. Water Poll. Control Fed. 48: 77. (1976)

48. Jelinek, C.F. and G. L. Braude. "Management of Sludge Use on Land: FDA Considerations," in Sludge Management, Disposal, and Utilization, Information Transfer, Inc., Rockville, Maryland. pp. 35-38. (1977)

49. Johnson, W.L. "Nutritional Aspects of Refeeding Cattle Manure to Ruminants." Agricultural and Food Chemistry 27: 690-694. (1979)

50. Joseph, S.W., O.P. Dailey, W.S. Hunt, R.J. Seidler D.A. Allen, and R.R. Colwell. "Aeromonas Primary Wound Infection of a Diver in Polluted Waters." J. of Clinical Microbiology 10: 46-49. (1979)

51. Kammerer, J.C. "Water Quantity Requirements for Public Supplies and Other Uses," in Handbook of Water Resources and Pollution Control, H.W. Gehm and J.I. Bregman, Eds. (Van Nostrand Reinhold Company, New York). (1976)

52. Kardos, L.T. "Crop Response to Sewage Effluent," in Municipal Sewage Effluent for Irrigation, C.W. Wilson and F.E. Beckett, Eds. (Louisiana Polytechnic Institute, Ruston, Louisiana). (1968)

53. Kardos, L.T. and W.E. Sopper. "Renovation of Municipal Wastewater through Land Disposal by Spray Irrigation," in Recycling Treated Municipal Wastewater and Sludge through Forest and Cropland, W.E. Sopper and L.T. Kardos, Eds. (Pennsylvania State University Press, University Park). (1973)

54. Kienholz, E., G.M. Ward, D.E. Johnson and J.C. Baxter. "Health Considerations Relating to Ingestion of Sludge by Farm Animals," in Sludge Management, Disposal, and Utilization, Information Transfer, Inc., Rockville, Maryland. pp. 128-134. (1977)

55. King, L.D. and H.D. Morris. "Land Disposal of Liquid Sewage Sludge: I. The Effect on Yield, in Vivo Digestibility, and Chemical Composition of Coastal Bermuda Grass (Cynodon dactylon L. pers). J. Environ. Qual. 1: 325-329. (1972)

56. Knittel, M.D., R.J. Seidler, C. Eby, and L.M. Cabe. "Colonization of the Botanical Environment by Klebsiella Isolates of Pathogenic Origin." Applied and Environmental Microbiology 34: 557-563. (1977)

57. Lance, J.C., C.P. Gerba, and J.L. Melnick. "Virus Movement in Soil Columns Flooded with Secondary Sewage Effluent." Appl. Environ. Microbiol. 32: 520. (1976)

58. Larkin, E.P., J.T. Tierney, and R. Sullivan. "Persistence of Virus on Sewage-Irrigated Vegetables." J. Environ. Eng. Div. A.S.C.E. 102(EE1): 29. (1976)

59. Lejcher, T.R. and S.H. Kunkle. "Restoration of Acid Spoil Banks with Treated Sewage Sludge," in Recycling Treated Municipal Wastewater and Sludge through Forest and Cropland, W.E. Sopper and L.T. Kardos, Eds. (Pennsylvania State University, University Park). (1973)

60. Linnman, L., A. Anderson, K.O. Nilsson, B. Lind, T. Kjellström, L. Friberg. "Cadmium Uptake by Wheat from Sewage Sludge Used as a Plant Nutrient Source." Arch. Environ. Health 27: 45-47. (1973)

61. Lowrance, W.W. "Of Acceptable Risk." William Kaufmann, Inc., Los Altos, CA. (1976)

62. Lynam, B.T., B. Sosewitz, and T.D. Hinesly. "Liquid Fertilizer to Reclaim Land and Produce Crops." Water Res. 6: 545-549. (1972)

63. McGauhey, P.H., R.B. Krone, and J.H. Winneberger. "Soil Mantle as a Wastewater Treatment System." SERL Report No. 66-7, University of California, Berkeley. (1966)

64. Metcalf, T.G. and W.C. Stiles. "The Accumulation of Enteric Viruses by the Oyster, Crassostrea virginica." J. Infectious Diseases 115: 68-78. (1965)

65. Miller, R.H. "The Soil as a Biological Filter," in Recycling Treated Municipal Wastewater and Sludge through Forest and Cropland, W.E. Sopper and L.T. Kardos, Eds. (Pennsylvania State University Press, University Park). (1973)

66. Moore, B.E., C.A. Turk, C.A. Sorber, and B.P. Sagik. "Recovery of Indigenous Viruses from Wastewater." Abstracts, ASM. Q 38. (1979)

67. Moore, B.E., B.P. Sagik, C.A. Sorber. "Land Application of Sludges: Minimizing the Impact of Viruses on Water Resources," in Risk Assessment and Health Effects

of Land Application of Municipal Wastewater and Sludges, B.P. Sagik and C.A. Sorber, Eds. (Center for Applied Research and Technology, The University of Texas at San Antonio). (1978)

68. Morris, G.K. and J.C. Feeley. "Yersinia enterocolitica: A Review of Its Role in Food Hygiene." Bulletin of the World Health Organization 54: 79-85. (1976)

69. Murphey, W.K., R.L. Brisbin, W.J. Young, and B.E. Cutter. "Anatomical and Physical Properties of Red Oak and Red Pine Irrigated with Municipal Wastewater," in Recycling Treated Municipal Wastewater and Sludge through Forest and Cropland, W.E. Sopper and L.T. Kardos, Eds. (Pennsylvania State University Press, University Park). (1973)

70. Packer, R.A. "Bovine Mastitis Caused by Pseudomonas aeruginosa." J. American Veterinary Medicine Assn. 170: 1166. (1977)

71. Parizek, R.P., L.T. Kardos, W.E. Sopper, E.A. Myers, D.E. Davis, M.A. Farrell, and J.B. Nesbitt. "Wastewater Renovation and Conservation." Pennsylvania State University, University Park. (1967)

72. Pipes, W.O. "Workshop Proceedings Water Quality and Health Significance of Bacterial Indicators of Pollution." Drexel University and National Science Foundation. (1978)

73. Reynolds, Reginald. "Cleanliness and Godliness." Doubleday and Company, Inc., Garden City, New York. (1946)

74. Ramirez, A. and J.F. Malina, Jr. "Chemicals Disinfect Sludge." Water and Sewage Works 127: 52-54. (1980)

75. Rudolfs, W., L. Falk, and R.A. Ragotzkie. "Contamination of Vegetables Grown on Polluted Soil. I. Bacterial Contamination." Sew. Works J. 23: 253. (1951a)

76. Rudolfs, W., L. Falk, and R.A. Ragotzkie. "Contamination of Vegetables Grown on Polluted Soil. II. Field and Laboratory Studies on Endamoeba cysts." Sew. Works J. 23: 498. (1951b)

77. Rudolfs, W., L. Falk, and R.A. Ragotzkie. "Contamination of Vegetables Grown in Polluted Soil. III. Field Studies on Ascaris eggs." Sew. Works J. 23: 656. (1951c)

78. Saari, T.N. and G.P. Jansen. "Waterborne Yersinia enterocolitica in the Midwest United States." Contributions to Microbiology and Immunology 5: 185-196. (1979)

79. Schiemann, D.A. "Association of Yersinia enterocolitica with the Manufacture of Cheese and Occurrence in Pasteurized Milk." Applied and Environmental Microbiology 36: 274-277. (1978)

80. Schiemann, D.A. "Enrichment Methods for Recovery of Yersinia enterocolitica from Foods and Raw Milk." Contributions to Microbiology and Immunology 5: 212-227. (1979)

81. Smith, B.P., F. Habasha, M. Reina-Guerra, A.J. Hardy. "Bovine Salmonellosis: Experimental Production and Characterization of the Disease in Calves, Using Oral Challenge with Salmonella typhimurium." American J. Veterinary Research 40: 1510-1513. (1979)

82. Smith, G.S., H.E. Kiesling, V.M. Cadle, C. Staples, L.B. Bruce, and H.D. Sivinski. "Recycling Sewage Solids as Feedstuffs for Livestock," in Sludge Management, Disposal, and Utilization, Information Transfer, Inc., Rockville, Maryland. pp. 119-127. (1977)

83. Sopper, W.E. "Revegetation of Strip Mine Spoil Banks through Irrigation with Municipal Sewage Effluent and Sludge." Compost Sci. 11(6): 6-11. (1970)

84. Sopper, W.E. and L.T. Kardos. "Vegetation Responses to Irrigation with Municipal Wastewater," in Recycling Treated Municipal Wastewater and Sludge through Forest and Cropland, W.E. Sopper and L.T. Kardos, Eds. (Pennsylvania State University Press, University Park). (1973)

85. Sorber, C.A. and K.J. Guter. "Health and Hygiene Aspects of Spray Irrigation," in Wastewater Management by Disposal on the Land, Cold Regions Research and Engineering Laboratory, Special Report 171, S. Reed, Ed. (Hanover, New Hampshire). (1972)

86. Sorber, C.A., S.A. Schaub, and H.T. Bausam. "An Assessment of a Potential Virus Hazard Associated with Spray Irrigation of Domestic Wastewaters," in Virus Survival in Water and Wastewater Systems, J.F. Malina, Jr., and B.P. Sagik, Eds. (Center for Research in Water Resources, The University of Texas, Austin). (1974)

44   HEALTH RISKS OF SPECIFIC AGENTS

87. Spite, G.T., D.F. Brown, and R.M. Tweat. "Isolation of an Enteropathogenic, Kanagana - Positive Strain of Vibrio parahae olyticus from Seafood Implicated in Acute Gastroenteritis." *Applied and Environmental Microbiology* 35: 1226-1227. (1978)

88. Sproul, Otis. "The Efficiency of Wastewater Unit Processes in Risk Reduction," in *Risk Assessment and Health Effects of Land Application of Municipal Wastewater and Sludges*, B.P. Sagik and C.A. Sorber, Eds. (Center for Applied Research and Technology, The University of Texas at San Antonio). (1978)

89. Steadham, J.E. "High-Catalase Strains of *Mycobacterium kansaii* Isolated from Water in Texas." *J. Clinical Microbiology*. In Press. (1980)

90. Steinmann, J. "Detection and Persistence of Human-Pathogenic Viruses in Chemical Sludges." *Zbl. Bakt. Hyg., I. Abt. Orig. B* 167: 470-477. (1978)

91. Subrahmanyan, T.P. "Persistence of Enteroviruses in Sewage Sludge." *Bulletin of the World Health Organization* 55: 431-434. (1977)

92. Tamminga, S.K., R.R. Beumer and E.H. Kampelmacher. "The Hygienic Quality of Vegetables Grown in or Imported into the Netherlands: A Tentative Survey." *J. Hyg., Camb.* 80: 143-154. (1978)

93. Taylor, K.C. "Persistent *Salmonella saintpaul* Excretion in a Small Dairy Herd." *Veterinary Record* 105: 35-36. (1979)

94. Thomas, R.E. "The Soil as a Physical Filter," in *Recycling Treated Municipal Wastewater and Sludge through Forest and Cropland*, W.E. Sopper and L.T. Kardos, Eds. (Pennsylvania State University Press). (1973)

95. Thunegard, E. "On the Persistence of Bacteria in Manure." *Acta. Vet. Scand., Suppl.* 56: 1-86. (1975)

96. Vassiliadis, P., A. Kalandidi, F. Zirouchaki, J. Papadakis, and D. Trichoponlas. "Isolement de Salmonellas à Partir de Saucisses de Porc en Utilisant un Nouveau Procédé D' Enrichissement (R10/43°)." *Réc. Méd. Vet.* 153: 489-494. (1977)

97. Vassiliadis, P., G. Papoutsakis, D. Avramidis, D. Trichopoulis, J. Papadakis. "Recherche de Salmonella sur des Carcasses de Poulets à Athènes en 1976." Archives de l' Institut Pasteur Hellénique XXII: 23-28. (1976)

98. Vassiliadis, P., D. Trichopoulos, E. Pateraki, and N. Papaiconomou. "Isolation of Salmonella from Minced Meat by the Use of a New Procedure of Enrichment." Zbl. Bakt. Hyg. I. Abt. Orig. B 166: 81-86. (1978)

99. Warner, D. and J.S. Dajani. "Water and Sewage Development in Rural America." C.C. Heath and Company, Lexington, Massachusetts. (1975)

100. Wauters, G. "Carriage of Yersinia enterocolitica Serotype 3 by Pigs as a Source of Human Infection." Contributions to Microbiology and Immunology 5: 249-252. (1979)

101. Way, J.T. "On the Power of Soils to Absorb Manure." J.R. Agric. Soc. Engl. 11: 313-379. (1850)

102. Wellings, F.M., A.L. Lewis, C.W. Mountain, and L.V. Pierce. "Demonstration of Virus in Groundwater after Effluent Discharge onto Soil." Appl. Environ. Microbiol. 29: 751. (1975)

103. Wellings, F.M., A.L. Lewis, and C.W. Mountain. "Demonstration of Solids Associated Virus in Wastewater and Sludge." Appl. Environ. Microbiol. 31: 354. (1976)

104. Winkelstein, W., Jr. and F.E. French. "Basic Readings in Epidemiology, 3rd Edition." MSS Educational Publishing Co., Inc., New York. (1972)

105. Winslow, C.-E.A. "Man and Epidemics." Princeton University Press, Princeton, New Jersey. (1952)

106. Wright, L. "Clean and Decent." Viking Press, New York. (1960)

107. Zenz, D.R., J.R. Peterson, D.L. Brooman, and C. Lue-Hing. "Environmental Impacts of Land Application of Sludge." J. Water Poll. Control Fed. 48: 2332. (1976)

CHAPTER 2

AGENTS OF HEALTH SIGNIFICANCE: PARASITES

<u>M. D. Little</u>. Department of Tropical Medicine, School of Public Health and Tropical Medicine, Tulane University, New Orleans, LA.

INTRODUCTION

The purpose of this presentation is to review the parasites that are known to occur in domestic waste sludges in the United States and, in regard to these parasites, to evaluate possible health risks to domestic animals and man that might result from the application of sludges to land or from the feeding of sludges directly to animals. Emphasis will be placed on how the basic life cycle pattern of a parasite relates to the possibility that it may be transmitted to animals or man through sludge.

Since this presentation is directed primarily to persons who are not microbiologists or parasitologists, it is perhaps necessary to indicate what types of organisms are usually included under the term "parasite" as it is used here. A parasite is often defined as an organism that lives at the expense of another organism. Stated in another way, a parasite is an organism that obtains its nutrition from the tissues or body fluids of another organism that is called the host [1]. By these definitions viruses and most pathogenic bacteria and fungi must be considered to be true parasites, and indeed they are. However, these organisms have traditionally been treated as separate groups, as they are in this symposium. The kinds of organisms that are generally included under the term parasites are the protozoa, the helminths, and the arthropods that infect or infest animals and plants. These are widely divergent groups. The protozoa are single-celled organisms that include the amoebae, the flagellates, and the sporozoa among others. The helminths include the nematodes, or roundworms; the trematodes, or flukes; and the cestodes, or tapeworms. Arthropod parasites include ticks, mites, fleas, lice, etc. While these groups contain large numbers of species, relatively few of them are of interest to us for the purpose of this discussion since we will only be concerned with those parasites that might possibly be transmitted to domestic animals or man through the use of sludges. This includes primarily those parasites that have stages in their life cycle that are adapted to long survival in the outside environment. Such stages must be relatively resistant to a wide range of environmental conditions. These include the eggs of helminths and the cysts of some protozoa. It is important to point out that these stages are microscopic in size; usually, helminth eggs are less than 100 μm in diameter and protozoan cysts are less than 30 μm. With few exceptions, these eggs and cysts reach the external environment by being passed in the feces of their host. Parasites that have stages passing in the feces of their hosts are mostly those that live in the gastrointestinal tract of the host or in organs such as the liver and lungs that communicate directly with it. Even in a relatively light infection literally millions of these eggs and cysts may be passed in the feces daily.

PARASITES IN SLUDGES

The presence of resistant stages of parasites in sludges in this country has been studied by relatively few workers [2-6]. In most of these studies, sludge samples from only one, or at most a few, waste treatment plants were examined. In 1977, our research team at Tulane University received a grant from the Environmental Protection Agency for a comprehensive study of domestic waste sludges in the southern part of the U.S. One purpose of this study was to determine the types and levels of parasites present in these sludges. Twenty-seven municipal waste treatment plants in five southern states, Texas, Louisiana, Mississippi, Alabama and Florida, were selected for study on the basis of their size, method of waste treatment, geographic location and type of population served. Primary and/or secondary and treated sludge samples were collected every three months from each of these plants over a period of one year. The parasites found in the samples from these plants are listed in Table I [7]. These results are similar to what other workers have found in sludges from other parts of the country, except that the numbers present in these sludges were perhaps higher and more parasites of animals were detected.

It can be seen that the eggs and cysts of numerous parasites were found and they came from a variety of host animals. It should be mentioned that none of the stages of protozoa found in any of the treated sludges appeared to be viable.

Table I.   Parasites Found in Domestic Waste Sludges in Southern U.S. (27 Treatment Plants Examined) [7]

Parasite	Host
Nematode Eggs	
Ascaris lumbricoides	Man
Ascaris suum	Pig
Toxocara canis	Dog
Toxocara cati	Cat
Toxascaris leonina	Dog, Cat
Trichuris trichiura	Man
Trichuris vulpis	Dog
Trichomosomoides-like	Rats
Capillaria, 3 species	Rats, Birds, Wild Mammals (?)
Cruzia-like	Opossum
Tapeworm Eggs	
Taenia sp.	Man, Dog, Cat (?)
Hymenolepis nana	Man, Rodent
Hymenolepis diminuta	Rat
Hymenolepis sp.	Birds (?)
Acanthocephala Eggs	
Macracanthorhynchus (?)	Pig
Protozoan Cysts	
Entamoeba coli-like	Man, Other ?
Giardia sp.	Man, Dog, Cat, Wild Mammals(?)
Coccidia (oocysts)	Birds, Dog, Cat, Other Mammals

From nearly all locations, the most commonly found stages were the eggs of Ascaris, Toxocara and Trichuris. The findings of eggs of Ascaris, Toxocara and Trichuris trichiura in sludges from the 27 plants are summarized in Table II.

Since the eggs of A. lumbricoides from humans and A. suum from pigs are indistinguishable, the relative occurrence of the individual species could not be determined. Almost all of the Toxocara eggs found were those of T. canis, the dog roundworm, but some were those of T. cati, the cat roundworm. It can be seen in Table II that viable eggs of Ascaris and Toxocara were found in one or more samples of some type of sludge from each of the 27 plants and that T. trichiura eggs, either viable or dead, were found in some sludge samples from all but one plant. Viable eggs of Ascaris and Toxocara were found in one or more samples of treated sludge from 26

Table II. Number of Southern Waste Treatment Plants in Which Eggs of <u>Ascaris</u>, <u>Toxocara</u> and <u>Trichuris trichiura</u> Were Found At Least Once During Year (27 Plants Studied) [7]

	Ascaris	Toxocara	Trichuris trichiura
Viable Eggs in Treated Sludges	26	23	12
Viable Eggs in Any Sludge Sample	27	27	15
Viable or Non-Viable Eggs in Any Sludge Sample	27	27	26

and 23 plants, respectively. Viable T. trichiura eggs were found in one or more samples of treated sludge from 12 plants and in one or more samples of any type sludge from 15 plants.

Helminth eggs, especially those of Ascaris, were present in relatively high numbers in some sludge samples. For example, in one drying bed sample from a plant receiving wastes from an abattoir that processed swine, over 200,000 viable Ascaris eggs/kg dry wt were recovered. These were undoubtedly mostly A. suum eggs. This represents about 40 viable Ascaris eggs/gm wet wt of the drying bed sludge. In another plant located in a small rural community, about 38,000 viable Ascaris eggs and 47,000 viable T. trichiura eggs/kg dry wt of drying bed sludge were found. This represents about 20 eggs of Ascaris and 25 eggs of Trichuris/gm wet wt of the sludge. There was no evidence of abattoir wastes entering this plant. The quantities of these eggs in sludges of other plants were usually much lower. The average number of viable and/or dead egg/kg dry wt found in samples of treated sludge from all plants was about 9000 for Ascaris, 500 for Toxocara and 1600 for T. trichiura. When only the samples of treated sludge in which eggs were found are considered, the average number of egg/kg dry wt recovered was about 13,000 for Ascaris, 800 for Toxocara and 3400 for T. trichiura.

While it is obvious how the resistant stages of human parasites enter domestic wastes, the manner by which the eggs and cysts of parasites of animals reach domestic sludges is not entirely known. It is known that some pet owners dispose of the feces of the cats and dogs by flushing them down the toilet. However, it is also likely that feces of dogs, cats and other animals that have been deposited on yards and streets are carried into the sewage system by way of surface water drainage. As indicated earlier, a potentially important source of parasites in sludge is the wastes from abattoirs and meat or poultry processing plants that pass directly into the municipal sewage system without prior treatment. The parasites of wild animals such as birds and rodents that are found in sludge probably come from animals living in or near the sewage system or the treatment plant.

Of the parasites that have been found in domestic sludges it is apparent that some more than others are likely to be transmitted through sludges to domestic animals and man. Based on our knowledge of their life cycles it would seem that those listed in Table III have the greatest potential for transmission through sludge.

Table III. Parasites of Veterinary and Medical Importance That Might Be Transmitted through Domestic Sludges.

Parasite	Stage	Definitive Host	Intermediate Host
Ascaris lumbricoides	Egg	Man	"Soil"
Ascaris suum	Egg	Pig	"Soil"
Toxocara canis	Egg	Dog	"Soil"
Toxocara cati	Egg	Cat	"Soil"
Trichuris trichiura	Egg	Man	"Soil"
Trichuris suis	Egg	Pig	"Soil"
Trichuris vulpis	Egg	Dog	"Soil"
Taenia saginata	Egg	Man	Cow
Taenia solium	Egg	Man	Pig (Man)
Echinococcus granulosus	Egg	Dog	Sheep, pig etc. (Man)
Toxoplasma gondii	Oocysts	Cat	"Soil" Mammals (Man)

The first seven parasites in this list, i.e., the species of Ascaris, Toxocara and Trichuris, are intestinal nematodes that belong to a group that is frequently referred to as the soil-transmitted parasites. In a sense, soil serves as an intermediate host for these parasites because their eggs must undergo a period of development in the external environment, usually the soil, before becoming infective for the next host. The length of this period of development varies depending upon the worm species and the environmental conditions; but it is usually a minimum of two to four weeks. The infective larva develops within the shell of egg and does not hatch from it until ingested by a suitable host. The eggs of these worms are resistant to a wide range of chemical and physical conditions. A Russian investigator has reported that some Ascaris eggs were still infective to animals after remaining in the soil for 15 years [8]. In a study carried out in Germany, Ascaris eggs were reported to remain infective in garden soil for 5 to 7 years [9]. Trichuris eggs have also been reported to survive for several years in the soil [10]. While the eggs of Ascaris, Trichuris and Toxocara are capable of surviving for several years in the soil when environmental conditions are ideal they undoubtedly survive for much shorter periods in most instances. The survival of these eggs in the environment depends on many factors including soil type, moisture, temperature, exposure to direct sunlight, etc. It should be mentioned that dewatered sludges, especially drying bed sludges, have characteristics similar to those of soil and in some circumstances may provide an ideal medium for these eggs to develop to the infective stage.

Ascaris lumbricoides and A. suum are the large roundworms of man and pigs, respectively. In their morphologic features and their life cycles, they are nearly identical. It has been demonstrated experimentally that it is possible to infect pigs with the human ascarid and to infect humans with the pig ascarid, but neither worm develops as well in the opposite host as it does in its own. However, both worms are capable of undergoing the initial tissue migration phase of their development in a wide variety of mammals. After infective eggs are ingested by these animals the larvae are able to hatch, enter the portal circulation and migrate through the liver and lungs just as they would in their normal definitive host. After completing the lung migration in an unnatural host the larvae are usually unable to establish themselves in the intestine and complete their development to the adult stage. Nevertheless, their migration through the liver and lungs may produce disease in these animals. This has been demonstrated in experimental infections in lambs [11] and in numerous cases of naturally acquired infections in cattle [10,12]. In helminthic infections, it is generally true that the more worms that are present the more it is likely that disease will be produced. With few exceptions helminths are unable to multiply in the body of the host; consequently, the number of worms developing in a host is directly related to the number of infective stages entering the body. However, repeated exposure to low numbers of infective stages of helminths may also result in the production of disease. In such cases, the host becomes sensitized and subsequent exposure to even a few larvae may result in a highly altered immune response. The so called "milk spots" in livers of pigs are manifestations of the pig's immune response to repeated exposure to infective eggs of Ascaris [13].

Ascaris pneumonitis (Loeffler's Syndrome) in humans may result from the ingestion of only a few eggs by a person sensitized by one or more previous Ascaris infections [14]. A similar type of pneumonitis due to Ascaris has also been reported in cattle [10,13].

The life cycles of Toxocara canis and T. cati, the roundworms of dogs and cats, respectively, are similar in many ways to those of the two Ascaris species. When the infective eggs of these worms are ingested by the definitive host the larvae must also migrate through the lungs before they are able to develop to adult worms in the host's intestine. However, these Toxocara parasites are additionally able to utilize several other routes for the transmission of infection. The mother may pass infective larvae transplacentally to the offspring before birth or through the milk after birth. Infections may also be acquired by eating paratenic hosts. When the infective eggs of Toxocara are ingested by a wide variety of mammals, the larvae can migrate into the tissues and persist there for a long period of time. They have been shown to persist there for at least 10 years in the tissues of a monkey [15]. Toxocara larvae behave in humans as they do in mammalian paratenic hosts. If a child eats dirt containing infective Toxocara canis eggs, the larvae enter the tissues and a condition known as visceral larva migrans (VLM) may result [16]. Severe cases of VLM are primarily due to the ingestion of large numbers of infective eggs, but it is also possible that one of the more serious complications of these infections, disease of the eye, may result from the ingestion of relatively few eggs. In numerous human cases, a lesion in the retina due to the presence of a T. canis larvae has resulted in the surgical removal of the eye [17]. Since Toxocara larvae may accumulate and persist in the tissues of animals, human consumption of raw or undercooked tissues from these animals could result in human infection [15].

The eggs of the three tapeworms included in Table III, i.e., Taenia saginata, Taenia solium, and Echinococcus granulosus, are infective for their intermediate host at the time they are discharged from the gravid segments of the worm. Consequently, unlike the eggs of the soil-transmitted

nematodes, when these tapeworm eggs are present in raw
sewage they are already infective for the next host. Cattle,
which serve as the intermediate hosts of T. saginata, become
infected by ingesting eggs while grazing on pasture. The
larvae which hatch from eggs in the duodenum migrate via the
blood stream to the skeletal muscles and heart where they
develop to the cysticercus stage in about 2 months. The
presence of cysticerci in cattle can, of course, result in
the carcasses being condemned and in economic loss to the
owner. Humans serve as host for the adult stage and become
infected by eating raw or undercooked beef containing
cysticerci. Human T. saginata infections are relatively rare
in this country, but infections are occasionally seen in
immigrants and travelers returning from endemic areas in
foreign countries.

Taenia solium is potentially more dangerous to humans
than is T. saginata since the egg stage of T. solium is infective to man as well as to the pig, the normal intermediate
host of the parasite. Human ingestion of these eggs can
result in a serious, even fatal, case of cysticercosis.
Human infection with the adult worm results from eating raw
or undercooked pork that contains cysticerci. While there
are no recent reports of autochthonous cases of human
cysticercosis or adult T. solium infections in this country,
such infections are occasionally seen in immigrants or
visitors from foreign countries, including those from Central
and South America.

The distribution in this country of Echinococcus
granulosus, the tapeworm that causes hydatid disease in animals and man, appears to be restricted to the sheep raising
areas of the western states [18]. Dogs are hosts for the
adult tapeworms and sheep, primarily, but other domestic
animals as well, may serve as the intermediate host. The
hydatid cyst is the larval stage that develops in the intermediate host and it may grow to a relatively large size in
the liver, lung or other organ. Infections in domestic animals are of importance for two reasons. First, there is
the economic loss that can result if the organs of the animals are condemned, and second, there is the role that these
animals play as reservoirs of infections for dogs which may
serve as a source of subsequent transmission of infection
to humans. An important point to remember is that a large
hydatid cyst can result from the ingestion of a single egg
of this parasite.

It should be mentioned that a related tapeworm,
Echinococcus multilocularis, might also be transmitted
through the land use of sludges. However, the possibilities
of this occurring are slight since the worm is primarily a
parasite of wild canines and its distribution in North
America is apparently restricted to North Dakota, South

Dakota, Iowa, Minnesota, Montana, Wyoming, Alaska and Canada. Wild rodents serve as intermediate host for the parasites and farm dogs and cats may become infected with the adult stages by preying on these rodents. The parasite produces alveolar hydatid disease in humans, but only one human case has been reported from the U.S., a woman in Minnesota [19].

The oocysts of Toxoplasma are not infective when passed in the feces of the parasite's definitive host, the cat, but require a period of a few days to reach infectivity in the soil. These oocysts are reported to be relatively resistant to a wide range of environmental conditions. Toxoplasma gondii oocysts have been reported to survive for at least 18 months in the soil under natural conditions [20]. Whether the oocysts of T. gondii are able to survive passage through various waste treatment processes is not known. Toxoplasma infections in domestic animals result from the animals ingesting oocysts from the soil or in the feed. Infections in these animals are apparently common but asymptomatic in most cases. However, cases of acute toxoplasmosis in cattle, sheep and swine have been reported [21]. Humans may become infected by eating improperly cooked meat containing the intracellular forms of these parasites, or by ingesting the infective oocysts in food or water contaminated by cats.

HEALTH RISKS

An evaluation of the health risks to domestic animals and man associated with the land application of municipal sludges is very difficult since many factors must be considered. These include the types and numbers of viable stages of parasites present in the sludge that is applied, the amount of sludge applied, the manner in which it is applied, characteristics of the soil, topography of the land, climatic conditions including rainfall and temperature, and the subsequent use of the land. The possibility of transmission to animals is greater, of course, if they graze directly on land to which sludge has been applied than if they are fed hay, corn, grain or other crops that have been grown on such land. Because of the number of factors involved, it is nearly impossible to make general conclusions regarding the health risks to animals resulting from the land application of domestic sludges. It would appear that each situation must be evaluated separately.

There are relatively few published reports of domestic animals acquiring parasitic infections as a result of sludge being applied to the land, and these are mostly from other countries. Investigators in Australia recently reported a relatively high prevalence of cysticerci of Taenia saginata in cattle reared on pastures irrigated with sewage effluent [22]. Apparently treated sludges were not applied to any of these pastures. Cysticercosis due to T. saginata was also recently reported in cattle raised on pastures in southern Virginia that had received applications of domestic sludges [23]. However, there seems to be a possibility that at least some of the sludges applied to these pastures were untreated.

The paucity of reports of domestic animals acquiring parasitic infections as a direct result of the land application of sludges would seem to indicate that the use of sludges in this way does not present any major health risks to these animals. However, it should be mentioned that light infections in domestic animals are difficult to detect and are likely to go unreported unless special efforts are made to look for them. While light infections would not ordinarily endanger the health of these animals, they could serve as sources for the dissemination of infections to other animals or to people.

One would have to conclude that in most circumstances domestic sludges should not be fed directly to animals without first having been treated to kill parasites and other infectious agents that might be present. Treatment processes in which temperatures of 55°C or above are reached uniformly throughout the sludge for at least two hours are effective in killing Ascaris eggs [24], but higher temperatures may be necessary to kill certain pathogenic organisms [25]. Irradiation, although not widely used, has also been demonstrated to be effective in inactivating parasites in sludges [26].

REFERENCES

1. Sprent, J.F.A. Parasitism (St. Lucia, Queensland, Australia: University of Queensland Press, 1963), p.12.
2. Hays, D.B. "Is There a Potential for Parasitic Disease Transmission from Land Application of Sewage Effluents and Sludges?" J. Environ. Hlth. 39: 424-426 (1977).
3. Fox, J.C. and P.R. Fitzgerald, "Parasitic Organisms Present in Sewage Systems of a Large Metropolitan Sewage District," Program and Abstracts, 51st Annual Meeting, American Society of Parasitologists, San Antonio, TX, (1976), pp. 28-29.
4. Fox, J.C. and P.R. Fitzgerald, "Parasite Content of Municipal Wastes from the Chicago area," Program and Abstracts, 52nd Annual Meeting, American Society of Parasitologists, Las Vegas, NV, (1977), pp. 68-69.
5. Jackson, G.L., J.W. Bier and R.A. Rude. "Recycling of Refuse into the Food Chain: The Parasite Problem", in

Proceedings of the Conference on Risk Assessment and Health Effects of Land Applications of Municipal Wastewater and Sludges, B.P. Sagik and C.A. Sorber, Eds. (San Antonio, TX: Center for Applied Research and Technology, The University of Texas at San Antonio, 1978), pp. 116-127.

6. Theis, J.H., V. Bolton and D.R. Storm. "Helminth Ova in Soil and Sludge from Twelve U.S. Urban Areas," J. Water Pollut. Contr. Fed. 50: 2485-2493 (1978).

7. Reimers, R.S., M.D. Little, D.B. Leftwich, D.D. Bowman, A.J. Englande and R.F. Wilkinson. "Investigation of Parasites in Southern Sludges and Disinfection by Standard Sludge Treatment Processes," Draft Report (Grant R805107) to Municipal Environmental Research Laboratory and Health Effects Research Laboratory, U.S. Environmental Protection Agency, Cincinnati, OH (1979).

8. Krasnonos, L.I. "Many-year Viability of Ascarid Eggs (Ascaris lumbricoides) in Soil of Samarkand," Medskaya Parazit. 47: 103-106 (1978). Abstract in Trop. Dis. Bull. 75: 991-992 (1978).

9. Müller, G. "Untersuchungen über die Lebensdauer von Askarideneirer in Gartenerde," Zbl. Bakteriol. 159: 377-379 (1953).

10. Levine, N.D. Nematode Parasites of Domestic Animals and of Man, (Minneapolis: Burgess Publishing Co., 1968), pp. 327, 526-539.

11. Fitzgerald, P.F. "The Pathogenesis of Ascaris lumbricoides var. suum in Lambs," Am. J. Vet. Res. 23: 731-736 (1962).

12. Morrow, D.A. "Pneumonia in Cattle Due to Migrating Ascaris lumbricoides Larvae," J. Am. Vet. Med. Ass. 153: 184-189 (1968).

13. Soulsby, E.J.L. Textbook of Veterinary Clinical Parasitology, Vol. 1, Helminths., (Philadelphia: F.A. Davis Company, 1965) pp. 184-208, 748.

14. Gelpi, A.P. and A. Mustafa. "Ascaris Pneumonia," Am. J. Med. 44: 377-389 (1968).

15. Beaver, P.C. "Zoonoses, with Particular Reference to Parasites of Veterinary Importance," in Biology of Parasites, E.J.L. Soulsby, Ed. (New York: Academic Press, Inc., 1966) pp. 215-227.

16. Beaver, P.C. "Toxocarosis (Visceral Larva Migrans) in Relation to Tropical Eosinophilia," Bull. Soc. Path. Exot. 55: 555-576 (1962).

17. Wilkinson, C.P. and R.B. Welch. "Intraocular Toxocara," Am. J. Ophthalmol. 71: 921-930 (1971).

18. Pappaioanou, M., C.W. Schwabe and D.M. Sard. "An Evolving Pattern of Human Hydatid Disease Transmission in the United States," Am. J. Trop. Med. Hyg. 26: 732-

742 (1977).
19. Gamble, W.G., M. Segal, P.M. Schantz and R.L. Rausch. "Alveolar Hydatid Disease in Minnesota," J. Am. Med. Ass. 241: 904-907 (1979).
20. Frenkel, J.K., A. Ring and M. Chinchilla. "Soil Survival of Toxoplasma Oocysts in Kansas and Costa Rica," Am. J. Trop. Med. Hyg. 24: 439-443 (1975).
21. Levine, N.D. Protozoan Parasites of Domestic Animals and Man (Minneapolis: Burgess Publishing Co., 1961), pp. 325-337.
22. Rickard, M.D. and A.J. Adolph. "The Prevalence of Cysticerci of Taenia saginata in Cattle Reared on Sewage-irrigated Pasture," Med. J. Australia, 1: 525-527 (1977).
23. Hammerberg, B., G.A. MacInnis, T. Hyler. "Taenia saginata Cysticerci in Grazing Steers in Virginia," J. Am. Vet. Med. Ass. 173: 1462-1464 (1978).
24. Brandon, J.R. "Parasites in Soil/Sludge Systems," in Proceedings of Fifth National Conference on Acceptable Sludge Disposal Techniques (Rockville, MD: Information Transfer, Inc., 1978), pp. 130-133.
25. Stern, G. "Pasteurization of Liquid Digested Sludges," in Proceedings of the National Conference on Municipal Sludge Management (Pittsburgh, PA, June, 1974), pp. 163-169.
26. Stern, G.S. and J.B. Farrell. "Sludge Disinfection Techniques," in Proceedings of the 1977 National Conference on Composting of Municipal Residue and Sludges (Silver Springs, MD: Information Transfer, Inc., 1978), pp. 142-148.

CHAPTER 3

HEALTH RISKS ASSOCIATED WITH TOXIC METALS IN
MUNICIPAL SLUDGE.

Rufus L. Chaney, Biological Waste Management and Organic
Resources Laboratory, USDA-SEA-AR, Beltsville, MD 20705.

INTRODUCTION

The potential for health effects in humans and domestic animals from toxic elements in municipal sludges utilized in agriculture depends on sludge, soil, plant, and animal characteristics. An individual may consume sludge, sludge-soil mixture, plant foliage, plant grain or fruit, or organ or skeletal muscle of a domestic animal thereby having great or negligible exposure. The sludge may have little or much industrial metal contribution. Soils can bind an element so strongly that plants do not absorb that element; alternatively, an element can be freely mobile even to plant grain and fruit tissues. Another important interaction results from other constituents in the sludge, soil, or animal diet which can affect absorption of a toxic element by an animal.

Risk assessment for potential health effects of sludge metals has progressed rapidly during the last decade [1]. This chapter attempts to summarize the present understanding. Much of this chapter focuses on Cd, an element which is mobile in the food chain, remains plant available for a long period after addition to soil [2], and which has caused health effects in humans from crops grown on Cd-contaminated soils [3-6].

TOXIC ELEMENTS IN SLUDGE

Sludges vary widely in composition. Table I shows the reported range and typical median levels based on numerous surveys of sludge composition [7, 8, 9, 10, 11, 12]. It is now clear that domestic use of water adds significant amounts of some toxic elements to sewage; however, domestic contribution of Cd is quite limited [13]. The discussion in this paper will consider both median quality and high-metal sludges.

Table I. Concentration of Trace Elements in Digested Sewage Sludges [7-12]

Element	Reported Range ppm	Typical Median ppm	"Maximum Domestic" ppm
As	1.1 - 230	10	--
B	4 - 1,000	33	--
Cd	1 - 3,410	15	25.
Co	1 - 260	10	--
Cu	84 -17,000	800	1,000.
Cr	10 -99,000	500	--
F	2 - 739	86	--
Fe, %	0.1 - 15.4	1.7	4
Hg	0.6 - 56.	6	10
Mn	32 - 9,870	260	--
Mo	1.2 - 40	10	--
Ni	2 - 5,300	80	200
Pb	13 -26,000	500	1,000
Se	1.7 - 17.2	5	--
Sn	40 - 700	150	--
V	15 - 400	36	--
Zn	101 -49,000	1,700	2,500.

TRANSFER OF METALS FROM SEWAGE SLUDGE TO THE FOOD-CHAIN

Fluid sewage sludge can be spray-applied to cropland and tilled into the soil, or to forage or pasture land where it can contact plants or remain on the soil surface. Dewatered or dried sludge, or sludge compost, can similarly be mixed with the soil or remain on the soil surface. These sludge management alternatives allow several routes for metals to enter the food-chain. Some routes involve direct consumption of sludge, while others involve plant absorption of metals from the sludge-soil mixture or sludge on the soil surface.

Adherence of Sludge to Forage Crops

During studies of land application of high copper pig manure, high Cu levels were found in harvested grasses due to adhering manure [14]. Because of this early observation, we began studies of the fate of spray-applied sludge. We found that once sludge dried on tall fescue forage it was not washed off by subsequent rainfall [14]; further growth of the tall fescue diluted the sludge percentage of the standing crop; Jones et al. [16] found that sludge could be washed off before it dried, and further, that the quantity of adhering sludge was approximately a linear function of the %-solids of the sprayed sludge. Sludge has adhered to all crops studied. Identification of potential sludge adherence is easily made since plant uptake or translocation to shoots for some metals (Cu, Pb, Cr, Fe) is exremely limited. Excessive forage levels of these elements indicates direct sludge contamination [15]. Many early publicatons on uptake of metals from surface applied sludges need to be reinterpreted based on these observations of sludge adherence [17, 18]. Studies with manure also indicate that organic wastes on the soil surface can be transferred into baled hay by hay making equipment [19].

Ingestion of Soil or Sludge on the Soil Surface.

Over a period of years scientists have established that grazing cattle, sheep, horses, etc., ingest soil as part of the grazing process. Most have measured Ti in forage and feces because Ti present in soil is not absorbed and translocated appreciably by plants. Mealy et al. [20] have characterized the role of weather and livestock management on the rate of soil ingestion. Some grasses are pulled from the soil with roots holding soil; soil comprised up to 20% of ingested diet when cattle grazed dryland crested wheatgrass [21].

Livestock have suffered Pb poisoning after grazing pastures established on soils naturally high in Pb or comprised of ore tailings [22, 23]. Similarly, Pb-contaminated organic sward thatch had to be removed before Pb poisoning of livestock could be corrected even though the stack emission Pb source had ceased emissions [24].

Sewage sludge or sludge compost can be ingested from the soil surface unless it is tilled into the soil. Decker et al. [25] found 6.5% (1977) and 2.0% (1978) compost in feces of cattle grazing sludge compost fertilized pastures even though the compost did not adhere to the tall fescue forage.

Another soil ingestion process of importance in assessing risk of sludge-borne heavy metals is soil consumption by children. Some children and adults deliberately consume soil in a practice called pica. If the soil is high in Pb (over 500-1,000 ppm), individuals may absorb excessive Pb [26, 27]. Others ingest soil and dust due to hand-to-mouth play activities and mouthing of toys, etc. [28, 29]. High Pb soils can be tracked into homes and cause high Pb housedust [30]. Several studies have shown that soil Pb contributes to Pb exposure of children [31, 32, 33]. Soil Pb, air Pb, housedust Pb, and paint Pb comprise a multiple source model.

Lead in soil was found to be nearly as bioavailable to rats as Pb acetate added to the same diet [34, 35]. Perhaps the clearest example of the possibility of health effects from inadvertent ingestion of high Pb housedust can be found in recent cases where the children of Pb industry workers experienced excessive blood Pb [36, 37, 38]. The only indentifiable source of Pb in most cases was in the industrial dusts carried into the house on the parent's clothing and shoes. A new National Research Council review modelled the evaluation of the importance of ingestion of soil Pb by children [39].

Soil ingestion can be an important process allowing entry of sludge-borne metals to the food chain when the metal is normally not absorbed by plants such that plant levels $\leq$ soil levels. When plant levels exceed soil levels, plant uptake becomes the more important process. Soil ingestion is a potential route for allowing excessive Pb, Fe, Cu, F, Mo, and other elements into the food chain.

Absorption of Toxic Metals From Soil and Translocation to Plant Shoots.

Plant absorption of toxic metals is affected by both soil and plant processes. The chapter by Sommers [40] in this book describes plant absorption of sludge-borne metals. In some of my earlier papers, I have also discussed this process [41, 42, 43]. The focus of the present discussion is soil and plant processes as limitations on toxic metal movement into the food-chain. The concepts summarized in this section have been developed over several decades as soil, plant, and animal scientists worked to understand natural animal poisoning incidents. Work by Underwood [44], Allaway [45], Bowen [46] and others [47, 48, 49, 50] evaluated plant uptake as compared to plant and animal tolerance of specific toxic elements. Models such as these have to include chemistry of metals in soils, plant uptake and translocation processes, and nutritional interactions influencing metal toxicity in animals.

Toxic Elements in Soils.

A metal can be insoluble in the soil, or be so strongly sorbed to the soil that plants absorb only small amounts of that metal. Lindsay [51] described the chemical equilibria of ions in soil, considering that crystalline inorganic phases can be the limit on chemical activity of an ion in soil solution. Activity of all metals is affected by soil pH, while redox status of the soil strongly influences only a few elements (theoretically, redox influences all equilibria; nearly all food crops require aerobic soils for growth, limiting the practical range of redox). Examples of "insoluble" elements include Pb, Hg, Cr^{3+}, F, Ag, Au, Ti, Sn, Si, and Zr.

Plant Absorption and Translocation of Toxic Elements.

Generally, plants absorb elements in proportion to their activity in the soil solution. Added chelating agents can increase the concentration of soluble metal and promote uptake by plants through increased rate of diffusion [43, 52, 53, 54]. After absorption by the root, a metal can be strongly sorbed on surfaces in the plant, or strongly chelated in the root cells, limiting that element's translocation to plant shoots. Tiffin [55] reviewed processes in microelement translocation, and this was recently updated [43]. This insolubility in the root or limited translocation to shoots keep several toxic elements (Fe, Pb, Hg, Al) from reaching excessive foliar levels in nearly all crops. Although Pb and Hg were listed above as insoluble in soil, healthy plant roots retain these elements even when they are supplied in solution. Because these soil and plant processes are so effective in protecting plants from these groups of elements, they do not cause phytotoxicity under conditions of municipal sludge utilization in agriculture.

Phytotoxicity as a Limit on Toxic Element Entry to Food-Chain.
Some elements (Zn, Cu, Ni, Co, Mn, As, B, and Cd) are known to be able to cause phytotoxicity if applied to cropland in excess [41, 42, 49, 56, 61].. Only Zn, Cu, and Ni are consistently increased in soil by utilization of "Domestic" sludge [42, 49]. In the absence of monitoring and source control programs, the potentially phytotoxic elements could also be important. If severe, obvious, yield reduction results from phytotoxicity, farmers generally are unable to use the injured crop and attempt to identify the cause. Thus, phytotoxicity becomes a natural limit on metal entry to the food-chain. Crops not harvested are not consumed by man or domestic animals, although they could affect wildlife [62, 63]. Both quantity and concentration of phytotoxic elements entering the food chain are limited by this natural process. If the levels of an element in a plant suffering phytotoxicity due to that element in soil are still low enough to be safe for the food-chain, then phytotoxicity has protected the food-chain.

"Soil-Plant Barrier" to Toxic Element Entry to Food-Chain.
During the last 30 years, the above concepts were developed by many researchers. Principal among them were Underwood [44], Allaway [45, 64], and Bowen [46]. These authors have not named this process, yet it is so important in considering soil-plant-animal relationships of toxic elements that I will call it the "Soil-Plant Barrier." By this I mean that either due to insolubility of an element in soil, or to its immobility in plant roots, or due to phytotoxicity limiting maximum plant shoot levels of the element to levels safe for animals, the food-chain is protected.

Unfortunately, the Soil-Plant Barrier does not protect animals from toxicities of all elements. The exceptions important to assessing risk from utilization of municipal sludge are Cd, Se, and Mo. Fortunately, Se and Mo have rarely been found at excessive levels in sludges; routine sludge monitoring would identify potentially excessive levels of these elements in sludge. Of course, soil or sludge ingestion completely circumvents the Soil-Plant Barrier.

The important exception is Cd [59, 65, 66, 67]. Soil Cd has caused health effects in humans through plant uptake.

Much of the rest of this chapter will be devoted to assessing risk to the food-chain of sludge applied Cd.

INTERACTIONS WITH DIETARY AND SLUDGE CONSTITUENTS LIMIT TOXICITY OF SLUDGE TOXIC ELEMENTS.

Consideration of the potential toxicity to animals from consumption of sludge, sludge-amended soil, or crops grown on sludge-amended soil is very complex. Animal species differ in tolerance of metals. Crops absorb different amounts and ratios of toxic metals from a soil. Uptake is affected by soil pH and other variables. Sludges differ in concentrations and ratios of metals. Further, individual potentially toxic elements interact with other dietary constituents.

As noted in the chapter by Kienholz [68], scientists expected to see some heavy metal toxicity in animals fed sludge. Because so few indications of toxicity have been observed, something may have reduced the toxicity of sludge-borne metals. This section reviews dietary metal interactions and evaluates potential toxicity of specific sludge-borne toxic metals.

Review of Interactions on Toxic Metal Bioavailability.

Perhaps the first observed dietary interaction on bioavailability of a microelement of practical significance was Mo-induced Cu deficiency of ruminants. The interaction involves dietary Cu, Mo, and S; sulfate is reduced to sulfide in the rumen and forms oxythiomolybdate; this forms an insoluble Cu compound, and leads to depletion of reserve liver Cu and if prolonged, clinical deficiency [69]. The overall animal Cu deficiency is confounded by Cu deficiency in the soil selecting for grasses which happen to be low in foliar Cu. The Cu in these grasses is less bioavailable if the grass is consumed fresh; drying the forage to make hay leads to increase in the portion of total Cu which is bioavailable. Further, soil consumption reduces Cu availability to sheep [70], perhaps due to soil Mo or Zn, but probably due to Cu sorption.

Later it was discovered that Zn, Cd, and Fe also interact with Cu to decrease Cu bioavailability to ruminants and monogastric animals [71 -78]. And, reciprocally, high dietary Cu interacts to reduce the absorption and potential toxicity of Zn, Fe, and Cd [79-81]. Many other metal interactions have been studied [44, 73, 82-86].

Thus, when one metal is so increased that the ratio of it to another is great enough to induce a deficiency of another, then animal performance declines and one observes a toxicity. However, "Domestic Sludge" (Table I) contains a mixture of potentially toxic metals. Consumption of "Domestic sludge" or "Domestic-sludge-amended" soil then represents a special case. Increased levels of dietary Zn (due to sludge) are balanced by increased levels of Cu and Fe. On the other hand, sludges contaminated by industrial discharges or chemicals added to facilitate sewage treatment or sludge processing may be unusually rich in one element (see ranges in Table I). These sludges high in one element can more easily cause health problems in animals.

Recent research has shown that interactions are especially important in understanding absorption of Cu, Cd, and Pb [76, 83, 85, 86]. These will be considered below under specific elements.

Considerations for Specific Elements.

In this section, I will summarize the concentrations of individual elements which can be reached in plants in relation to tolerance of that element in plants and tolerance of that element by animals. Interactions from other constituents in these plants will be considered as often the interaction produces or prevents a health effect in animals. Table II shows the element levels in feeds above which health effects on domestic animals could occur [83]. Unfortunately, the NRC committee considered increased levels of only one element at a time, and these tolerance levels may not be valid for sludge-fertilized crops or for ingestion of sludge.

Zinc phytotoxicity (ca. 25% yield reduction) occurs in most plants at about 500 ppm dry weight [48, 87, 88]. However, leafy vegetables such as lettuce and chard do not experience phytotoxicity in acid soils until foliar Zn is about 1500 ppm dry weight [89]. Zn toxicity in domestic animals occurs at 300 to 1,000 ppm (Table II) in diet. Zn toxicity in animals appears to result from Zn-induced Cu deficiency [79, 81, 90]; dietary Zn can also interact with dietary Fe, Mn, and Cd. Sheep were unusually sensitive to dietary Zn if feed Cu was marginal to deficient [90].

Sludge grown crops are seldom as high in Zn as 500 ppm; most forage crops are not Zn accumulators. The crops evaluated as the worst case for cattle and sheep would be enriched in Cu, not Cu-deficient. Cattle and sheep would not likely be continuously exposed to this worst case. If these animals consumed sludge, dietary Cu and Fe would be substantially increased along with Zn; only with high-Zn/normal-Cu sludges would one expect any possibility of Zn-induced Cu deficiency in domestic animals. The studies of Ott et al. [91] and Campbell and Mills [90] strongly support this conclusion. Because some wildlife would chronically consume Zn-phytotoxic plants as most of their diet, high Zn sludges could cause health effects in wildlife (e.g. voles, shrews).

Although humans could consume appreciable amounts of high Zn leafy vegetables (acid soil worst case), it is very

Table II. Maximum Tolerable Levels of Dietary Minerals for Domestic Animals [83][a]

Element	Cattle	Sheep	Swine
	---------ppm dry diet---------		
As, inorganic	50	50	50
B	150	(150)	(150)
Cd	0.5	0.5	0.5
Cr, oxide	(3000)	(3000)	(3000)
Co	10	10	10
Cu	100	25	250
Fe	1000	500	3000
Pb	30	30	30
Mn	1000	1000	400
Mo	10	10	20
Ni	50	(50)	(100)
Se	(2)	(2)	(2)
V	50	50	(10)
Zn	500	300	1000

[a] Continuous long-term feeding of minerals at the maximum tolerable levels may cause adverse effects. Levels in parentheses were derived by interspecific extrapolation.

unlikely that Zn toxicity would occur in even the worst case. The leafy vegetable would be high in Cu, counteracting the worst case. Individuals are unable to consume high Zn leafy vegetables as the bulk of their diet year around. Many individuals presently consume low or deficient amounts of Zn [92]; increased crop Zn might even be beneficial.

In summary, use of sludge is extremely unlikely to cause negative health effects in domestic animals or humans even under worst case conditions. Increased dietary Zn may be a benefit of sludge utilization.

Copper phytotoxicity occurs in most plants at about 25-40 ppm dry weight [42, 48, 93]. Copper toxicity to sensitive sheep and cattle occurs at 25-100 ppm in diet. Cu toxicity can be counteracted by high dietary Zn, Cd, Fe, or Mo [69, 71, 76, 77, 84, 90].

Crops with 25-40 ppm Cu could be grown only on strongly acid soils amended with Domestic Sludges. These crops would have high Zn and normal Fe. Copper toxicity to domestic animals from sludge-fertilized crops appears extremely unlikely. In the one case where Cu toxicity was expected (direct consumption of sludge or sludge-contaminated forages or sludge from the soil surface), the interactions of other dietary constituents with Cu was so pronounced that liver Cu was depleted rather than rising to toxic levels [25, 68].

High Cu sludges may reverse these interactions. It may be possible to have Cu toxicity expressed as induced-Mo deficiency if forages are already marginal or deficient in Mo and the sludge is very low in Mo. This seems most likely for wildlife.

Because monogastric animals tolerate even more Cu than ruminants, and because the level of Cu in grains and fruits is not increased appreciably by excess soil Cu, humans are at no risk from Cu even under the conditions of the worst case acid garden scenario.

Nickel phytotoxicity becomes apparent (interveinal chlorosis) in most plants at about 50-100 ppm dry weight in leaves [48, 57, 94]. This holds for grasses, legumes, and leafy vegetables. Ni toxicity in domestic animals occurs at 50-100 ppm of added soluble salt Ni [83]. Greater tolerance may occur in green forage or hay grown on domestic sludge fertilized land. Little study has been done with foods rich in intrinsic Ni. The study by Alexander et al. [95] showed no health effect and no bioaccumulation of Ni from sludge-fertilized soybeans (30 ppm Ni). When $NiCO_3$ was added to cattle diets, no toxicity was observed at 250 ppm Ni [96].

Based on these results, it seems very unlikely that Ni toxicity would occur in ruminants grazing worst case, acidic, Ni-phytotoxic, sludge fertilized pastures. While forage crops may include some Ni-accumulators [57], soil Ni will seldom reach excessive levels unless high Ni sludges are used at high cumulative rates. Grains and garden crops do not accumulate Ni to levels dangerous to monogastric animals. It may be possible for wildlife to be injured if high Ni sludges remain on the soil surface where they spend their lives. A few non-economic plant species accumulate Ni and or tolerate Ni [43, 97].

Lead in sewage sludge has never been observed to cause phytotoxicity, partially due to the luxury supply of phosphate from the sludge. Plants do not normally translocate high amounts of Pb to their shoots. Collard greens growing in a fertile urban garden soil containing 10,800 ppm Pb were only 9.4 ppm Pb dry weight. (Chaney et al, unpublished). On the other hand, high Pb has been observed in grasses growing on infertile, strongly acidic Pb-mine spoils [98, 99].

Lead toxicity to animals occurs at about 30 ppm in diet [83], and interacts strongly with dietary Fe, Ca, Zn, P, etc. [82, 85]. Thus, animals appear to be protected from excess Pb in crops grown on sludge fertilized cropland. A seasonal increase in foliar Pb occurs in forages growing during a mild winter [106]; however, this was not exaggerated in crops grown on soil enriched in Pb from sludge [101].

As noted previously, consumption of high Pb soil has caused Pb toxicity in domestic animals [22, 23, 24]. The Pb level is low enough in Domestic Sludges that sludge factors can counteract the toxicity of sludge Pb. Sludges over 1,000 ppm Pb, remaining on the soil surface, may be a risk for domestic animals or children. Sludge feeding trials showed little indication of continuing Pb accumulation by cattle [68].

The home garden case is difficult to evaluate, as sludge use is unlikely to contribute large amounts of Pb to the soil. When garden soils are very high in Pb, lettuce and some root crops can have increased Pb [102, 103]. In one study, we found 70 ppm Pb in a Pb-accumulator lettuce cultivar growing in pots of an unfertilized urban garden soil with over 5,000 ppm Pb: application of composted limed-raw-sludge and NPK fertilizer caused plant Pb to drop to below 10 ppm. Further, increased Pb in plants on unfertilized soil was not pronounced below 1,000 ppm Pb in soil. Kneip [104] and Kneip and Lowenberg [105] recently evaluated risks to children from Pb in crops grown in urban gardens; there was little risk of increase in children's dietary Pb due to consumption of urban garden grown foods; soil consumption was the important risk.

<u>Iron</u> phytotoxicity is not an agronomic problem except for rice where soil Fe is reduced to the ferrous form [43]. Leaves of most crops contain 50-300 ppm Fe (dry weight). Fertilization with sludge Fe does not cause Fe to rise above these normal levels, but may raise Fe in deficient plants from 30-40 ppm (chlorotic, Fe deficient leaves) to normal 50-300 ppm. Sludges may be a valuable Fe fertilizer for sorghum and other crops [106].

Plant Fe can reach unusually high levels due to interactions with other plant micronutrients [107]. Deficiencies of Cu [108], Mn, or Zn [109, 110] can cause foliar Fe to reach or exceed 1,000 ppm. When excessive Fe reaches leaves, it is stored in phytoferritin [111, 112].

The Fe levels shown above are for washed leaves. Soil dust on leaves may supply more total Fe than internal plant Fe. Trampled forages become "soiled" [20]. Spray-applied sludge sharply increases Fe in/on forages [15, 16]. Forage Fe in one sludge field trial exceeded 10,000 ppm [25]. This external Fe comprises a risk to domestic animals; intrinsic plant Fe has shown no evidence of toxicity to animals.

Animals tolerate higher levels of Fe than normally occur in feedstuffs [83]. However, chronic Fe toxicity is complex, and toxicity of sludge Fe especially complex. Dietary Fe interacts with Cu and Zn; 1,000 ppm Fe reduced liver Cu in cattle and sheep [113, 114, 115]. On low Cu diets, Fe toxicity is manifested as induced-Cu deficiency. When sludge supplies the Fe, it also supplies Zn, Cd, and sulfide which would further interfere with Cu utilization [116]; and sludge supplies Cu along with Cu-antagonistic and Cu sorbing constituents. As noted above, adding soil to diets reduced Cu bioavailability [70].

Further, the chemical form of Fe influences bioavailability. Iron in FeSO$_4$ is 3 to 5 times more usable by cattle than iron as ferric oxide [117]. Digested sludges contain large proportions of their Fe as ferrous; however, ferrous rapidly oxidizes as the sludge becomes aerobic.

Iron toxicity was suspected in cattle grazing pastures sprayed with a digested sludge high in Fe (11%). That sludge was also low in Cu. Liver Cu was at marginal or deficient levels in the cattle on sludge or sludge compost fertilizer treatments [25]. Liver Cu has been low in most studies of sludge feeding. One study at Denver, with sludge Cu at 1,000 ppm, reported Cu balance or slight Cu increase [68].

Thus, high Fe sludges which are simultaneously low in Cu can allow induced-Cu deficiency and extreme Fe accumulation in the spleen, liver, and duodenum. Even the erosion of joint cartilage may have been due to the induced Cu deficiency. Zn-induced Cu deficiency in horses [116, 118, 119] caused erosion of cartilage very similar to that observed in cattle grazing pastures fertilized with high-Fe sludge [25]. Sludges high in Fe (>4%) should not be surface-applied to fertilize forage crops. Supplemental copper may be needed for pasture use of even low Fe sludges.

Arsenic is 1 to 230 ppm in sludge, with median levels of 10 ppm dry sludge (Table I). Use of median As sludges at high cumulative loadings thus would not increase soil arsenic to phytotoxic levels even on sandy soils [120, 121]. Plants can experience phytotoxicity when soil As is high, with greater toxicity possible on sandier soils since they have fewer As sorbing Fe and Al surfaces. Because As is so phytotoxic, severe yield reduction occurs in most plants before As is appreciably increased in edible plant tissues [120, 121]. However, As is increased in the peel of root crops when As is increased in soil, apparently due to soil contamination of the peel. Because sludges apply so little As, the food chain is protected from sludge-applied As. Domestic animals are similarly protected against As in ingested sludge. Sludges should be routinely analyzed to assure that industrial As sources have not caused unacceptable sludge As levels.

Fluorine uptake is strongly controlled by precipitation of CaF$_2$ in the soil. Plant F is increased by addition of F to soils only where Ca activity is very low in the soil, e.g. strongly acid soils [48]. Sludges are not high in F, and increased F in crops due to sludge use has not been reported.

Direct ingestion of sludge circumvents the "Soil-Plant Barrier" on F. Kienholz et al. observed significant increase in bone F in cattle fed Denver sludge [122]. Sludge F should be CaF$_2$, a form of relatively low bioavailability [83]. Avoidance of sludge ingestion, or of sludges high in F, should protect domestic animals from health effects of sludge F.

Monogastric animals are protected by the "Soil-Plant Barrier."

Molybdenum uptake by plants depends strongly on soil pH. Uptake is greatest in calcareous soils since Mo sorption is weak. Leaching of Mo from the root zone is also promoted by higher soil pH.

Plants tolerate very high levels of Mo, and translocate Mo into edible plant tissues [48, 123]. Thus, the "Soil-Plant Barrier" does not protect the food chain from excessive soil Mo. Ruminant animals are susceptible to Mo toxicity because Mo interacts to deplete liver Cu (see above) [69, 75].

Sewage sludge contains 1-40 ppm Mo, with median levels of 10 ppm (Table I). One sludge enriched (> 1,000 ppm) in Mo due to industrial Mo discharges has been reported (Bates, T. E., Univ. Guelph, personal communication). Thus, to avoid Mo problems with ruminants, sludges should be monitored for Mo. Only substantially Mo-polluted sludges should be able to cause toxicity since sludges also contain Cu. No occurrence of Mo toxicity due to sludge use has been reported. Mo-toxicity would not be a persistent problem in regions with adequate rainfall to leach Mo through the soil.

Selenium is readily absorbed from soil by plants, although selenate is more plant available than selenite which is more strongly sorbed by soils [125]. Once absorbed, Se is freely translocated to edible plant tissues [45, 126]. Some areas of the U.S. have naturally Se toxic soils [124]. Because plants can accumulate toxic levels of Se without phytotoxicity, the food chain is not protected from excessive Se by the "Soil-Plant Barrier." Plant uptake is greater at higher soil pH, and neutral soil pH is recommended for sludge use.

Fortunately, sewage sludges are generally low in Se; the reported range is 1.7 to 17 ppm, with median levels of 5 ppm dry sludge (Table I). Because sludge contains only low levels of Se, the food chain is protected from plant absorbed Se. Ingestion of 5% sludge with median Se levels by grazing ruminants does not add excessive Se to diets [83]. Sludges should be routinely monitored for Se to assure that Se remains acceptably low.

Recent studies on potential environmental consequences of landspreading flyash have indicated that high plant levels of Se can result from present disposal practices [127]. Some Western U.S. coal is particularly high in Se. Lower risk of high plant Se results when flyash is used at low rates as a fertilizer or liming material.

Cadmium phytotoxicity and associated tissue Cd levels vary among plant species [2, 40]. At 50% yield reduction due to Cd, leafy vegetables contain ca. 200 ppm Cd in calcareous soils but ca. 500 ppm Cd in acid soils [128]. If all plant foods an individual consumed had phytotoxic Cd levels, chronic

Cd health effects would rapidly result. The food chain is clearly not protected from excessive Cd by a "Soil-Plant Barrier."

However, interactions can reduce the potential for health effects of sludge-borne Cd. NRC recommended that animal feeds not exceed 0.5 ppm Cd in order to protect the food chain [38]. But cattle exposed to sludges with 10 ppm Cd had normal kidney Cd [68]. It appears that sludge Fe, Zn, Ca, etc. must act to reduce Cd absorption. These cattle retained only a very low percentage of their dietary Cd [68]; sludge use in animal agriculture is perhaps the lowest Cd-food-chain risk of all agriculture sludge use systems. The next section summarizes Cd health effects and risks from sludge Cd to populations.

POTENTIAL FOR HUMAN HEALTH EFFECTS FROM SLUDGE-BORNE CADMIUM.

The first human health effects from increased cumulative Cd ingestion is Renal Tubular Dysfunction (Franconi Syndrome) resulting in low-molecular-weight proteinuria, glucosuria, aminoaciduria, phosphaturia, etc. [3-6, 67, 129, 130]. With further Cd ingestion, the severity of the proteinuria, etc., increases. Subsequently, bone problems can develop with repeated fractures; the bone disease (Itai-itai) brought attention to the Cd-exposed Japanese farmers [3, 4]. Severely exposed industrial Cd-workers have experienced increased incidence of kidney stones. Lab animals also show enteropathy, anemia, and teratogenesis (from Cd-induced Zn and Cu deficiencies in the fetus). Humans do not experience greater incidence of cancer or hypertension due to Cd ingestion.

Dose-Effect Relationships For Ingested Cd

Kjellstrom and Nordberg [129] developed a sophisticated multi-compartmental model for Cd metabolism in humans: "The present model predicted that a daily intake corresponding to 440 µg at age 50 would give 200 µg Cd/g of (wet) kidney cortex at age 45-50." The model calculated (best fit) 4.8% absorption of dietary Cd, and a 12 year biological half-life for Cd. Cd intake was assumed to vary with age in the same way as calorie intake [4].

It should be noted that Cd ingestion in this model is 589 µg Cd/d for teenage persons at the greatest calorie intake period (based on Figure 4.32 in [4]). FDA's teenage male Cd ingestion estimate would be even higher. Although the model [129] is presently generally accepted, a number of specific values used are not well supported experimentally [67, 130].

Dose-Response Relationships for Ingested Cd.

Extending the 440 µgCd/d "Average Human" Cd-effect

model to a population has involved further assumptions. Individuals vary widely in self-selected dietary Cd [131], Cd absorption rate, and sensitivity to absorbed Cd. These phenomena are generally assumed to vary in a log-normal fashion within a population. Kjellstrom [132] extended the 440 µg/d value to a population by arbitrarily using a geometric standard deviation of 2.35 based on studies of Cd in autopsy tissues (see [67] for details). This Dose-Response model is shown in Figure 1. His model appears to require greater than 100% absorption of dietary Cd for the most sensitive individuals. Another line extends to 25% absorption by the most sensitive individuals; this is the highest result reported for a human subject [133] who had mild anemia. Fe-deficiency strongly enhanced Cd absorption by humans [133].

It seems unreasonable to extrapolate the 440 "Average Human" model result to an assumed maximum sensitivity group with much greater absorption than ever observed in humans. Further, individuals are unlikely to be in this greatest risk

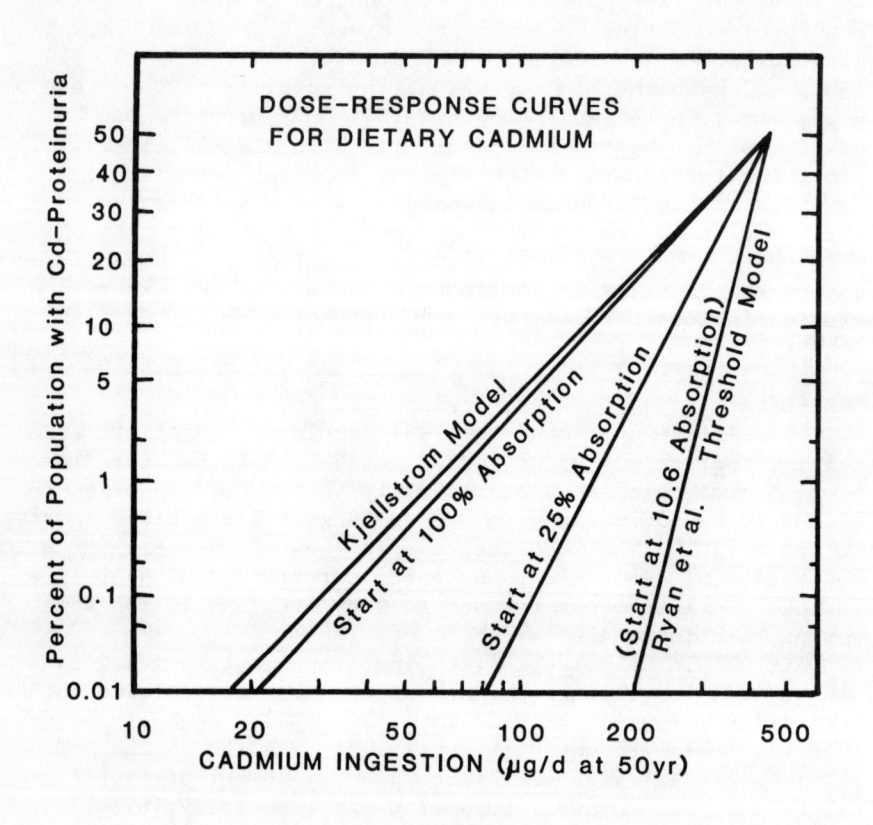

Figure 1. Plots of Several Dose-Response Models For Ingested Cd. See Text For Details.

group for their whole life. Ryan et al. [67] concluded that a 200 µg Cd/d threshold model (Figure 1) was most appropriate, as did a CEC Workgroup [130]; this value corresponds to a 10.6%-absorption rate for the most sensitive individuals. Individuals with other nutritional deficiencies (Zn, Ca, protein, vitamin D) could also be in the sensitive group [85, 86].

In the third section of this paper I repeatedly noted that interactions from other toxic elements, nutrients, etc., applied in a sludge could limit the toxicity from a specific element. Interactions with Cd are no less important in assessing human health risks from sludge Cd. The "worst case" for human exposure to sludge Cd usually evaluated is consumption of crops grown in sludge-amended, strongly acidic gardens. When low-Cd, low-Cd:Zn-ratio sludges are applied to these gardens, the crops grown after high cumulative sludge applications are only somewhat increased in Cd [2, 57, 65]; further, these crops have normal or increased levels of Zn, Fe, and other nutrients which can inhibit Cd absorption. For this low Cd, low Cd:Zn sludge scenario, the very act of consuming enough Cd in leafy and root vegetable crops to be of concern for susceptible individuals would shift the individual into a less susceptible population!

Fe-sufficient humans absorbed only 2.3% of dietary Cd compared to average absorption of 4.8% [133]. Adjusting the "Average Human" Cd-health-effect dose of 440 µg/d at age 50 from average Fe status to adequate Fe status would increase the limit from 440 to 918 µg/d.

Leafy vegetables grown with "Domestic Sludge" have been fed to lab animals to evaluate bioavailable Cd in sludge fertilized crops. Although dietary Cd was increased up to 5 fold by sludge, kidney Cd did not increase [134, 135]. In other studies with high metal sludges kidney Cd has been increased [135, 136, 137]. Sludge and crop Zn may be important in reducing absorption of crop Cd in the intestine [38, 139]. More sludge-soil-plant-animal studies are needed to firmly establish the bioavailability of Cd in sludge-fertilized crops. The recommendations of Fox [83, 140] should be carefully followed in conducting these studies.

The special protection afforded for low-Cd sludges can not be extrapolated to industrially polluted high Cd sludges. Another consideration is age; the model [129] for age 50 should probably be adjusted for longer lived populations [141]. Although unusual consumption rates for garden vegetables now appear not to comprise a high risk case for Domestic Sludges, dietary choice effects on Cd-risk need further evaluation [131].

Based on the considerations in this paper it now seems increasingly likely that low-risk systems can be developed for utilization of the beneficial aspects of sludges.

REFERENCES

1. Sagik, B. P. and C. A. Sorber, Eds. *Risk Assessment and Health Effects of Land Application of Municipal Wastewater and Sludges* (San Antonio, TX: Center for Applied Research and Technology, University of Texas, 1978), 329 pp.
2. "Sewage Sludge Applications on Agricultural Soils: Effect of Annual and Cumulative Additions of Cadmium and Zinc on Plants," Council for Agricultural Science and Technology (1980). In press.
3. Tsuchiya, K., Ed. *Cadmium Studies in Japan: A Review* (New York: Elsevier/North-Holland Biomedical Press, 1978), 376 pp.
4. Friberg, L., M. Piscator, G. Nordberg and T. Kjellstrom. *Cadmium in the Environment*, 2nd Ed. (Cleveland, OH: CRC Press, 1974), 248 pp.
5. Fulkerson, W., and H. E. Goeller. "Cadmium - The Dissipated Element," NTIS No. ORNL-NSF-EP-21 (1973), 473 pp.
6. Hammons, A. S., J. E. Huff, H. M. Braunstein, J. S. Drury, C. R. Shriner, E. B. Lewis, B. L. Whitfield and L. E. Towill. "Reviews of the Environmental Effects of Pollutants: IV. Cadmium", EPA-600/1-78-026 (1978).
7. Berrow, M. L. and J. Webber. "Trace Elements in Sewage Sludges," *J. Sci. Food Agr.* 23:93-100 (1972).
8. Blakeslee, P. A. "Monitoring Considerations for Municipal Wastewater Effluent and Sludge Application to the Land," in *Proc. Jt. Conf. Recycling Municipal Sludges and Effluents on Land* (Washington, D.C.: Nat. Assoc. St. Univ. Land-Grant Coll., 1973). pp.183-198.
9. Chaney, R. L., S. B. Hornick, and P. W. Simon. "Heavy Metal Relationships During Land Utilization of Sewage Sludge in the Northeast," in *Land as a Waste Management Alternative*, R. C. Loehr, Ed. (Ann Arbor, MI: Ann Arbor Science Publishers, Inc., 1977). pp. 283-314.
10. Furr, A. K., A. W. Lawrence, S. S. C. Tong, M. C. Grandolfo, R. A. Hofstader, C. A. Bache, W. H. Gutenmann, and D. J. Lisk. "Multielement and Chlorinated Hydrocarbon Analyses of Municipal Sewage Sludges of American Cities," *Environ. Sci. Technol.*, 10:683-687 (1976).
11. LaConde, K. V., R. J. Lofy, and R. P. Stearns. *Municipal Sludge Agricultural Utilization Practices. An Environmental Assessment, Vol. 1.* EPA Solid Waste Management Series SW-709, (1978). 150 pp.
12. Sommers, L. E. "Chemical Composition of Sewage Sludges and Analysis of Their Potential Use as Fertilizers," *J. Environ. Qual.* 6:225-232 (1977).
13. Gurnham, C. F., B. A. Rose, H. R. Ritchie, W. T. Fetherston, and A. W. Smith. "Control of Heavy Metal

Content of Municipal Wastewater Sludge," Report to Natl. Sci. Found. (NTIS)-PB-295917, (1979). 144 pp.
14. Batey, T., C. Berryman, and C. Line. "The Disposal of Copper-Enriched Pig-Manure Slurry on Grassland," J. Br. Grassland Soc. 27:139-143 (1972).
15. Chaney, R. L. and C. A. Lloyd. "Adherence of Spray-Applied Liquid Digested Sewage Sludge to Tall Fescue," J. Environ. Qual. 8:407-411 (1979).
16. Jones, S. G., K. W. Brown, L. E. Deuel, and K. C. Donnelly. "Influence of Rainfall on the Retention of Sludge Heavy Metals by the Leaves of Forage Crops," J. Environ. Qual. 8:69-72 (1979).
17. Boswell, F. C. "Municipal Sewage Sludge and Selected Element Applications to Soil: Effect on Soil and Fescue," J. Environ. Qual. 4:267-273 (1975).
18. Fitzgerald, P. R. "Toxicology of Heavy Metals in Sludges Applied to the Land." in Proc. Fifth Natl. Conf. Acceptable Sludge Disposal Techniques (Rockville, Md., Information Transfer, Inc., 1978). pp. 106-116.
19. Dalgarno, A. C. and C. F. Mills. "Retention by Sheep of Copper from Aerobic Digests of Pig Faecal Slurry," J. Agric. Sci 85:11-18 (1975).
20. Healy, W. B., P. C. Rankin, and H. M. Watts. "Effect of Soil Contamination on the Element Composition of Herbage," N.Z.J. Agr. Res. 17:59-61 (1974).
21. Mayland, H. F., G. E. Schewmaker, and R. C. Bull. "Soil Ingestion by Cattle Grazing Crested Wheatgrass," J. Range Mgmt. 30:264-265 (1977).
22. Egan, D. A., and T. O'Cuill. "Cumulative Lead Poisoning in Horses in a Mining Area Contaminated with Galena," Vet. Rec. 86:736-738 (1970).
23. Harbourne, J. F., C. T. McCrea, and J. Watkinson. "An Unusual Outbreak of Lead Poisoning in Calves," Vet. Rec. 83:515-517 (1968).
24. Edwards, W. C., and B. R. Clay. "Reclamation of Rangeland Following a Lead Poisoning Incident in Livestock from Industrial Airborne Contamination of Forage," Vet. Human Toxicol. 19:247-249 (1977).
25. Decker, A. M., R. L. Chaney, J. P. Davidson, T. S. Rumsey, S. B. Mohanty, and R. C. Hammond. "Animal Performance on Pastures Topdressed with Liquid Sewage Sludge and Sludge Compost," in Proc. Natl. Conf. Municipal and Industrial Sludge Utilization and Disposal (Silver Spring, Md: Information Transfer, Inc., 1980). In press.
26. Wedeen, R. P., D. K. Mallik, V. Batuman, and J. D. Bogden. "Geophagic Lead Nephropathy: Case Report," Environ. Res. 17:409-415 (1978).
27. Shellshear, I. D., L. D. Jordan, D. J. Hogan, and F. T. Shannon. "Environmental Lead Exposure in Christchurch Children: Soil Lead a Potential Hazard," N. Z. Med. J. 81:382-386 (1975).

28. Lepow, M. L., L. Bruckman, M. Gillette, S. Markowitz, R. Robino, and J. Kapish. "Investigations into Sources of Lead in the Environment of Urban Children," Environ. Res. 10:415-426 (1975).
29. Sayre, J. W., E. Charney, J. Vostal, and I. B. Pless. "House and Hand Dust as a Potential Source of Childhood Lead Exposure," Am. J. Dis. Child. 127:167-170 (1974).
30. Jordan, L. D. and D. J. Hogan. "Survey of Lead in Christchurch Soils," N.Z.J. Sci. 18:253-260 (1975).
31. Galke, W. A., D. I. Hammer, J. E. Keil, and S. W. Lawrence. "Environmental Determinants of Lead Burdens in Children," in Proc. Int. Conf. Heavy Metals in the Environment. III: 53-74 (1977).
32. Angle, C. R. and M. S. McIntire. "Environmental Lead and Children: the Omaha, Nebraska, USA Study," J. Toxicol. Environ. Health 5:855-870 (1979).
33. Charney, E., J. W. Sayre, and M. Coulter. "Increased Lead Absorption in Inner City Children: Where does the Lead Come From?" Proc. Second Intern. Symp. Environ. Lead Research (1979). In press.
34. Dacre, J. C. and G. L. Ter Haar. "Lead Levels in Tissues from Rats Fed Soils Containing Lead," Arch. Environ. Contam. Toxicol. 6:111-119 (1977).
35. Stara, J., W. Moore, M. Richards, N. Barkley, S. Neiheisel, and K. Bridbord. "Environmentally Bound Lead. III. Effects of Source on Blood and Tissue Levels of Rats." Environ. Health Effects Res. Ser. A-670/1-73-036 (1973), pp. 28-29.
36. Baker, E. L., Jr., D. S. Folland, T. A. Taylor, M. Frank, W. Peterson, G. Lovejoy, D. Cox, J. Housworth, and P. J. Landrigan. "Lead Poisoning in Children of Lead Workers. Home Contamination with Industrial Dust," New Engl. J. Med. 296:260-261 (1977).
37. Giguere, C. G., A. B. Howes, M. McBean, W. N. Watson, and L. E. Witherell. "Increased Lead Absorption in Children of Lead Workers -- Vermont," Morbidity Mortality Weekly Report 26(8):61-62 (1977).
38. Rice, C., A. Fischbein, R. Lilis, L. Sarkozi, S. Kon, and I. J. Selikoff. "Lead Contamination in the Homes of Employees of Secondary Lead Smelters," Environ. Res. 15:375-380 (1978).
39. Lead in the Human Environment, National Research Council (Washington, D.C.: National Academy of Sciences, 1980), 525 pp.
40. Sommers, L. E. "Toxic Metals in Agricultural Crops," in Proc. Conf. Evaluation of Health Risks Associated with Animal Feeding and/or Land Application of Municipal Sludge, G. Bitton et al., Eds. (Ann Arbor, MI, Ann Arbor Sci. Publ., 1980). In press.

41. Chaney, R. L. "Crop and Food Chain Effects of Toxic Elements in Sludges and Effluents," in <u>Recycling Municipal Sludges and Effluents on Land</u> (Washington, D.C.: Natl. Assoc. St. Univ. and Land-Grant Coll., 1973). pp. 129-141.
42. Chaney, R. L., and P. M. Giordano. "Microelements as Related to Plant Deficiencies and Toxicities," in <u>Soils for Management of Organic Wastes and Waste Waters</u>, L. F. Elliot and F. J. Stevenson, Eds. (Madison, WI: Am. Soc. of Agr., 1977), pp. 234-279.
43. Foy, C. D., R. L. Chaney, and M. C. White. "The Physiology of Metal Toxicity in Plants," <u>Ann. Rev. Plant Physiol</u>. 29:511-566 (1978).
44. Underwood, E. J. <u>Trace Elements in Human and Animal Nutrition</u>, 4th ed., (New York: Academic Press, Inc., 1978), 545 pp.
45. Allaway, W. H. "Agronomic Controls over the Environmental Cycling of Trace Elements," <u>Adv. Agron</u>. 20:235-274 (1968).
46. Bowen, H. J. M. <u>Environmental Chemistry of the Elements</u>, (New York: Academic Press, 1979), 334 pp.
47. Hodgson, J. F. "Chemistry of the Micronutrient Elements in Soil," <u>Adv. Agron</u>. 15:119-159 (1963).
48. Chapman, H. D. <u>Diagnostic Criteria for Plants and Soils</u> Riverside, CA: Univ. Calif. Dir. Agr. Sci., 1966), 793 pp.
49. Page, A. L. "Fate and Effects of Trace Elements in Sewage Sludge when Applied to Agricultural Lands. A Literature Review Study." U.S. Environ. Prot. Agency Rept. No. EPA-670/2-75-005 (1974), 108 pp.
50. Lonegran, J. F. "The Availability and Absorption of Elements in Soil-Plant Systems and their Relation to Movement and Concentrations of Trace Elements in Plants," in <u>Trace Elements in Soil-Plant-Animals Systems</u>, D.J.D. Nicholas and A. R. Egan, Eds. (New York: Academic Press, Inc. 1975), pp. 109-134.
51. Lindsay, W. L. <u>Chemical Equilibria in Soils</u>, (New York: Wiley-Interscience, 1979), 449 pp.
52. Barber, S. A. "Influence of the Plant Root on Ion Movement in Soil," in <u>The Plant Root and its Environment</u>, E. W. Carson, Ed. (Charlottesville, Va.: Univ. Press of Virginia, 1974), pp. 525-564.
53. Lindsay, W. L. "Role of Chelation in Micronutrient Availability." in <u>The Plant Root and its Environment</u>, E. W. Carson, Ed. (Charlottesville, Va.: Univ. Press of Virginia, 1974), pp. 508-524.
54. Wallace, A., E. M. Romney, G. V. Alexander, S. M. Soufi, and P. M. Patel. "Some Interactions in Plants Among Cadmium, Other Heavy Metals, and Chelating Agents," <u>Agron. J</u>. 69:18-20 (1977).

55. Tiffin, L. O. "The Form and Distribution of Metals in Plants: An Overview." in Biological Implications of Metals in the Environment, Proc. Fifteenth Annu. Hanford Life Sciences Symposium. ERDA-TIC-Conf. No. 750929, (NTIS) (1977). pp. 315-334.
56. Antonovics, J., A. D. Bradshaw, and R. G. Turner. Heavy Metal Tolerance in Plants," Adv. Ecol. Res. 7:1-85 (1971).
57. Chaney, R. L., P. T. Hundemann, W. T. Palmer, R. J. Small, M. C. White, and A. M. Decker. "Plant Accumulation of Heavy Metals and Phytotoxicity Resulting from Utilization of Sewage Sludge and Sludge Composts on Cropland," in Proc. Natl. Conf. Composting Municipal Residues and Sludges (Rockville, Md.: Information Transfer, Inc., 1987). pp. 86-97.
58. Mitchell, G. A., F. T. Bingham, and A. L. Page. "Yield and Metal Composition of Lettuce and Wheat Grown on Soils Amended with Sewage Sludge Enriched with Cadmium, Copper, Nickel, and Zinc," J. Environ. Qual. 7:165-171 (1978).
59. Walsh, L. M., M. E. Sumner, and R. B. Corey. "Consideration of Soils for Accepting Plant Nutrients and Potentially Toxic Nonessential Elements," in Land Application of Waste Materials (Ankeny, IA.: Soil Conserv. Soc. Am., 1976), pp. 22-47.
60. Webber, J. "Effects of Toxic Metals in Sewage on Crops," Water Poll. Contr. 71:404-413 (1972).
61. Williams, J. H. "Use of Sewage Sludge on Agricultural Land and the Effects of Metals on Crops," Water Pollut. Contr. 74:635-644 (1975).
62. Martin, M. H., and P. J. Coughtrey. "Comparisons Between the Levels of Lead, Zinc, and Cadmium within a Contaminated Environment," Chemosphere 5:15-20 (1976).
63. Roberts, R. D., and M. S. Johnson. "Dispersal of Heavy Metals from Abandoned Mine Workings and Their Transference through Terrestrial Food Chains," Environ. Pollut. 16:293-310 (1978).
64. Allaway, W. H. "Food Chain Aspects of the Use of Organic Residues," in Soils for Management of Organic Wastes and Wastewaters, L. F. Elliott and F. J. Stevenson, Eds. (Madison, WI.: Amer. Soc. Agron., 1977). pp. 282-298.
65. Chaney, R. L. and S. B. Hornick. "Accumulation and Effects of Cadmium on Crops," in Proc. First International Cadmium Conference (London: Metals Bulletin Ltd., 1978). pp. 125-140.
66. Pahren, H. R., J. B. Lucas, J. A. Ryan and G. K. Dotson. "Health Risks Associated with Land Application of Municipal Sludge," J. Water Pollut. Contr. Fed. 51:2588-2601 (1979).

67. Ryan, J. A., L. D. Grant, J. B. Lucas, R. E. Marland, H. R. Pahren, W. A. Galke, and D. J. Ehreth. "Cadmium Health Effects: Implications for Environmental Regulations." EPA External Review Draft (1979), 121 pp.
68. Keinholz, E. W. "Effect of Toxic Chemicals Present in Sewage Sludge on Animal Health," in *Proc. Conf. Evaluation of Health Risks Associated with Animal Feeding and/or Land Application of Municipal Sludge*, G. Bitton et al., Eds. (Ann Arbor, MI.: Ann Arbor Science Publishers, Inc., 1980). In press.
69. Mills, C. F., I. Bremner, T. T. El-Gallad, A. C. Dalgarno, and B. W. Young. "Mechanisms of the Molybdenum-Sulfur Antagonism of Copper Utilization by Ruminants," in *Trace Element Metabolism in Man and Animals* -3, M. Kirchgessner, Ed., pp. 150-158.
70. Suttle, N. F., B. J. Alloway, and I. Thornton. "An Effect of Soil Ingestion on the Utilization of Dietary Copper by Sheep," *J. Agr. Sci.* 84:249-254 (1975).
71. Suttle, N. F. and C. F. Mills. "Studies of the Toxicity of Copper to Pigs. I. Effects of Oral Supplements of Zinc and Iron Salts on the Development of Copper Toxicosis," *Br. J. Nutr.* 20:135-148 (1966).
72. Hill, C. H., G. Matrone, W. L. Payne, and C. W. Barber. "*In vivo* Interactions of Cadmium with Copper, Zinc, and Iron," *J. Nutr.* 80:227-235 (1963).
73. Matrone, G. "Chemical Parameters in Trace-element Antagonisms," in *Trace Element Metabolism in Animals-2* W. G. Hoekstra et al., Eds. (Baltimore, MD.: University Park Press, 1974), pp. 91-103.
74. Mills, C. F. in *Trace Element Metabolism in Animals-2*, W. G. Hoekstra et al., Eds., (Baltimore: University Park Press, 1974). pp. 79-90.
75. Bunn, C. R., and G. Matrone. *J. Nutr.* 90:395-399 (1966).
76. Mills, C. F. "Heavy Metal Toxicity and Trace Element Imbalance in Farm Animals," *World Cong. Anim. Feed*, 3rd, 7:275-281 (1978).
77. McGhee, F., C. R. Creger, and J. R. Couch. "Copper and Iron Toxicity," *Poult. Sci.* 44:310-312 (1965).
78. Standish, J. F., C. B. Ammerman, A. Z. Palmer, and C. F. Simpson. "Influence of Dietary Iron and Phosphorus on Performance, Tissue Mineral Composition and Mineral Absorption in Steers," *J. Anim. Sci.* 33:171-178 (1971).
79. Grant-Frost, D. B. and E. J. Underwood. *Aust. J. Exp. Biol. Med. Sci.* 36:339-346 (1958).
80. Cox, D. H., and D. L. Harris. "Effect of Excess Dietary Zn on Fe and Cu in the Rat," *J. Nutr.* 70:514-520 (1960).
81. Lee, D., and G. Martrone. "Fe and Cu Effects on Serum Ceruloplasmin Activity of Rats with Zn-induced Cu Deficiencies," *Proc. Soc. Exp. Biol. Med.* 130:1190-1194 (1969).

82. Levander, O. A. "Lead Toxicity and Nutritional Deficiencies," Environ. Health Perspect. 29:115-125 (1979).
83. Mineral Tolerance of Domestic Animals, NRC (Washington, D.C.: National Academy of Sciences, 1980), 577 pp.
84. Mills, C. F., and A. C. Dalgarno. "Copper and Zinc Status of Ewes and Lambs Receiving Increased Dietary Concentrations of Cadmium," Nature 239:171-173 (1972).
85. Mahaffey, K. R., and J. E. Vanderveen. "Nutrient-Toxicant Interactions: Susceptible Populations," Environ. Health Perspect. 29:81-87 (1979).
86. Fox, M. R. S. "Effect of Essential Minerals on Cadmium Toxicity," J. Food Sci. 39:321-324 (1974).
87. Boawn, L. C., and P. E. Rasmussen. "Crop Response to Excessive Zinc Fertilization of Alkaline Soil," Agron. J. 63:874-876 (1971).
88. Walsh, L. M., D. R. Stevens, H. D. Seibel, and G. G. Weis, "Effects of High Rates of Zinc on Several Crops Grown on an Irrigated Plainfield Sand," Commun. Soil Sci. Plant Anal. 3:187-195 (1972).
89. Baxter, J. C., R. L. Chaney, and K. S. Kinlaw. "Reversion of Zn and Cd in Sassafras Sandy Loam As Measured by Several Extractants and By Swiss Chard," Agron. Abstr. 1974:23 (1974).
90. Campbell, J. K. and C. F. Mills. "The Toxicity of Zinc to Pregnant Sheep," Environ. Res. 20:1-13 (1979).
91. Ott, E. A., W. H. Smith, R. B. Harrington, and W. M. Beeson. "Zinc Toxicity in Ruminants. II. Effect of High Levels of Dietary Zinc on Gains, Feed Consumption, and Feed Efficiency of Beef Cattle," J. Anim. Sci. 25:419-423 (1966).
92. Hambridge, K. M., C. Hambridge, M. Jacobs, and J. D. Baum. "Low Levels of Zinc in Hair, Anorexia, Poor Growth, and Hypogeusia in Children," Pediat. Res. 6:868-874 (1972).
93. Walsh, L. M., W. H. Erhardt, and H. D. Seibel. "Copper Toxicity in Snapbeans (Phaseolus vulgaris L.)," J. Environ. Qual. 1:197-200 (1972).
94. Hunter, J. G. and O. Vergnano. "Nickel Toxicity in Plants," Ann. Appl. Biol. 39:279-284 (1952).
95. Alexander, J., R. Koshut, R. Keefer, R. Singh, O. J. Horvath, and R. L. Chaney. "Movement of Nickel from Sewage Sludge into Soil, Soybeans, and Voles," in Trace Substances in Environ. Health -12, D. D. Hemphill, Ed. (Columbia, MO.: Univ. Missouri, Columbia, 1979), pp. 377-388.
96. O'Dell, G. D., W. J. Miller, W. A. King, S. L. Moore, and D. M. Blackmon. "Nickel Toxicity in the Young Bovine," J. Nutr. 100:1447-1453 (1970).
97. Jaffre, T., R. R. Brooks, J. Lee, and R. D. Reeves. "Sebertia acuminata: A Hyperaccumulator of Nickel From New Caledonia," Science 193:579-580 (1976).

98. Johnson, W. R., and J. Proctor. "A Comparative Study of Metal Levels in Plants From Two Contrasting Lead-Mine Sites," Plant Soil 46:251-257 (1977).
99. Johnson, M. S., T. McNeilly, and P. D. Putwain. "Revegetation of Metalliferous Mine Spoil Contaminated by Lead and Zinc," Environ. Pollut. 12:261-277 (1977).
100. Mitchell, R. L., and J. W. S. Reith. "The Lead Content of Pasture Herbage," J. Sci. Food Agr. 17:437-440 (1966).
101. Haye, S. N., D. J. Horvath, O. L. Bennett, and R. Singh. "A Model of Seasonal Increase of Lead in a Food Chain," in Trace Substances in Environmental Health -9, D. D. Hemphill, Ed. (Columbia, MO., Univ. Missouri, Columbia, 1976), pp. 387-393.
102. Davies, B. E. "Plant-Available Lead and Other Metals in British Garden Soils," Sci. Total Environ. 9:243-262 (1978).
103. Preer, J. R., H. S. Sekhon, B. R. Stephens, and M. S. Collins. "Factors Affecting Heavy Metal Content of Garden Vegetables," Environ. Pollut. pp. 95-104 (1980).
104. Kneip, T. J. "Concentrations of Lead and Cadmium in Garden Vegetables Grown in New York City," Environ. Health Perspect. In press.
105. Lowenberg, R. and T. J. Kneip. "Dietary Intakes of Lead and Cadmium in Vegetables Grown in New York City," Environ. Health Perspect. In press (1980).
106. McClaslin, B. D. and V. L. Rodriguez. "Gamma Irradiated Digested Sewage Sludge as Micronutrient Fertilizer on Calcareous Soil," Agron. Abstr. 1978:30-31 (1978).
107. Olsen, S. R. "Micronutrient Interactions," in Micronutrients in Agriculture, J. J. Mortvedt, P. M. Giordano, and W. L. Lindsay, Eds. (Madison, WI: Soil Sci. Soc. Am., 1972), pp. 243-264.
108. DeKock, P. C., M. V. Cheshire and A. Hall. "Comparison of the Effect of Phosphorus and Nitrogen on Copper-deficient and -sufficient Oats," J. Sci. Fd. Agric. 22:437-440 (1971).
109. Ambler, J. E. and J. C. Brown. "Cause of Differential Susceptibility to Zinc Deficiency in Two Varieties of Navy Beans (Phaseolus vulgaris L.)," Agron. J. 61:41-43 (1969).
110. Warnock, R. E. "Micronutrient Uptake and Mobility Within Corn Plants (Zea mays L.) in Relation to Phosphorus-induced Zinc Deficiency," Soil Sci. Soc. Amer. Proc. 34:765-769 (1970).
111. Seckbach, J. "Studies on the Deposition of Plant Ferritin as Influenced by Iron Supply to Iron-deficient Beans," J. Ultrastruct. Res. 22:413-423 (1968).
112. Seckbach, J. "Iron Content and Ferritin in Leaves of Iron Treated Xanthium pensylvanicum Plants," Plant Physiol. 44:816-820 (1969).

113. Standish, J. F., C. B. Ammerman, C. F. Simpson, F. C. Neal and A. Z. Palmer. "Influence of Graded Levels of Dietary Iron, as Ferrous Sulfate, on Performance and Tissue Mineral Composition of Steers," J. Anim. Sci. 29:496-503 (1969).
114. Grun, M., M. Anke, A. Hennig, W. Seffner, M. Partschefeld, G. Flachowsky and B. Groppel. "Excessive Oral Iron Application to Sheep. 2. The Influence on the Level of Iron, Copper, Zinc, and Manganese in Different Organs," (IN GERMAN), Arch. Tierernahr 28:341-347 (1978).
115. Standish, J. F. and C. B. Ammerman. "Effect of Excess Dietary Iron as Ferrous Sulfate and Ferric Citrate on Tissue Mineral Composition of Sheep," J. Anim. Sci. 33:481-484 (1971).
116. Anke, M. "Disorders Due to Copper Deficiency in Sheep and Cattle," Monat. Vet. Med. 28:294-298 (1973).
117. Ammerman, C. B., J. M. Wing, B. G. Dunavant, W. K. Robertson, J. P. Feaster and L. R. Arrington. "Utilization of Inorganic Iron by Ruminants as Influenced by Form of Iron and Iron Status of the Animal," J. Anim. Sci. 26:404-410 (1967).
118. Willoughby, R. A., E. MacDonald, B. J. McSherry and G. Brown. "Lead and Zinc Poisoning and the Interaction Between Pb and Zn Poisoning in the Foal," Can. J. Comp. Med. 36:348-359 (1972).
119. Willoughby, R. A. and W. Oyaert. "Zinc Poisoning in Foals," (IN DUTCH), Vlaams Diergeneeskd. Tijdschr. 42:134-143 (1973).
120. Walsh, L. M. and D. R. Keeney. "Behavior and Phytotoxicity of Inorganic Arsenicals in Soils," in Arsenical Pesticides ACS Symp. Ser. 7, E. A. Woolson, Ed. (Washington, D. C.: American Chemical Soc., 1975). pp. 35-52.
121. Woolson, E. A. "Arsenic Phytotoxicity and Uptake in Six Vegetable Crops," Weed Sci. 21:524-527 (1973).
122. Kienholz, E. W., G. M. Ward, D. E. Johnson, J. Baxter, G. Braude and G. Stern. "Metropolitan Denver Sewage Sludge Fed to Feedlot Steers," J. Anim. Sci. 48:734-741 (1979).
123. Hornick, S. B., D. E. Baker and S. B. Guss. "Crop Production and Animal Health Problems Associated with High Soil Molybdenum," in Molybdenum in the Environment-2, W. R. Chappell and K. K. Peterson, Eds. (New York: Marcel Dekker, 1977), pp. 665-684.
124. Kubota, J. and W. H. Allaway. "Geographic Distribution of Trace Element Problems," in Micronutrients in Agriculture, J. J. Mortvedt, P. M. Giordano and W. L. Lindsay, Eds. (Madison, WI: Soil Sci. Soc. Amer., 1972). pp. 525-554.

125. Cary, E. E. and W. H. Allaway. "The Stability of Different Forms of Selenium Applied to Low-selenium Soils." Soil Sci. Soc. Amer. Proc. 33:571-574 (1969).
126. Hamilton, J. W. and O. A. Beath. "Selenium Uptake and Conversion by Certain Crop Plants," Agron. J. 55:528-531 (1963).
127. Stoewsand, G. S., W. H. Gutenmann and D. J. Lisk. "Wheat Grown on Fly Ash: High Selenium Uptake and Response when Fed to Japanese Quail," J. Agr. Food Chem. 26:757-759 (1978).
128. Mahler, R. J., F. T. Bingham, and A. L. Page. "Cadmium-Enriched Sewage Sludge Application to Acid and Calcareous Soils: Effect on Yield and Cadmium Uptake by Lettuce and Chard," J. Environ. Qual. 7:274-281 (1978).
129. Kjellstrom, T., and G. F. Nordberg. "A Kinetic Model of Cadmium Metabolism in the Human Being," Environ. Res. 16:248-269 (1978).
130. Criteria (Dose/Effect Relationships) for Cadmium, Commission of the European Communities, (New York: Pergamon Press, 1978), 202 pp.
131. Yost, K. J., L. J. Miles, and T. A. Parsons. "A Methodology for Estimating Dietary Intake of Environmental Trace Contaminants: Cadmium, a Case Study," Environ. Intern. In press.
132. Kjellstrom, T. "Comparative Study on Itai-itai Disease," in Proc. First International Cadmium Conference (London: Metal Bulletin, 1978), pp. 224-231.
133. Flanagan, P. R., J. S. McLellan, J. Haist, M. G. Cherian, M. J. Chamberlain, and L. S. Valberg. Gastroenterol. 74: 841-876 (1978).
134. Chaney, R. L., G. S. Stoewsand, A. K. Furr, C. A. Bache, and D. J. Lisk. "Elemental Content of Tissues of Guinea Pigs Fed Swiss Chard Grown on Municipal Sewage Sludge-Amended Soil." J. Agr. Food Chem. 26:994-997.
135. Chaney, R. L., G. S. Stoewsand, C. A. Bache, and D. J. Lisk. J. Agr. Food Chem. 26:992-994 (1978).
136. Miller, J., and F. C. Boswell. J. Agr. Food Chem. 27: 1361-1365 (1979).
137. Williams, P. H., J. S. Shenk, and D. E. Baker. J. Environ. Qual. 7:450-454 (1978).
138. Welch, R. M., W. A. House, and D. R. VanCampen. Nutr. Rep. Intern. 17:35-42, (1978).
139. Jacobs, R. M., A. O. L. Jones, B. E. Fry, Jr., and M. R. S. Fox. J. Nutr. 108:901-910 (1978).
140. Fox, M. R. S., R. M. Jacobs, A. O. L. Jones, and B. E. Fry, Jr. Environ. Health Perspect. 28:107-114 (1979).
141. Nogawa, K. in Proc. First International Cadmium Conference, (London: Metal Bulletin, 1978), pp. 213-231.

CHAPTER 4

POTENTIAL HEALTH HAZARDS OF TOXIC ORGANIC RESIDUES IN SLUDGE

Jack C. Dacre. U.S. Army Medical Bioengineering Research and Development Laboratory, Fort Detrick, Frederick, Maryland

INTRODUCTION

Sludges tend to concentrate the most unwanted materials. This is, of course, an advantage in obtaining pure clean water, but it makes disposal a more difficult and perhaps dangerous operation [1]. These toxic unwanted materials have been identified as highly pathogenic microorganisms, some highly toxic trace metals and some residual organochlorine pesticides and other related compounds. Pathogenic bacteria, viruses, and parasite eggs are all to be found in raw sludges and to a lesser extent in digested sludges. Heavy metals, such as tin, zinc, mercury and cadmium, are to be found in much higher concentrations in municipal sludges than in the wastewater from which they are produced. Most sludges contain small amounts of chlorinated pesticides as well as the highly toxic polychlorinated biphenyls.

Traces of organic compounds are known to be present in municipal wastewater and sludges and much recent research has been directed to the ultimate identification of these compounds. Some 700 or more organic compounds have already been identified in waters in the United States, many of them being highly toxic to mammalian systems [2,3]. Table I summarizes these identified compounds according to their chemical groups. They range from simple hydrocarbons to complex polynuclear aromatic hydrocarbons and many of them are chlorinated. The last two groups only -- the organochlorine pesticides and the polyaromatic hydrocarbons -- will be considered here since it is the compounds in these two groups that have been identified in municipal sludges. Compounds within all the other groups, of course, are removed in most water treatment processes.

Table I. Toxic Pollutants
in Water

Purgeable Halocarbons
Purgeable Aromatics
Acrolein/acrylonitrile
Phenols/benzidines
Phthalate esters
Nitroaromatics and Isophorone
Nitrosamines
Haloesters
Chlorinated Hydrocarbons
Organochlorine Pesticides/PCBs
Polynuclear Aromatic Hydrocarbons

ORGANIC COMPOUNDS IN SLUDGES

Organic compounds in sludges, with particular reference to the health risks associated with land application of municipal sludge, have been reviewed in the literature by Kover [4], Dacre [5], Kowal and Pahren [6], Pahren, Lucas, Ryan and Dotson [7], and briefly by Connery [8].

An extensive search of the literature on sludges has shown many organochlorine insecticides to occur in sludges throughout the world. These compounds are listed in Table II together with their reported levels in sludge. (See Appendix for the chemical name of all the compounds discussed.) All these compounds are highly chlorinated and complex molecules. In the case of dieldrin in the Netherlands, an adjacent wool textile industry contributed this insecticide to the wastewaters and hence to the sludges. The high levels of aldrin and chlordane should be noted.

The related polyhalogenated biphenyl compounds and their levels found in sludge in many countries of the world are given in Table III. Compared with the organochlorine pesticides, the levels reported for PCBs are very high, especially in sludge from the Midwest area, with values ranging from 240 to 1700 ppm, and an average of 765 ppm [19]. The PCBs and PBBs have been identified extensively in the environment -- the PCBs as a result of accidental spillage and

Table II. Organochlorine Insecticides in Municipal Sewage Sludge

Insecticide	Country	ppm (dry wt)	Reference
DDT (+ DDD, DDE)	Sweden	0.24	9
DDT	USA	0.35 (5)[a]	10
DDD	USA	0.25 (10)	10
DDE	USA	+	11,12
Dieldrin	Netherlands	7.5 (12)	13
Dieldrin	USA	0.31 (14)	14
Dieldrin	USA	0.91 (11)	10
Aldrin	USA	16.2 (1)	10
Chlordane	USA	16.04 (11)	10

a. Number of sludge samples analyzed.

Table III. Polyhalogenated Compounds in Municipal Sewage Sludge

Compound	Country	ppm (dry wt)	Reference
PCBs	USA	9.11 (21)[a]	10
PCBs	USA (MA)	10.8	11,12
PCBs	USA	5.15 (14)	14
PCBs	USA (WI)	20-69 (2)	15
PCBs	Japan	6.35 (7)	16
PCBs	Sweden	0.97	9
PCBs	Canada (Ontario)	0.074-1.122	17
PCBs	Canada (Southern Ontario)	0.141 (4)	18
PCBs	USA (Midwest)	765 (6)	19
PCBs	Canada (Ontario)	0.6-76.6 (33)	20
PCTs	Japan	0.15 (7)	16
PBBs	USA (NJ)	431	21
PBT/TBT	Sweden	8-180	22

a. Number of sludge samples analyzed.

resultant soil contamination from transformer fluids -- and the PBBs from animal food contamination. Reports show that nearly everyone has ingested some PCBs and a recent government study show PCBs in the sperm of every male tested.

Many other chlorinated compounds have been identified in sewage sludge and they are listed in Table IV. The origins and toxic properties of most of these compounds are not well known. There are also many other chlorinated residual organopesticides that have been identified in water and wastewater. It is almost certain that many of these compounds will be accumulated in municipal sludges. Some of them are listed in Table V. The insecticide kepone will without a doubt be found in sludges in Virginia, while mirex should be found extensively in sludges in Louisiana, Mississippi, and Alabama. The chlorinated dibenzo-p-dioxins and dibenzo-furans have been identified as contaminants in fungicides (e.g. pentachlorophenol), the herbicides (e.g. 2,4,5-T) and the PCBs. TCDD (dioxin) is reported to be one of the most toxic chemicals known. It is teratogenic and carcinogenic in experimental animals and has most of the toxic properties of the organochlorine insecticides [23,24].

Table IV. Other Chlorinated Compounds Found in Sewage Sludge[a]

Dichlorobenzene

Trichlorobenzene

Tetrachlorobenzene

Chloroaniline

Dichloroaniline

Trichloroaniline

Dichloronaphthalene

Tetrachloronaphthalene

Trichlorophenol

Chlorobiphenyl

Dichlorobiphenyl

Trichlorobiphenyl

Tetrachlorobiphenyl

Pentachlorobiphenyl

---

a. Erickson and Pellizzari [11,12].

Table V. Other Chlorinated Compounds
Probably Present in Sewage Sludge

Endrin

Heptachlor

Endosulfan

Kepone

Mirex

Lindane (γ-BHC)

2,4-D

2,4,5-T

Dioxins

Pentachlorophenol

The polynuclear aromatic hydrocarbons have been shown to occur in air, food and water -- both ground and wastewater. They are formed via automobile and engine exhausts, the burning of fossil fuels, oil spills, food processing, tobacco smoking and from many other natural sources. Table VI lists the hydrocarbon compounds and the levels found in samples of municipal sludge in Germany [25], Czechoslovakia [26], and England [27]. The relatively high levels of the two proven carcinogens -- benzo(b)fluoranthene and the more common benzo(a)pyrene should be noted.

POTENTIAL HEALTH HAZARDS OF ORGANIC COMPOUNDS IN SLUDGES

Why then are these organic compounds potentially hazardous to the ultimate health of animals and man? The five principal reasons are summarized in Table VII. All the compounds have very low solubility in water, and hence they do not move readily in soil. They are relatively stable in soil and do not degrade readily in microbiological systems. In the case of the PCBs it has been shown that the less chlorinated materials degrade in activated sludge more rapidly than the more highly chlorinated ones [28]. Consequently, the more highly chlorinated PCBs may have extremely long biological half-lives. Table VIII sets out the approximate half-lives of many of these residual organochlorine compounds. This is the time it takes for one-half of the compound to degrade or disappear in soil. So they are around for a long time. All these compounds have a very high affinity for lipids -- that is, they are soluble in fat and

Table VI. Polynuclear Aromatic Hydrocarbons in Municipal Sewage Sludge

Compound	Grimmer et al. [25] ppm[a]	Hotař et al. [26] ppm[b]	Nicholls et al. [27] ppm[c]
Chrysene	3.57[d]	0.11 - 1.31	
Fluoranthene	4.28		0.91
Benzo(b)fluoranthene	⎱3.22	1.0 - 5.78	0.30
Benzo(k)fluoranthene	⎰		0.27
Benzo(e)pyrene	1.40	0.02 - 0.08	
Benzo(a)pyrene	1.70	0.01 - 0.27	0.42
Perylene	0.51	0.05 - 0.10	
Dibenz(a,j)anthracene	0.26	0.03 - 0.15	
Indeno(1,2,3,c,d)pyrene	1.34	0.38 - 1.87	0.30
Benzo(g,h,i)perylene	1.10	0.02 - 0.10	0.23
Country:	Germany	Czechoslovakia	England

a. Mean of five sludge samples analyzed.
b. Ten sludge samples analyzed.
c. Mean of three sludge samples analyzed.
d. Mixture of chrysene and benzo(a)anthracene (1:1).

Table VII. Potential Health Hazards of Organic Compounds in Sludge

1. Very low water solubility
2. Do not degrade readily - long half-lives in soil
3. Very high affinity for lipids - bioaccumulate in tissue
4. All accumulate and translocate in the food chain (soil → plants → animals → man)
5. All are highly toxic to mammals - many are carcinogenic

Table VIII. Approximate Half-Lives of Insecticides in Soil[a]

Insecticide	$t_{1/2}$ (years)
DDT	3 - 10
Aldrin	1 - 4
Dieldrin	1 - 7
Chlordane	2 - 4
Endrin	4 - 8
Heptachlor	7 - 12
Toxaphene	10
Lindane	2
PBBs	4+
TCDD	1+

a. In part from Menzie [29].

hence will accumulate in fatty tissues. They have been found to accumulate in animal and human fat and milk and all the analytical results of this work are well documented, e.g., TCDD has recently been shown to accumulate in the fat and tissues of beef cattle and sheep exposed to the compound in their diet. Another study reported residues in the milk and cream from cows fed TCDD. The only class of compound in sludge for which there is any data is PCB, and this has been summarized in Table IX. Many plants bioaccumulate PCBs especially the root vegetables. However, a recent study on the long-term uptake of PCBs and their conversion products by Spruce trees from soil treated with sewage sludge, indicated very low uptake levels, of the order of 0.8% PCBs [30].

There are numerous pathways for all these compounds to bioaccumulate and translocate up the food chain from the soil to eventually reach man (Figure 1). The pathways for the polynuclear aromatic hydrocarbons have not been studied in any detail; some studies show that despite their high lipid solubility, they show little tendency for bioaccumulation in the fatty tissues of animals or man. This observation is not unexpected, in light of convincing evidence to show that they are rapidly and extensively metabolized. Finally, all these compounds in both groups are highly toxic to mammalian systems; many of them are carcinogenic, mutagenic, and teratogenic.

92   HEALTH RISKS OF SPECIFIC AGENTS

Table IX. Uptake of PCBs from Sludge Treated Soils

Country	Plant → Animal	Uptake	Reference
Canada	Grain/Corn	>50%	2
Canada	Grass/Corn	High	16
Germany	Spruce Trees	0.8%	30
USA (Midwest)	Corn → Swine	No effects	31
USA (Midwest)	Soil/Grass: Milk of Grazing Cow	Low	17
	Fish in Adjacent Stream	Very high	
USA (NY)	Cabbage: Fed to Guinea Pigs	275% High (in liver)	32

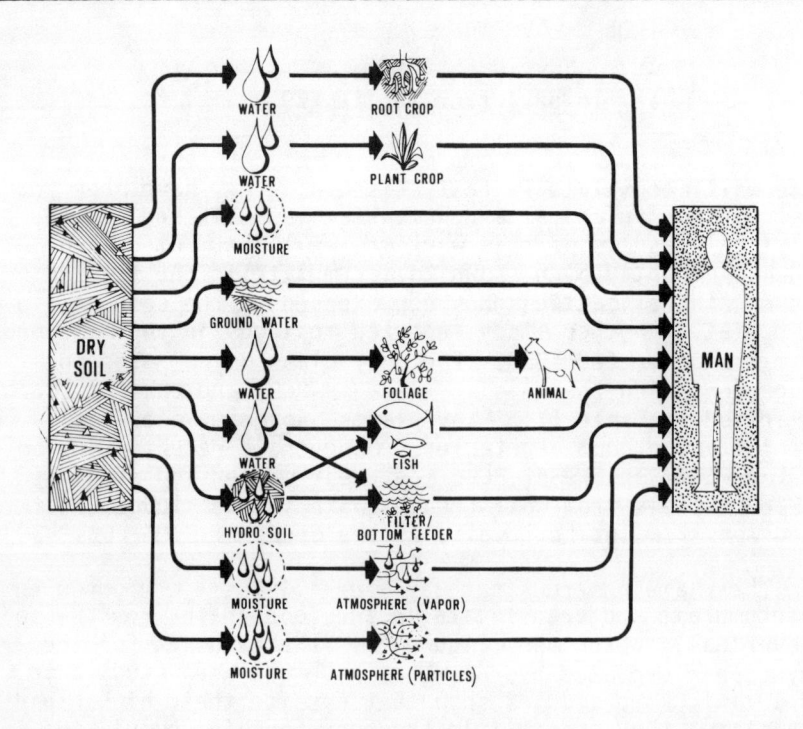

Figure 1. Pollutant Pathways from Soil to Man.

All these organic compounds, especially the synthetic insecticides, have been the subject of extensive toxicological investigations. A toxicological evaluation of chemicals found in the environment however must not be simplified to cover only acute or subacute effects. Today, such an evaluation includes all studies listed in Table X. The acute toxicity of some of the chlorinated insecticides is given in Table XI.

Table X. Toxicological Tests

Acute Ingestion Toxicity
    Acute oral toxicity to rodents
    Acute oral toxicity to wildlife

Acute Inhalation Toxicity to Mammals

Acute Dermal Toxicity to Mammals

Toxicity to Eyes of Mammals

Subacute Effects on Mammals

Chronic (life-time) Effects on Mammals
    Behavioral/reproductive
    Carcinogenic/mutagenic/teratogenic

Pharmacokinetic Effects

Metabolism in Mammals

Table XI. Acute Toxicity of Chlorinated Insecticides in Rats

Insecticide	Oral LD_{50}, mg/kg Male	Female
Aldrin	39	60
Dieldrin	46	46
Photodieldrin	9.6	
Endrin	17.8	7.5
Chlordane	335	430
Heptachlor	100	162
Heptachlor Epoxide	46.5	61.3
Endosulfan	43	18

The carcinogenicity of most of these organochlorine insecticides has been studied and in most cases confirmed by the National Cancer Institute under their Carcinogenesis Bioassay Program. This program calls for life-time feeding studies in both rats and mice. A list of the compounds is given in Table XII and the detailed reports are available from the National Cancer Institute. From the toxicological data it is now possible to estimate the potential carcinogenic risk to man. Some of these risk estimates are given in Table XIII. The figures are expressed as a probability of cancer after a life-time consumption of 1 liter of water per day containing 1 ppb or 1 µg/L of the compound.

Table XII. NCI Carcinogen Bioassay Program Carcinogenicity Evaluations of Pesticides

Pesticide	Report Number NCI-CG-TR
DDT, DDD, DDE	131
Dieldrin	22
Aldrin/Dieldrin	21
Photodieldrin	17
Chlordane	8
Heptachlor	9
Endrin	12
Lindane	14
Toxaphene	37
Endosulfan	62
PCB (Aroclor 1254)	38

Even if the specific toxicity of a certain chemical is established, the possible toxic effects of their breakdown products may be more difficult to determine. Natural biodegradation or specific wastewater treatment processes may lead to the development of new compounds or metabolites. For example, many insecticide degradation products such as photodieldrin are highly active neurotoxins. Table XI compares the oral LD_{50} of dieldrin and photodieldrin and heptachlor and heptachlor epoxide.

Table XIII. Lifetime Cancer Risk to Man
of Known Organochlorine
Animal Carcinogens

Insecticide	µg/L
DDT	1.2×10^{-5}
Dieldrin	2.6×10^{-4}
PCB (Aroclor 1260)	3.1×10^{-6}
Chlordane	1.8×10^{-5}
Lindane (γ-BHC)	9.3×10^{-6}
Heptachlor	4.8×10^{-5}
Kepone	4.4×10^{-4}

Many of the polynuclear aromatic hydrocarbons have been similarly evaluated for toxicity and carcinogenicity. They range from the highly carcinogenic benzo(a)pyrene to the moderately carcinogenic dibenzanthracene, to the inactive perylene.

Because of their toxic properties, these compounds have naturally been subject to regulation by the Environmental Protection Agency. Water quality criteria have been established [33-35], toxic pollutant effluent standards have been promulgated [36,37] and finally the banning of some seven of the residual organochlorine insecticides and a suspension of the herbicide 2,4,5-T (Table XIV).

Table XIV. Pesticides Banned by EPA

Pesticide	Date Banned
DDT	1973
Aldrin/Dieldrin	1974
Endrin	1979
Chlordane	1975
Heptachlor	1975
Kepone	1978
2,4,5-T (suspended)	1979

SUMMARY

Two groups of highly toxic organic compounds -- the organohalogen pesticides and the polynuclear aromatic hydrocarbons -- occur in municipal sludge. The properties of these organochlorine compounds are such that they resist degradation in the soil environment; they are retained in fat and fatty tissues and hence will bioaccumulate as they pass up the food chain to eventually reach man himself; and all of them are highly toxic and proven to be carcinogenic in extensive animal studies. None of them, with the possible exception of benzopyrene, has so far, been implicated as a human carcinogen. The importance of the potentially carcinogenic polynuclear aromatic hydrocarbons, however, can be minimized since they all appear to be rapidly and extensively metabolized in mammalian systems. Further research is needed to identify and confirm the presence in sludges of other known residual chlorinated pesticides, especially of the highly toxic TCDD. It is essential that the identified compounds be quantified. The organics and their resultant animal and human health hazards certainly warrant as much attention and research as has already been given to the trace metals in sludge, especially cadmium.

REFERENCES

1. Reuse of Effluents: Methods of Wastewater Treatment and Health Safeguards (Report of a WHO Meeting of Experts) (Geneva: World Health Organization, 1973), WHO Tech. Ref. Ser. No. 517, p. 37.

2. Health-Effects Relating to Direct and Indirect Re-Use of Waste Water for Human Consumption (Report of an International Working Meeting) (The Hague, Netherlands: Nw. Havenstraat 6, Voorburg, 1975), WHO International Reference Centre for Community Water Supply, Technical Paper No. 7.

3. Drinking Water and Health, Washington, DC: National Academy of Sciences, 1977, Chapter VI, Organic Solutes, pp. 489-856.

4. Kover, F.D. "Considerations Relating to Toxic Substances in the Application of Municipal Sludge to Cropland and Pastureland (A Background Summary)," EPA 560/8-76-004. EPA, Washington, DC (November 1976).

5. Dacre, J.C. "The Potential Health Hazards and Risk Assessment of Toxic Organic Chemicals in Municipal Wastewaters and Sludges with Special Reference to Their Effects on the Food Chain," in Proceedings of the Conference on Risk Assessment and Health Effects of Land Application of Municipal Wastewaters and Sludges (San Antonio, TX: The University of Texas at San Antonio, 1978), pp. 141-148.

6. Kowal, N.E. and H.R. Pahren. "Health Effects Associated with Wastewater Treatment and Disposal," J. Water Pollut. Control Fed. 51(6):1301-1315 (1979).

7. Pahren, H.R., J.B. Lucas, J.A. Ryan, and G.K. Dotson. "Health Risks Associated with Land Application of Municipal Sludge," J. Water Pollut. Control Fed. 51(11): 2588-2601 (1979).

8. Connery, J.L. "Health Implications of Sludge Disposal Alternatives: Ocean Disposal, Incineration, Land Application, and Land Filling," in Workshop on the Health and Legal Implications of Sewage Sludge Composting, Vol. 2, Position Papers (Cambridge, MA: Energy Resources Co. Inc., 1979), Chapter 9, pp. 9-1 to 9-53.

9. Mattsson, P.E. and S. Nygren. "Gas Chromatographic Determination of Polychlorinated Biphenyls and Some Chlorinated Pesticides in Sewage Sludge Using a Glass Capillary Column," J. Chromatog. 124:265-275 (1976).

10. Farrell, J.B. and B.V. Salotto. "The Effect of Incineration on Metals, Pesticides and Polychlorinated Biphenyls in Sewage Sludge," in Proceedings of a National Symposium on Ultimate Disposal of Wastewaters and Their Residuals, F.E. McJunkin and P.A. Vesilind, Eds. (Raleigh, NC: Water Resources Research Institute, 1973).

11. Erickson, M.D. and E.D. Pellizzari. "Identification and Analysis of Polychlorinated Biphenyls and Other Related Chemicals in Municipal Sewage Sludge Samples," EPA 560/6-77-021. EPA, Washington, DC (August 1977).

12. Erickson, M.D. and E.D. Pellizzari. "Analysis of Municipal Sewage Sludge Samples by GC/MS/Computer for Polychlorinated Biphenyls and Other Chlorinated Organics," Bull. Environ. Contam. Toxicol. 22:688-694 (1979).

13. de Haan, F.A.M. "The Effects of Long Term Accumulation of Heavy Metals and Selected Compounds in Municipal Wastewater on Soil," in <u>Wastewater Renovation and Reuse</u> Vol. 3, D'Itri F.M. Ed. (New York: Marcel Dekker, Inc. 1977), Chapter 9, pp. 283-303.

14. Furr, A.K., A.W. Lawrence, S.S.C. Tong, M.C. Grandolfo, R.A. Hofstader, C.A. Bache, W.H. Gutenmann, and D.J. Lisk. "Multielement and Chlorinated Hydrocarbon Analysis of Municipal Sewage Sludges of American Cities," <u>Environ. Sci. Technol.</u> 10(7):683-687 (1976).

15. Dube, D.J., G.D. Veith, and G.F. Lee. "Polychlorinated Biphenyls in Treatment Plant Effluents," <u>J. Water Pollut. Control Fed.</u> 46(5):966-972 (1974).

16. Kowase, T., Y. Tsuchiya, Y. Okamoto, K. Yamazaki, and S. Mimura. "Research on Pollution with Polychlorinated Biphenyls and Polychlorinated Terphenyls in Sewage Plants," <u>Ann. Rep. Tokyo Metr. Res. Lab. P. H.</u> 25: 411-416 (1974).

17. Liu, D. and V.K. Chawla. "Polychlorinated Biphenyls (PCBs) in Sewage Sludges," in <u>Proc. Univ. Mo. Ann. Conf. Trace Subs. Environ. Hlth.</u> 10:247-250 (1976).

18. Lawrence, J. and H.M. Tosine. "Polychlorinated Biphenyl Concentrations in Sewage and Sludges of Some Waste Treatment Plants in Southern Ontario," <u>Bull. Environ. Contam. Toxicol.</u> 17:49-56 (1977).

19. Bergh, A.K. and R.S. Peoples. "Distribution of Polychlorinated Biphenyls in a Municipal Wastewater Treatment Plant and Environs," <u>Sci. Total Environ.</u> 8:197-204 (1977).

20. Shannon, E.E., F.J. Ludwig, and I. Valdmanis. "Polychlorinated Biphenyls (PCB's) in Municipal Wastewaters: An Assessment of the Problem in the Canadian Lower Great Lakes," Research Report No. 49. Ontario Ministry of the Environment, Pollution Control Branch, Toronto, Canada (1978).

21. DeCarlo, V.J. "Studies on Brominated Chemicals in the Environment," <u>Ann. N.Y. Acad. Sci.</u> 320:678-681 (1979).

22. Mattsson, P.E., A. Norström, and C. Rappe. "Identification of the Flame Retardant Pentabromotoluene in Sewage Sludge," <u>J. Chromatog.</u> 111:209-213 (1975).

23. Blair, E.H. Ed. <u>Chlorodioxins - Origin and Fate</u> (Washington, DC: American Chemical Society, 1973) Advances in Chemistry Series 120.

24. Cattabeni, F., A. Cavallaro, and G. Galli, Eds. <u>Dioxin: Toxicological and Chemical Aspects.</u> (New York: SP Medical & Scientific Books, 1978).

25. Grimmer, G., H. Böhnke, and H. Borwitzky. "Profile-Analysis of Polycyclic Aromatic Hydrocarbons in Sewage Sludge by Gas Chromatography," <u>Fresenius Z. Anal. Chem.</u> 289:91-95 (1978).

26. Hotař, Z., J. Šula, J. Křemen, E. Břízová, Z. Vozňáková, and J. Vencl. "Carcinogenic Polycyclic Aromatic Hydrocarbons in Digested Sludge from Wastewater Treatment Plants," <u>Čas. Lék. česk.</u> 118:110-114 (1979).

27. Nicholls, T.P., R. Perry, and J.N. Lester. "The Influence of Heat Treatment on the Metallic and Polycyclic Aromatic Hydrocarbon Content of Sewage Sludge," <u>Sci. Total Environ.</u> 12:137-150 (1979).

28. Tucker, E.S., V.W. Saeger, and O. Hicks. "Activated Sludge Primary Biodegradation of Polychlorinated Biphenyls," <u>Bull. Environ. Contam. Toxicol.</u> 14:705-713 (1975).

29. Menzie, C.M. "Fate of Pesticides in the Environment," <u>Ann. Rev. Entomol.</u> 17:199-222 (1972).

30. Moza, P.N., I. Scheunert, W. Klein, and F. Korte. "Long-Term Uptake of Lower Chlorinated Biphenyls and Their Conversion Products by Spruce Trees (<u>Picea abies</u>) from Soil Treated with Sewage Sludge," <u>Chemosphere</u> 8: 373-375 (1979).

31. Hansen, L.G., J.L. Dorner, C.S. Byerly, R.P. Tarara, and T.D. Hinesly. "Effects of Sewage Sludge-Fertilized Corn Fed to Growing Swine," <u>Amer. J. Vet. Res.</u> 37:711-714 (1976).

32. Babish, J.G., G.S. Stoewsand, A.K. Furr, T.F. Parkinson, C.A. Bache, W.H. Gutenmann, P.C. Wszolek, and D.J. Lisk. "Elemental and Polychlorinated Biphenyl Content of Tissues and Intestinal Aryl Hydrocarbon Hydroxylase Activity of Guinea Pigs Fed Cabbage Grown on Municipal Sewage Sludge," <u>J. Agric. Food Chem.</u> 27:399-402 (1979).

33. "Water Quality Criteria," Federal Register, Vol. 44, No. 52, 19526-15981 (15 March 1979).

34. "Water Quality Criteria; Availability," Federal Register, Vol. 44, No. 144, 43660-43697 (25 July 1979).

35. "Water Quality Criteria; Availability," Federal Register, Vol. 44, No. 191, 56628-56657 (1 October 1979).

36. "Water Program Proposed Toxic Pollutant Effluent Standards for Polychlorinated Biphenyls," Federal Register, Vol. 41, No. 143, 30468-30476 (23 July 1976).

37. "Toxic Pollutant Effluent Standards," Federal Register, Vol. 42, No. 8, 2588-2621 (12 January 1977).

APPENDIX

Chemical Name of Compounds Mentioned in the Text

Aldrin	1,2,3,4,10,10-Hexachloro-1,4,4a,5,8,8a-hexahydro-1,4-endo, exo-5,8-dimethanonaphthalene 95% and related compounds 5%
Aroclor 1254	PCB, approximately 54% chlorine
Aroclor 1260	PCB, approximately 60% chlorine
Chlordane	1,2,4,5,6,7,8,8-Octachloro-2,3,3a,4,7,7a-hexahydro-4,7-methano-1H-indene. The technical product is a mixture of several compounds including heptachlor, chlordane, and two isomeric forms of chlordane.
2,4-D	2,4-Dichlorophenoxyacetic acid
DDD (TDE)	2,2-Bis(p-chlorophenyl)-1,1-dichloroethane (including isomers and dehydrochlorination products)
DDE	Dichlorodiphenyldichloroethylene (degradation product of DDT); p,p'-DDE: 1,1-Dichloro-2,2-bis(p-chlorophenyl) ethylene; o,p'-DDE: 1,1-Dichloro-2-(o-chlorophenyl)-2-(p-chlorophenyl) ethylene

DDT	Main component (p,p'-DDT): α-Bis(p-chlorophenyl) β,β,β-trichloroethane. Other isomers are possible and some are present in the commercial product. o,p'-DDT: 1,1,1-Trichloro-2-(o-chlorophenyl)-2-(p-chlorophenyl)ethane
Dieldrin	Not less than 85% of 1,2,3,4,10,10-hexachloro-6,7-epoxy-1,4,4a,5,6,7:8,8a-octahydro-1,4-endo-exo-5,8-dimethanonaphthalene
Dioxin (TCDD)	2,3,7,8-Tetrachlorodibenzo-p-dioxin
Endosulfan	6,7,8,9,10,10-Hexachloro-1,5,5a,6,9,9a-hexahydro-6,9-methano-2,4,3-benzodioxathiepin-3-oxide
Endrin	1,2,3,4,10,10-Hexachloro-6,7-epoxy 1,4,4a,5,6,7,8,8a-octahydro-endo,endo-1,4,5,8-dimethanonaphthalene
Heptachlor	1,4,5,6,7,8,8-Heptachloro-3a,4,7,7a-tetrahydro-4,7-endo-methanoindene
Heptachlor epoxide	1,4,5,6,7,8,8-Heptachloro-2,3-epoxy-3a,4,7,7a-tetrahydro-4,7-methanoindene
Kepone	1,1a,3,3a,4,5,5,5a,5b,6-Decachlorooctahydro-1,3,4-metheno-2H-cyclobuta[cd]pentalene-2-one
Lindane	1α,2α,3β,4α,5α,6β-Hexachlorocyclohexane
Mirex	1,1a,2,2,3,3a,4,5,5,5a,5b,6-Dodecachlorooctahydro-1,3,4-metheno-1H-cyclobuta[cd]pentalene
PBBs (Polybrominated Biphenyls)	Mixtures of brominated biphenyl compounds having various percentages of bromine
PCBs (Polychlorinated Biphenyls)	Mixtures of chlorinated biphenyl compounds having various percentages of chlorine

PCTs (Polychlorinated Terphenyls)	Mixtures of chlorinated terphenyl compounds having various percentages of chlorine
Photodieldrin	1,1,2,3,3a,7a-Hexachloro-5-6-epoxy-decahydro-2,4,7-metheno-1H-cyclopenta[a]pentalene
2,4,5-T	2,4,5-Trichlorophenoxyacetic acid
Toxaphene	Chlorinated camphene (67-69% chlorine). Product is a mixture of polychlor bicyclic terpenes with chlorinated camphenes predominating.

NOTE ADDED IN PROOF

A paper by Liu, Chawla, and Chau [38] describes the levels of chlorinated hydrocarbon insecticides in four digested sewage sludges. Three sludges were from chemical industry and the fourth from a residential community in Ontario, Canada. The following insecticides were identified in the municipal sludge: DDT (mean level, wet wt., 2.27 ppb); DDD (1.60); DDE (2.81); Dieldrin (1.65); Aldrin (0.70); Chlordane (8.53); Heptachlor (1.28); Lindane (0.63). Much higher levels of all these pesticides were found in the three chemical sewage sludges.

38. Liu, D., V.K. Chawla, and A.S.Y. Chau. "Chlorinated Hydrocarbon Pesticides in Chemical Sewage Sludges," in <u>Proc. Univ. Mo. Ann. Conf. Trace Subs. Environ. Hlth.</u> 9:189-196 (1975).

SECTION II

HEALTH RISKS ASSOCIATED WITH USING MUNICIPAL SLUDGE FOR PRODUCTION OF FOOD CHAIN SUBSTANCES

CHAPTER 5

TOXIC METALS IN AGRICULTURAL CROPS

Lee E. Sommers. Department of Agronomy, Purdue University, West Lafayette, Indiana.

INTRODUCTION

The application of municipal sewage sludges on agricultural land is receiving increased emphasis because of environmental and economic constraints associated with alternative disposal methodologies. The most beneficial use of sewage sludge involves the application to soils at a rate consistent with nutrient utilization by the crop grown. Even though sewage sludges contain essentially all macro- and micro-nutrients needed by agronomic crops, the utilization of sludge on cropland will not affect the overall need for fertilizer in the U.S.A. because the total amount of sludge produced could supply the annual N needed for only 1 to 2% of the cropland [1, 2]. In addition to essential plant nutrients, sludges contain elements which are non-essential or potentially toxic not only to crops but also to animals or man consuming the plant product. Included in this category are heavy metals (e.g., Hg, Cd, Ni, and Pb), trace elements (e.g., As, Be), and persistent organics (e.g., polychlorinated biphenyls, chlorinated hydrocarbon pesticides). The primary focus of this paper will be to discuss the toxic metals present in sludges and their fate in soil-plant systems.

COMPOSITION OF SEWAGE SLUDGE

The composition of sewage sludge is influenced by many factors, including the type of treatment process used within the sewage treatment plant and the composition of the sewage influent. In many cases, the degree and type of industrialization will significantly affect the input of metals into a sewage treatment plant, realizing that metals are also being contributed from other sources such as urban run-off, leaching from plumbing fixtures and human wastes. As shown by the data in Table I, the metal concentrations found in sewage sludges cover a wide range. Data have been published on the metal concentrations found in sewage sludges from several states in the U.S.A. [5-10], Sweden [11], and Great Britain [12]. It is readily apparent from the available data that a "typical" sewage sludge does not exist from the standpoint of metal content. It has been suggested [13] that a sludge derived from primarily domestic sewage would contain no more than the following metal levels (in mg/kg): Zn, 2,500; Cd, 25; Cu, 1,000; Ni, 200; and Pb, 1,000. To characterize the metal content of a specific sludge, a sound sludge sampling program is needed because the metal content of sludge at a treatment plant can vary with time [6, 10].

The chemical species of metals present in sewage sludge applied to soils undoubtedly influence subsequent availability of metals to plants. Unfortunately, the metal species existing in sludges have not been adequately characterized, but several studies indicate that variations in the chemical forms of metals occur in different sludges. For example, anaerobic incubation of sludge-amended soils resulted in decreased extractability (EDTA or acetic acid) of Pb, Zn, Cu, Ni, and Cd, while aerobic incubation caused an increase in metal extractability [25]. In a soil incubation study, the extractability of Pb, Zn, Cu, and Cd with DTPA changed not only with time, but also with the type of sludge added, suggesting the presence of different forms of metals in the three sludges used [26]. This hypothesis is supported by data obtained from a sequential extraction procedure designed to remove sludge-borne metals present in soluble, exchangeable, sorbed, organic, carbonate, and sulfide forms [27]. Even though such procedures are semi-quantitative at best, the results indicated that metal forms vary with sludge source. The analysis of sludges by X-ray diffraction procedures has not shown the presence of crystalline metal precipitates [28], while infrared spectroscopy has indicated that the peptide bond in residual proteins may be a potential site for metal binding in the organic fraction of sludge [29]. Additional research has indicated that metals interact with various functional groups (e.g., carboxyl) in sludge organic matter [30,31]. In addition to organic matter, metals contained in sludges at low concentrations (i.e., <50 mg/kg) may be largely co-precipitated during the formation of Fe, Al or Mn hydrous oxides, Ca phosphates and carbonates or Fe sulfides. Preliminary results suggest that this co-precipitation mechanism may account for the relative availability of Cd to plants when sludges containing similar Cd concentrations are added to soils [32]. Additional research is needed to define the metal species, both soluble and insoluble, existing in different sewage sludges and their effect on controlling the relative plant availability of metals following application to soils.

Table I. Concentration of Selected Elements in Sewage Sludge (mg/kg)

Element	Sommers [2][a] Range	Median	Furr et al.[b]	Chaney et al. [4][c] Range	Median
As	6–230	10	3–30		
B	4–760	33	16–90		
Cd	3–3,410	16	7–444	1–970	13
Co	1–18	4	4–18		
Cr	10–99,000	890	169–14,000		
Cu	84–10,400	850	458–2,890	240–3,490	790
Hg	1–10,600	5	3–18		
Mn	18–7,100	260	32–527		
Mo	5–39	30	1–40		
Ni	2–3,520	82	36–562	10–1,260	42
Pb	13–19,700	500	136–7,627	52–4,900	500
Se			2–9		
V			15–92		
Zn	101–27,800	1,740	560–6,390	228–6,430	1,430

[a] Summary of data from approximately 200 treatment plants in 8 states.
[b] Sludges from 16 treatment plants.

The metals of primary concern when considering application of sewage sludge to cropland are Pb, Zn, Cu, Ni, and Cd. In certain instances, other elements may be of importance, e.g., B toxicity to plants in irrigated soils and Mo toxicity to animals grazing pastures with elevated Mo. Either the relatively low concentrations found in sludges or the lack of significant effects on plant composition have resulted in minimal concern over the presence of As, Co, Cr, Hg, Mo, Se, and V in sewage sludges [1]. Even though one study has reported uptake of Hg into tomato fruit [14], it is generally concluded that application of Hg to soils in sewage sludges will not result in significant changes in the Hg content of crops [1, 15-18]. Additional information on the effects of metals and other trace elements on soils and crops can be obtained from several reviews [1, 19-24].

As stated above, this review will focus on the effect of Pb, Zn, Cu, Ni, and Cd on agricultural crops following sludge application to soils. One goal of sludge management is to utilize sludges on cropland in such a manner that the risks to human and animal health are minimized. It is generally concluded that phytotoxicities will result from excessive soil levels of Zn, Cu, and Ni before the concentration of these metals in plant tissues reaches a level that is detrimental to animals or man. For Pb, the greatest threat to animal health is from direct ingestion of soils amended with sludge or sludge adhering to forages. Because minimal plant uptake of Pb occurs, humans, except for infants who may ingest soil or dust, will not be exposed to elevated levels of Pb as a result of sludge application on cropland. The greatest concern from a human health standpoint is the accumulation of Cd in various crops. Even though Cd can be phytotoxic when large amounts are added to soils (e.g., >100 kg/ha), total Cd applications are currently limited to 5 to 20 kg/ha by the U.S. EPA because excessive Cd concentrations are found in many crops before toxicity symptoms develop. Thus, in contrast to Zn, Cu, and Ni, an inherent plant control mechanism does not exist to protect animals or humans from elevated levels of Cd in their food supply.

EFFECT OF PLANT, SOIL AND ENVIRONMENTAL FACTORS ON METAL UPTAKE

Plant Factors

Several plant variables influence the metal concentrations found in the leaf, grain, tuber, or fruit of crops grown after application of sludge to soils. These variables include species, cultivar, maturity and plant part. Background levels have been determined for Cd, Cr, Cu, Ni, Pb, and Zn in corn grain collected at various locations in Illinois (Table II).

Table II. Metal Concentrations for Corn Grain Samples Collected from Various Locations in Illinois [33].

Location[a]	Soil pH	Cd	Cr	Cu	Ni	Pb	Zn
				mg/kg[b]			
Research Fields							
Aledo	6.3	0.055e	0.094c	2.69e	0.91d	0.191cd	16.4ab
Brownstown	6.8	0.080f	0.110d	2.68e	0.78c	0.170bc	18.8c
Carthage	6.0	0.010a	0.059a	2.60de	0.49a	0.192cd	20.1d
DeKalb	6.3	0.037c	0.073b	2.18b	0.79c	0.141a	20.7e
Dixon	6.0	0.034c	0.068ab	2.49d	0.78c	0.149ab	16.9b
Elwood	6.3	0.026b	0.118d	3.22f	1.43f	0.133a	19.0c
Hartsburg	6.2	0.051e	0.094c	1.67a	1.03e	0.213d	17.1b
Kewanee	6.3	0.046d	0.193e	4.41g	0.58b	0.184c	15.7a
Urbana	6.2	0.012a	0.059a	2.29c	<0.80	0.243c	18.6c
Soil Association							
Clinton-Hickory-Keomah	5.6	0.043cd	0.107d	2.73a	0.83c	0.204a	21.4a
Fayette-Hickory-Rozetta	5.5–6.0	0.032c	0.078c	3.95b	0.80c	0.226a	24.3b
Ipava-Sable	6.0–6.5	0.042cd	0.099cd	2.82a	0.80c	0.230a	21.7ab
Lawson-Titus Beaucoup	6.7–6.8	0.053d	0.078c	3.76b	0.80c	0.204a	21.6ab
Tama-Ipava	6.0	0.015a	0.047a	3.64b	0.62a	0.180a	22.5ab
Mine-Spoil	7.4	0.023b	0.065b	3.76b	0.73b	0.230a	22.0ab

[a] Corn was grown on soils that were not amended with sewage sludge.
[b] Means in the same column followed by the same letter are not significantly different at the 5% level using the Student-Newman-Keul's test.

The range in metal concentrations of corn grain found were as follows (in mg/kg): Cd, 0.01-0.08; Cu, 1.67-4.41; Ni, <0.8-1.43; Pb, 0.133-0.230; and Zn, 15.7-24.3. Undoubtedly, part of the variability shown for the metal content of corn grain is due to the cultivar grown, soil properties, and management practices (e.g., N fertilization) at the various sampling locations. For all metals, considerable variability exists for the various sampling sites, especially when expressed on a percentage basis. For example, Cd ranged from 0.01 to 0.08 mg/kg, an 800% change from the minimum to maximum value. The concept of relative increases in plant metal concentrations is pertinent when evaluating crop composition data obtained after amending soils with sewage sludge. Even though percentage increases in crop Cd of several hundred percent may be caused by sludge addition, the absolute value of the increase may be quite small (<1 mg/kg). Therefore, it is essential to evaluate whether or not a statistically significant increase in plant Cd levels from 0.01 to 0.08 mg/kg is also biologically significant when predicting the impact of sludge use on human health. Similar considerations apply to interpreting the significance of sludge-induced increases in the concentrations of all metals and trace constituents in crops.

Plant species differ markedly in their ability to assimilate metals from soil and translocate them to various plant parts. The data presented in Table III show that leafy vegetables (lettuce, spinach, and curlycress) tend to accumulate significantly greater amounts of Cd than most forages, corn, soybeans, or wheat when crops are grown under constant conditions [34-36]. The Cd concentrations in crops shown in Table III are considerably greater than those obtained in practice when sludge is applied to soils, but they are included to indicate the relative response of different crops to soil-borne Cd. The response of most plant species to increased soil concentrations of Zn and Ni is essentially linear whereas plant concentrations of both Cu and Pb tend to show a non-linear response to the amount of sludge-borne Cu and Pb applied. Typically, concentrations of Cu and Pb in various plant parts are <50 mg/kg irrespective of the amount of Cu and Pb applied to soil in sewage sludge. Experiments conducted under field conditions indicate increased uptake of Zn and Cd following sludge additions; however, the increases are relatively small compared to greenhouse studies (Table IV). The data in Table IV also show that leafy vegetables and the vegetative part of most plants (e.g., pea, corn, radish, carrot) tend to accumulate more Zn and Cd than the edible or reproductive parts. Numerous studies have been conducted evaluating the uptake of metals by various species of plants from soil amended with sewage sludge in either field or greenhouse experiments [see 1, 13].

Table III. Concentration of Cd in Plant Tissues Associated with a 25% Yield Decrement (YD)[a].

Crop Species	Cd added to soil for 25% YD[a]	Tissue Cd at 25% YD Diagnostic Leaf	Edible[b]	Edible part Cd at 5 μg Cd/g soil [36]
		mg/kg		
Spinach	4	75	75	91
Soybean	5	7	7	6.8
Curlycress	8	70	80	55
Lettuce	13	48	70	27
Corn	18	35	2	0.5
Carrot	20	32	19	8.2
Turnip	28	121	15	4.9
Field Bean	40	15	2	0.4
Wheat	50	33	12	2.8
Radish	96	75	21	2.5
Tomato	160	125	7	1.6
Zucchini Squash	160	68	10	0.4
Cabbage	170	160	11	1.3
Rice	>640	3	2	0.2
Sudangrass	15	9	--	--
Alfalfa	30	24	--	--
White Clover	40	17	--	--
Tall Fescue	95	37	--	--
Bermudagrass	145	43	--	--

[a] Greenhouse experiment using a Domino soil (pH 7.5) amended with 1% sewage sludge containing varying amounts of $CdSO_4$ [34, 35].

[b] Refers to leaf, bean, fruit, grain or tuber.

Table IV. Uptake of Zn, Cu, Cd, Ni, and Pb by Several Crops Grown Under Field Conditions.

Crop	Zn[a] Ck	Zn SS	Cu Ck	Cu SS	Cd Ck	Cd SS	Pb Ck	Pb SS	Ni Ck	Ni SS
					mg/kg					
Carrots[b]	23	103	0.3	1.5	0.48	1.15	<0.4	0.9	--	--
Radishes	37	98	<0.3	<0.3	0.13	0.31	0.5	0.7	--	--
Potatoes	24	53	8.6	19.0	0.12	0.23	<0.4	0.6	--	--
Pea Fruit	70	130	6.0	7.0	<0.03	0.04	<0.1	<0.1	--	--
Pea Vine	49	327	5.9	18.1	0.02	0.20	0.7	1.2	--	--
Tomato Fruit	9	31	<0.3	11.5	0.08	0.33	<0.4	<0.4	--	--
Corn Grain	41	65	2.0	1.9	<0.02	0.05	<0.2	<0.2	--	--
Corn Leaf	22	293	8.7	8.5	0.26	1.32	3.5	4.0	--	--
Lettuce	21	225	1.6	11.9	0.61	2.67	1.1	0.8	--	--
Rye[c]	21.9	55.8	3.9	11.7	0.18	0.40	--	--	0.9	2.5
Sorghum-Sudan	69.9	121.9	6.1	9.4	0.53	0.95	--	--	2.4	4.5
Corn Stover	19.2	52.5	2.0	3.3	0.08	0.27	--	--	0.6	1.1
Corn Grain	20.1	26.9	0.8	0.7	0.09	0.10	--	--	0.9	1.2
Soybean Leaf[d]	205	331	5	15	0.11	0.15	--	--	--	--
Soybean Grain	45	69	11	8	0.35	1.08	--	--	12.7	12.5
Radish Shoot[e]	70	164	14	22	3.2	3.7	--	--	20	22
Radish Root	53	73	7	13	1.3	1.2	--	--	6.9	10.7
Carrot Shoot	40	79	3	5	2.4	2.6	--	--	8	11
Carrot Root	39	59	2	4	2.3	2.4	--	--	13	11
Lettuce	46	133	12	17	1.4	3.9	--	--	5.6	9.6
Tomato Fruit	42	49	10	13	1.4	1.4	--	--	4.4	6.9

[a] CK = non-treated soil; SS = sludge-amended soil.
[b] Sludge added 842 kg Zn/ha, 110 kg Cu/ha, 3.3 kg Cd/ha and 232 kg Pb/ha; soil pH = 5.3 to 5.6. [37].
[c] Sludge added 180 kg Zn/ha, 86 kg Cu/ha, 4.3 kg Cd/ha and 42 kg Ni/ha; soil pH = 5.0 to 6.0. [38].
[d] Sludge added 420 kg Zn/ha, 400 kg Cu/ha, 2.4 kg Cd/ha, 420 kg Pb/ha and 88 kg Ni/ha; soil pH = 6.5. [39].
[e] Sludge added 99 kg Zn/ha, 150 kg Cu/ha, 1.2 kg Cd/ha and 33 kg Ni/ha; soil pH = 5.6 [40].

Different cultivars of the same crop species can exhibit wide variations in metal content. Field experiments with corn cultivars grown on soil treated with sewage sludge have indicated that corn leaf Zn and Cd ranged from 62 to 282 and 2.47 to 62.93 mg/kg, respectively, while corn grain Zn and Cd ranged from 34 to 40 and 0.08 to 3.87 mg/kg, respectively (Table V). In addition, the cultivar exhibiting maximum uptake of Zn did not always show the maximum uptake of Cd. For both Zn and Cd, concentrations in the leaf were always greater than those found in the grain. This study demonstrated that it is possible to select a corn inbred line which will show minimal increases in corn grain Cd when grown on soils amended with sewage sludge. Related studies have been conducted using $CdCl_2$ amended soils and growing soybean seedlings under greenhouse conditions [42]. The concentration of Cd in soybean seedlings varied by a factor of 3 to 4 for the 30 different cultivars studied. Growth of lettuce in solution culture with varying concentrations of Cd [43] and in sludge-amended soils [44] has also shown that Cd uptake is dependent on the cultivar used. In general, the effect of sludge-borne metals on plants is strongly influenced not only by the crop species, but also by the cultivar grown. In most cases, the addition of sludge to soil will result in an increase in the Zn and Cd concentration of the plant grown, while Ni and Cu concentrations may or may not be elevated. It must be realized that the absolute increases in concentrations of Zn and Cd in plants are directly related to the species and cultivar grown. In view of this, it is possible to only state a range for the background concentrations of Zn, Cu, Pb, Ni and Cd in crops grown on either non-amended or sludge-amended soils.

The metal levels in plants vary with the plant part analyzed (i.e., roots, stems, leaves, grain, fruit) and also with maturity. Several studies have concluded that the reproductive part (e.g., grain) of most plants contains lower metal concentrations than roots, leaves, or stems (e.g., corn, pea, soybean data in Table IV). The concentration of metals in various plant parts varies with the physiological maturity of the plant [39, 45]. Furthermore, differences exist between the distribution of Zn, Cd, Ni, Cu, and Pb in various plant parts. A study [39] of the metal concentrations in soybean plants at maturity indicated the following:

Zn: Leaf > Seed > Stem > Pod
Cd: Leaf > Stem > Seed > Pod
Cu: Leaf > Seed > Pod > Stem

Even though soybean leaves contained the highest concentrations of Cu, Zn, and Cd, the ratio of metal concentrations in leaf:grain was greater than that found in other crops such as

Table V. Concentrations of Zn and Cd in Leaves and Grain of Corn Inbred Lines [41].

Inbred Line[a]	Leaf Zn Ck	Leaf Zn SS	Grain Zn Ck	Grain Zn SS	Leaf Cd Ck	Leaf Cd SS	Grain Cd Ck	Grain Cd SS
	---------- mg/kg --------------							
Oh 43	14	282	21	52	0.38	11.33	0.09	0.19
R 806	18	268	29	50	0.40	11.38	0.07	0.40
H 98	18	217	28	54	1.49	48.84	0.11	2.43
A 619	16	193	30	70	0.41	8.06	0.10	0.28
Mo 17	28	189	23	38	0.59	27.06	0.08	1.20
Oh 454	15	171	28	52	0.22	9.57	0.12	0.16
B 37	22	164	27	64	0.92	62.93	0.12	3.87
R 805	18	148	34	44	<0.06	2.47	0.06	0.08
Va 26	17	144	24	40	0.16	3.71	0.09	0.18
H 100	20	140	26	47	0.25	15.40	<0.06	0.51
B 73	20	134	26	43	<0.06	3.74	0.09	0.15
B 14	17	130	22	34	0.32	19.75	0.11	0.73
A 632	17	113	21	46	0.14	11.45	0.11	0.37
N 28	17	109	26	47	0.39	23.41	0.08	0.38
W 64A	17	107	21	39	0.15	9.79	0.06	0.31
R 802A	19	105	25	40	0.16	10.74	<0.06	0.34
H 99	15	103	26	34	0.24	6.81	0.11	0.23
H 96	13	94	27	54	0.22	5.03	0.07	0.18
R 177	16	88	36	52	0.18	5.00	0.16	0.33
B 77	15	62	19	45	0.34	14.67	0.16	1.06

[a] Corn was grown on non-treated (Ck) and sludge-treated (SS) soils. Total Zn in soil: Ck, 68 mg/kg; SS, 454 mg/kg; Total Cd in soil: Ck, 0.3 mg/kg; SS, 21 mg/kg.

corn. For example, the ratio of leaf Cd to grain Cd was
~ 7:1 for soybeans, whereas this ratio for corn typically
ranges from 30 to 80:1 [18] indicating that soybeans transfer
a greater proportion of Cd accumulated from leaves to grain
than does corn.

To characterize the metal status of plants grown on soils
amended with sewage sludge, it is essential that species and
cultivar, maturity and plant part sampled be considered when
developing a plan for monitoring crops. The impact of sludge
application on metal concentrations in crops can be validly
assessed only when plant samples are collected from both
non-treated and sludge-treated soils and are consistent with
respect to: cultivar; degree of physiological maturity;
plant part (e.g., leaf, grain); and management practices (e.g.,
fertilizer use).

Soil Factors

The availability of metals to plants after application of
sewage sludge on soils is dependent on several soil factors,
including rate of metal addition, pH, cation exchange capacity,
and the amounts of organic matter and hydrous Fe, Al, and Mn
oxides present. Of these factors, soil pH has been evaluated
to the greatest extent and appears to be the most critical
factor in controlling metal accumulation by crops, at a
specific concentration of metals in soil. It is generally
recognized that the concentration of Zn and Cd in plant tissue
is directly proportional to the amount of metal applied to
soil in sewage sludge for a given set of soil-plant-
environmental conditions. Sludge applications also influence
Ni and Cu concentrations in crops but to a more limited extent
than Zn and Cd. The effect of metal application rates on
metal concentrations in plants will be discussed in more
detail in a later section of this review.

Soil pH

The addition of sewage sludges to soil can either increase or decrease pH, which subsequently alters the
availability of metals to plants. Sewage sludges may contain
alkaline-earth carbonates (e.g., $CaCO_3$), and consequently
increase soil pH values [10, 46]. Alternatively, the
microbial oxidation of NH_4^+ and reduced S in sludges generates
H^+, resulting in a decrease in soil pH. Short-term field
experiments have indicated that soil pH can increase [44],
decrease [38], or remain unchanged [40] following application
of sludge. Obviously, sludge applications have an insignificant effect on the pH of calcareous soils.

The effect of soil pH on metal uptake by plants has been
evaluated in both field and greenhouse experiments. Liming

an acid soil amended with sewage sludge from pH 5.1 to 6.8 decreased metal uptake by rye in the following order: Zn > Cd > Pb > Cu [47]. Other greenhouse studies have demonstrated the importance of pH in minimizing Zn, Cd, Ni, and Cu uptake by plants from soils amended with sewage sludge [48, 51]. Representative data from field plot studies evaluating the effect of pH on Cd uptake from sludge-amended soils are presented in Table VI. It is apparent from this data that a near neutral pH decreases the plant availability of Cd to a wide-variety of crops grown on soils treated with sludge. In addition, the plant availability of indigenous soil Cd is lowered although to a lesser extent than sludge-borne Cd. In general, the availability of Zn, Cu, Ni, and Pb is similarly decreased by liming acid soils treated with sludge [1]. The relationship between soil pH and the solubility of metal solid phases, sorption of metals by soil inorganic components and complexation of metals by soil organic matter are likely involved in controlling the plant availability of metals [1, 13, 54-59].

Soil Cation Exchange Capacity

Soil cation exchange capacity (CEC) has been proposed as a parameter that is directly proportional to the ability of a soil to minimize plant uptake of Cd [60] and decrease the potential for phytotoxicity from Zn, Cu, and Ni [61, 62] in soils amended with sewage sludge. Maximum Cd applications for agricultural cropland have been established by U.S. EPA as 5 kg/ha for a CEC of <5 meq/100g, 10 kg/ha for CEC's of 5 to 15 meq/100g, and 20 kg/ha for a CEC of >15 meq/100g. Insufficient data is available from field plots to indicate whether or not soil CEC influences plant uptake of Cd at a constant soil pH. Greenhouse experiments have indicated that Cd concentrations in oat shoots decrease with increasing CEC [63]; however, total uptake of Cd was approximately constant because total growth was increased with increasing CEC. Thus, the lower Cd concentration observed may be a result of dilution within the plant rather than lowered Cd availability in the soil. A similar experiment with soybean seedlings suggested that CEC did not affect plant uptake of Cd [64]. Since the soil pH may decrease following sludge applications, use of CEC to scale metal additions has an indirect benefit in that the pH buffering capacity of a soil is directly related to CEC, primarily the CEC contributed by soil organic matter. Thus, pH shifts following sludge application will be more gradual in soils with a CEC of 20 meq/100g than in those with a CEC of 3 meq/100g.

Additional soil properties (i.e., organic matter, clays, and hydrous oxides) are also likely involved in controlling the availability of sludge-borne metals to crops. Only pH

Table VI. Effect of Soil pH on Cd Concentration in Representative Crops

Crop	Control Soil[a] Soil pH L	Control Soil[a] Soil pH H	Control Soil[a] Tissue Cd L	Control Soil[a] Tissue Cd H	Sludge-amended Soil[a] Soil pH L	Sludge-amended Soil[a] Soil pH H	Sludge-amended Soil[a] Tissue Cd L	Sludge-amended Soil[a] Tissue Cd H
			- mg/kg -				- mg/kg -	
Lettuce (Bibb)[b]	4.6	6.3	1.18	0.78	6.0	6.7	8.40	4.18
Lettuce (Romaine)	4.6	6.3	0.88	0.78	6.0	6.7	2.25	1.78
Lettuce (Boston)	4.6	6.3	0.95	0.90	6.0	6.7	3.10	1.85
Cabbage	4.6	6.3	0.19	0.16	6.0	6.7	0.35	0.19
Carrot	4.6	6.3	0.96	0.71	6.0	6.7	2.29	1.25
Swiss Chard[c]	6.1	--	1.31	--	5.3	6.7	13.50	2.91
Corn Grain	6.1	--	0.04	--	5.3	6.7	0.29	0.12
Swiss Chard[d]	6.1	--	1.31	--	4.5	6.6	10.60	2.10
Corn Grain	6.1	--	0.04	--	4.5	6.6	0.52	0.29
Lettuce[e]	4.9	6.3	1.6	0.6	4.9	6.3	20.4	4.6
Swiss Chard	4.9	6.3	--	0.8	4.9	6.3	37.1	2.9
Soybean Grain	4.9	6.3	0.20	0.08	4.9	6.3	1.07	0.38
Oat Grain	4.9	6.3	0.22	0.04	4.9	6.3	2.12	0.38
Orchardgrass	4.9	6.3	0.34	0.17	4.9	6.3	1.67	0.66
Swiss Chard[f]	5.7	6.7	0.6	0.5	5.2	6.2	1.9	0.6
Oat Grain	5.7	6.7	0.05	0.04	5.2	6.2	0.23	0.07
Swiss Chard[g]	5.3	6.4	0.89	0.49	5.6	6.6	70.4	17.7
Oat Grain	5.3	6.4	0.11	0.07	5.6	6.6	3.38	0.54

[a] L = low pH; H = high pH
[b] pH adjusted by liming; 11.2 kg Cd/ha added in sludge [44]
[c] pH decreased (L) by adding S°; 2 kg Cd/ha added in sludge [52]
[d] as in c but 6 kg Cd/ha added in sludge
[e] pH adjusted by liming; old sludge disposal site - total Cd in soil was 0.25 and 2.8 µg/g for control and sludge treated soils, respectively [52]
[f] pH adjusted by liming; old sludge disposal site - DTPA extractable Cd was 0.13 and 0.55 µg/g for control and sludge-treated soils, respectively [53]
[g] as in f but DTPA extractable Cd was 0.94 and 6.3 µg/g for control and sludge-treated soils, respectively

and CEC were discussed above because of their demonstrated importance or current use in regulatory documents.

Environmental Factors

The concentration of metals found in crops following sludge application is modified by both climatic and management factors imposed during a given growing season. Yearly variations are found for the metal content of plants grown on sludge-amended soils even if all soil and plant variables remain constant. To simulate the effects of different climatic regimes on metal uptake by crops, soils have been amended with sewage sludge and maintained at temperatures ranging from 16 to 35° C. Corn grain yields were found to increase with increasing temperature and rate of sludge application [65]. In various corn tissues, concentrations of Zn, Cu, Ni, and Cd were elevated by application of sludge, but similar metal levels were found at all imposed soil temperatures [66]. In contrast, soil heating has been found to increase the concentrations of Zn and Cd in some vegetable species [44]. This study also indicated that cultivars of lettuce accumulate different levels of Zn and Cd in the presence and absence of soil heating.

One source of seasonal variation in the concentration of metals in crops is the various stresses imposed during the growth period. Potential stresses include nutrient deficiencies or toxicities, temperature, drought, and excess soil moisture. Many of these factors are interrelated, i.e., available N losses through denitrification and/or leaching may follow a period of excess soil moisture. In general, the concentrations of metals in vegetative parts can be increased when crops are grown under stress if uptake remains relatively constant but dry matter production decreases.

PHYTOTOXICITY FROM ADDITION OF SLUDGE-BORNE METALS TO SOILS

The historical aspects of phytotoxicity problems resulting from sludge additions to soils have been recently reviewed [1, 13]. In general, phytotoxicities have been caused by elevated soil concentrations of Zn and Cu in conjunction with an acidic pH and growth of metal sensitive crops such as vegetables. When a low metal content sludge is applied to soils at rates based on utilization of nutrients applied by the crop grown (e.g., 10-20 metric tons/ha/yr), potential phytotoxicity problems are minimal, especially if an upper limit on sludge additions is employed and the soil pH is maintained at 6.5 [61, 62].

Several greenhouse experiments have been conducted to evaluate the toxicity of Zn, Cu, Cd, and Ni to various crops.

Soil Cd additions causing a 25% yield decrement are shown in Table III. These data indicate that crop species differ markedly in their susceptibility to soil applied Cd. Similar studies have been conducted to determine the relative phytotoxicity of Cu, Zn, Ni, or Cd either present in sewage sludge or added to sludge as a metal salt for lettuce [51, 67, 68], wheat [67, 69], corn [46, 68, 70], rye [46, 70], barley [71], and soybeans [39, 52]. Greenhouse studies using sludge amended with a metal salt have shown that the relative toxicity of metals is dependent on the crop grown and soil pH. For example, at a soil pH of 5.7, the relative toxicity of Zn:Cu:Ni:Cd was 1:4:6:12 and 1:1:2:2 for wheat and lettuce, respectively [67]. For rye, the relative toxicity of Zn:Cu:Ni was 1.0:1.8:1.0 [46]. In addition, the absolute and relative toxicity of metals to plants is altered by varying soil pH. Zinc toxicity to soybean seedlings is observed at 262 μg Zn/g soil at pH <6.0 but not at pH >6.3 [72]. In an acid soil, Zn, Cu, Ni, and Cd are associated with yield reductions of wheat grain while liming to pH 6.7 results in only Cu and Cd affecting yield [69].

It should be noted that many phytotoxicity studies have been conducted with sludges amended with metal salts. The plant availability of metals added to soils as a soluble salt is significantly greater than the metal forms present in sludges [39, 70-73]. In fact, addition of sludge to soil reduced the toxicity of Cu and Ni to lettuce [68]. Nevertheless, the above cited studies demonstrate the importance of soil properties, mainly pH, and the interactions between metals when considering the potential for phytotoxicity arising from application of sludge to cropland. A summary of sludge application rates resulting in yield reductions for a variety of crops grown under greenhouse conditions are shown in Table VII.

In contrast to greenhouse studies, very few experiments conducted under field conditions have shown decreased crop yields because of metal toxicities. Yield reductions were shown to occur immediately after sludge applications and may have been caused by accumulation of soluble salts or NH_4^+ in the root zone [74]. In subsequent years of this study, all sludge application rates (0 to 60 metric tons/ha) resulted in increased corn yields. Yields of barley have not been significantly reduced for 3 years even though sludge applications annually added 876 kg Zn/ha, 244 kg Pb/ha, 40 kg Ni/ha, and 134 kg Cu/ha to an acid soil (pH ~4.5) [75]. Yield reductions of small grains have been observed for 3 years after a single large application of sludge containing relatively high metal concentrations; however, corn yields were increased while soybeans were unaffected (Table VIII) [87]. It should be noted that no yield reductions would have occurred if these metal-enriched sludges were applied to soil at rates consistent with current metal addition guidelines [61, 62]. Research in England has shown toxicities of Zn, Cu, and Ni to sensitive crops grown on soils (pH = 6.1 to 7.0) treated with sludges containing elevated metal concentrations [100]. The relative toxicity of Zn, Cu, and Ni was approximately 1:2:8. However, the relative toxicity was also related to the plant grown and soil properties [101]. In general, sludge applications within the 10 to 40 metric ton/ha/yr range do not result in metal phytotoxicities and more typically cause yield increases. The effect of sludge applications on yield has been studied for the following selected crops under field conditions: barley [75], corn [65, 74-81, 87], soybeans [39, 82, 83, 87], sorghum [82], wheat [84, 87], potatoes [16], vegetables [85] and Coastal bermuda grass [86]. Even though phytotoxicity does not commonly occur under field conditions at recommended rates of sludge application, yield depressions could readily occur if the soil pH decreased (e.g., from 6.5 to approximately 5.5) and a metal sensitive crop was grown. Maintenance of soil pH at 6.2 to 6.5 is probably the most critical factor in minimizing potential phytotoxicity problems following application of sludges to soil.

Table VII. Sludge Application Rates Associated With a 25% Yield Reduction and Interveinal Chlorosis for Crops Grown Under Greenhouse Conditions [52].

Crop	Yield Reduction pH 5.5	Yield Reduction pH 6.5	Chlorosis pH 5.5	Chlorosis pH 6.5
	----- metric tons sludge/ha[a] -----			
Swiss Chard	$\geq$ 110	$\geq$ 220	$\geq$ 110	$\geq$ 330
Soybeans	$\geq$ 110	$\geq$ 220	$\geq$ 110	--
Snapbean	$\geq$ 110	--	$\geq$ 110	--
Orchardgrass	$\geq$ 330	$\geq$ 330	$\geq$ 330	--
Oat	--	$\geq$ 550	$\geq$ 550	--
Corn	--	$\geq$ 440	--	--

[a] Metals applied at 110 metric tons/ha were (in kg/ha): Zn, 530; Cu, 240; Ni, 32; Cd, 2.3; Pb, 145.

Table VIII. Effect of Three Sludges on Grain Yields of Small Grains, Corn, and Soybeans [87].

Treatment[a]	Sludge Applied	Corn 1977	Corn 1978	Corn 1979	Soybeans 1977	Soybeans 1979	Oats 1977	Winter Wheat 1978	Winter Wheat 1979
		\-\-\-\-\-\-\-\-\-\-\-\-\- metric tons/ha \-\-\-\-\-\-\-\-\-\-\-\-\-							
None		7.44	4.20	3.56	3.34	3.25	2.77	2.32	5.33
Sludge A[b]	56	9.28	5.28	4.58	3.32	3.28	3.44	1.91	5.66
	112	9.79	5.98	5.97	3.40	3.13	2.84	1.29	4.45
	224	9.98	5.72	6.86	3.44	3.30	2.54	0.82	3.60
	448	8.26	8.52	8.90	3.11	3.10	1.26	0.95	3.82
Sludge B	56	10.23	--	--	3.32	3.18	4.53	1.50	4.43
	112	10.68	--	--	3.40	3.38	4.15	2.11	3.06
	168	--	--	--	3.26	2.79	3.26	2.04	2.97
Sludge C	56	9.60	--	--	3.43	3.28	3.78	2.52	5.22
	112	10.30	5.40	5.72	3.49	3.34	3.72	1.50	4.86
	168	--	--	--	3.20	3.36	5.00	2.25	3.94

[a] Sludge was applied in the fall of 1976 to Chalmers silt loam. Soil pH in 1979 ranged from 5.8 on the control to 7.4 for high application rate of sludge A and C.

[b] Metal concentrations in sludges were (in mg/kg):
  Sludge A - Zn, 6,800; Cu, 1,200; Ni, 2,040; and Cd, 284.
  Sludge B - Zn, 1,900; Cu, 1,330; Ni, 430; and Cd, 1,210.
  Sludge C - Zn, 5,200; Cu, 450; Ni, 215; and Cd, 247.

EFFECT OF SLUDGE APPLICATION RATE ON METAL UPTAKE BY CROPS

Other than soil pH, the concentration of metals in crops will be most strongly influenced by the amount of sludge-borne metals applied to soils. The majority of field experiments consist of multiple sludge application rates over a period of years in conjunction with growing one or more crops. Only representative data will be presented here to indicate typical metal concentrations in crops following sludge applications. The data presented are for a given combination of sludge-soil-plant variables, and it is likely that the absolute values of plant-metal concentrations will change if one or more of the above variables is altered.

The uptake of sludge-borne Pb, Zn, Cu, and Ni by a variety of crops has been described in several reviews [1, 13, 22] and numerous research articles [16-18, 32-39, 66-86]. The concentrations of Zn, Cu, and Ni in selected vegetable crops as a function of metals applied are shown in Table IX. The concentrations of Zn tend to increase to a greater extent than either Cu or Ni in sludge-amended soils because of more Zn being added to the soil and the general tendency for plants to accumulate minimal levels of Cu and Ni. Elevated Ni concentrations in plants are primarily obtained in greenhouse studies, especially when acid soils are used [50].

The Zn and Cd content of most crops are essentially a linear function of the amount of sludge-borne Zn and Cd applied to the soil. The effect of sludge applications on the Zn and Cd concentrations in the leaves and grain of corn and soybeans are presented in Tables X and XI, respectively. These data represent both low and high levels of Zn and Cd additions to soil. For both corn and soybeans, Zn and Cd concentrations are greater in the leaves (or stover) than in the grain. The importance of soil pH in controlling Zn and Cd availability is also apparent by comparing the plant composition data at **approximately** equivalent metal addition rates for studies 1 and 3. Typically, the amounts of Zn and Cd applied to soils are considerably less than those shown in Tables X and XI when sewage sludge application rates are based on nutrient utilization by the crop grown.

The relative importance of annual and cumulative application rates of Zn and Cd on crop composition remains to be established. Studies in Illinois [78, 83] suggest that the most recent application of sludge controls plant uptake of Zn and Cd to a greater extent than the cumulative amounts of Zn and Cd applied over a period of years. If the plant availability of sludge-borne metals does not change significantly with time after application, then comparable plant concentrations of metals should be obtained for 1) a single, large application of sludge, and 2) repeated annual applications to attain the same total amount applied (i.e., metal additions are additive). As reviewed by Chaney and Hornick [53], the effect of sludge metal additions on plant composition differs for a single and repeated sludge applications. Both annual and cumulative limits are being recommended for Cd additions to soils [61] to minimize both short- and long-term increases in the Cd concentrations of crops.

Table IX. Effect of Sludge Application Rate on Zn, Cu, and Ni Concentrations in Selected Vegetables.

Crop	Metal Applied			Plant Tissue Concentration		
	Zn	Cu	Ni	Zn	Cu	Ni
	--- kg/ha ---			--- mg/kg ---		
Lettuce[a]	0	0	0	21	1.6	--
	120	27	2.7	94	5.4	--
	241	55	5.4	155	8.1	--
	482	110	10.8	255	11.9	--
Lettuce[b]	0	0	0	46	12	5.6
	33	50	11	73	17	7.6
	99	150	33	133	17	9.6
Carrots[a]	0	0	0	23	<0.3	--
	120	27	2.7	53	<0.3	--
	241	55	5.4	72	<0.3	--
	482	110	10.8	103	1.5	--
Carrots[b]	0	0	0	39	2	13
	33	50	11	51	4	15
	99	150	33	59	4	11
Radishes[a]	0	0	0	37	<0.3	--
	120	27	2.7	46	<0.3	--
	241	55	5.4	50	<0.3	--
	482	110	10.8	98	<0.3	--
Radishes[b]	0	0	0	53	7	6.9
	33	50	11	61	9	10.5
	99	150	33	73	13	10.7

[a] Soil pH ranged from 5.3 for control to 6.5 for high sludge treatment [37].

[b] Soil pH ranged from 5.6 to 5.8 [40].

Table X. Effect of Sludge Application Rate on Zn and Cd Concentration in Corn Leaf and Grain.

Study	Metal Applied Zn	Metal Applied Cd	Metal in Leaf Zn	Metal in Leaf Cd	Metal in Grain Zn	Metal in Grain Cd	Soil pH
	-- kg/ha --		------ mg/kg ------				
1[a]	0	0	59	0.2	28	0.09	~6.2
	590	25	122	1.4	37	0.18	~6.2
	1,179	50	193	3.2	46	0.40	~6.2
	2,358	101	393	10.9	56	0.81	~6.2
2[b]	0	0	19[c]	0.08[c]	20	0.09	5.0
	90	2.2	41	0.23	26	0.09	5 4
	180	4.3	52	0.27	27	0.10	5.4
3[d]	0	0	37	0.42	13	<0.05	6.0
	381	16	40	1.07	14	<0.05	6.7
	762	32	62	1.55	19	<0.05	7.0
	1,523	64	69	2.04	19	<0.05	7.1
	3,046	128	77	1.66	16	<0.05	7.3
	0	0	37	0.42	13	<0.05	6.0
	106	68	50	5.08	21	0.06	6.5
	213	136	49	7.81	22	0.14	6.6
	0	0	37	0.42	13	<0.05	6.0
	291	14	44	1.13	23	<0.05	7.2
	582	28	37	1.62	18	<0.05	7.3
	1,164	42	53	0.92	12	<0.05	7.4

[a] Crop grown in 1974 on soil treated with sludge from 1968 to 1974 [78]

[b] Crop grown on Plano silt loam which received a single sludge application [38]

[c] Values for corn stover rather than leaf

[d] Single application of three sludges on Chalmers silt loam [87]

Table XI. Effect of Sludge Application Rate on Zn and Cd in Soybeans.

Study	Metal Applied Zn	Metal Applied Cd	Metal in Leaf Zn	Metal in Leaf Cd	Metal in Grain Zn	Metal in Grain Cd	Soil pH
	-- kg/ha --		------ mg/kg ------				
1[a]	0	0	42	0.05	47	0.07	6.5
	105	0.6	66	0.08	68	0.11	
	210	1.2	75	0.12	69	0.11	
	420	2.4	64	0.10	68	0.12	
2[b]	0	0	49	0.22	49	0.31	6.0-6.5
	408	19.4	103	0.89	63	0.31	
	816	38.9	132	2.73	65	0.57	
	1,632	77.6	143	5.78	78	0.92	
3[c]	0	0	42	1.59	39	0.41	6.0
	381	16	53	2.24	44	0.51	6.7
	762	32	67	1.78	47	0.75	7.0
	1,523	64	55	1.80	44	0.78	7.1
	3,046	128	63	2.42	59	0.93	7.3
	0	0	42	1.59	39	0.41	6.0
	106	68	54	4.62	42	2.07	6.5
	213	136	59	5.02	47	3.31	6.6
	320	204	63	5.97	46	3.36	6.6
	0	0	42	1.59	39	0.41	6.0
	291	14	44	2.10	40	0.48	7.2
	582	28	48	1.72	42	0.55	7.3
	1,164	42	51	2.13	43	0.52	7.4

[a] Single sludge application on a Waukegan silt loam [39]

[b] Crop grown in 1974 on soil treated with sludge from 1969 to 1974 [83]

[c] Single application of three sludges on a Chalmers silt loam [87]

The accumulation of Cd in crops is of concern not only during the time period of sludge application, but also following cessation of this practice. Two approaches can be used to evaluate the residual plant availability of sludge-borne metals, primarily Cd, in soils, namely monitoring of "old" sludge application sites and using experimental field plots. Limitations with old sludge sites include a lack of knowledge of sludge composition, application rates, and time of application. Since most experimental plots have been established since 1970, old sludge sites enable an examination of metal availability to plants in soils where longer time periods have elapsed allowing the soil-sludge system to approach equilibrium. Analyses of corn leaves (1973) grown on a soil treated with sludge from 1930 to 1965 indicated that both Zn and Cd were increased by a factor of three to four while only Zn was elevated in corn grain [92]. Chaney and co-workers have grown a variety of crops on old sludge sites in the northeastern U.S. [4, 53]. In essence, their studies show that a part of the Zn and Cd applied to soil in sewage sludge remains available for plant uptake and that liming acid soils results in significant reductions in Zn and Cd concentrations in crops. Relatedly, Cd applied to soils as a contaminant in superphosphate fertilizer can result in small increases in plant Cd concentrations [93-95], suggesting that the total Cd level in a soil is an important factor in regulating Cd uptake by plants. Furthermore, the Cd concentration of plants tends to parallel the native soil total Cd level at a constant pH.

The residual availability of sludge-borne Cd in soils is summarized for several field studies in Table XII. Corn leaf Cd concentrations tend to remain above the control for at least 3 to 4 years after sludge applications cease. The relative increases are greater at the higher Cd addition rates. In study 1, corn grain Cd concentrations were essentially the same for the control and sludge-amended plots after 4 years, whereas insignificant changes in grain Cd occurred in study 2 during and after sludge additions. The data for corn leaves in studies 1 and 2 and corn grain in study 3 suggest that the plant availability of Cd in sludge-amended soils definitely decreases with time, but the concentration of Cd in the vegetative part of plants will likely remain elevated over the normal background level. Maintaining soil pH at 6.5 or above is critical to minimize the uptake of Cd after terminating sludge applications.

Data obtained from both field and greenhouse experiments have been used to project the impact of applying sewage sludge on agricultural cropland on the Cd content of the human diet. Greenhouse studies are more manageable and allow the simultaneous evaluation of numerous sludge application rates, soil types and crop species, and thus, have been employed in Cd

Table XII. Plant Availability of Cd in Soils After Termination of Sludge Applications.

Study	Year	Sludge Applied	Corn Leaf Cd at Relative Cd Application Rate of 0	1	2	4	Corn Grain Cd at Relative Cd Application Rate of 0	1	2	4
			--- mg Cd/kg ---							
1[a]	1971	Yes	0.5	3.1	11.6	35.6	0.15	0.36	0.70	1.37
	1972	Yes	0.5	4.3	10.4	23.3	0.15	0.27	0.43	0.89
	1973	Yes	0.3	1.3	2.5	7.1	0.16	0.20	0.27	0.44
	1974	No	0.1	0.3	0.9	3.6	0.18	0.18	0.15	0.23
	1975	No	0.7	1.2	1.9	5.9	0.15	0.15	0.22	0.17
	1976	No	0.3	0.5	1.3	2.9	0.14	0.14	0.14	0.15
	1977	No	0.3	0.6	1.3	2.1	0.10	<0.06	<0.06	0.07
2[b]	1972	Yes	0.17	0.48	1.08	1.49	0.02	0.04	0.05	0.07
	1973	Yes	0.18	0.84	1.24	1.68	0.01	0.04	0.06	0.12
	1974	Yes	0.07	0.19	0.53	0.98	0.01	0.01	0.05	0.10
	1975	No	0.19	0.40	0.93	1.64	0.05	0.05	0.08	0.10
	1976	No	0.20	0.46	1.13	2.05	0.04	0.06	0.08	0.11
	1977	No	0.26	0.55	0.93	1.36	0.03	0.04	0.04	0.06
3[c]	1974	No	--	--	--	--	0.12	0.39	0.58	0.68
	1975	No	--	--	--	--	0.20	0.30	0.38	0.44
	1976	No	--	--	--	--	0.09	0.22	0.28	0.37

[a] Total Cd applied through 1973 was 0, 14.5, 29 and 58 kg/ha for the 0, 1, 2, and 4 relative rates [77]

[b] Total Cd applied through 1974 was 0, 4.8, 9.6, and 19.2 for the 0, 1, 2, and 4 relative rates [91]

[c] Sludge applied in 1972. Total Cd applied was 0, 1, 2, and 4 kg/ha for the 0, 1, 2, and 4 relative rates [52]

impact projections. A recent study has indicated that significant differences in Cd content are observed when lettuce and onions are grown in the greenhouse, in mini-plots outside the greenhouse, or in field plots (Table XIII) [96]. Under field conditions, sludge applications caused slight increases in lettuce Cd concentrations and had no effect on onions. The absolute concentrations of Cd in plants grown on control and sludge-treated soils were from 3 to 20 fold greater under greenhouse than field conditions. Similar differences were obtained for plant uptake of Cu, Ni, and Zn. Since both lettuce and onions are shallow-rooted plants and were grown in 16 kg of soil in the greenhouse, less root exploration of soil under field conditions does not appear to explain these results. The authors discussed several other factors to explain the differences but they did not state a definite explanation. It is also possible that differential changes in soil pH may explain some of the differences found. In view of this study, the absolute values of plant metal concentrations obtained under greenhouse conditions cannot be extrapolated to crops grown in the field. Nevertheless, greenhouse studies are valuable to examine the complexity of sludge, soil and plant factors governing metal concentrations in crops grown on sludge-amended soils.

Table XIII. Comparison of Cd Uptake by Lettuce and Onions in Various Experiments [96].

Crop	Sludge Applied (metric tons/ha)	Greenhouse (mg Cd/kg)	Mini-Plots[a] (mg Cd/kg)	Field (mg Cd/kg)
Lettuce	0	2.1	0.3	0.6
	1	3.0	0.3	0.8
	3	5.0	0.4	0.8
	9	6.0	0.8	0.9
Onion bulbs	0	1.1	0.15	0.15
	1	1.4	0.25	0.15
	3	2.1	0.40	0.15
	9	3.2	0.45	0.15

[a] 1-m x 1-m plots located outside greenhouse

CURRENT GUIDELINES FOR APPLICATION OF SEWAGE SLUDGE ON CROPLAND

The U.S. EPA along with regulatory agencies in numerous states either have developed or are in the process of drafting regulations concerning application of sewage sludge on agricultural cropland. The "Criteria for Classification of Solid Waste Disposal Facilities" [60] contains limits for Cd additions on soils used for growing food-chain crops and also for "dedicated sludge disposal sites" where the crop grown is solely used for animal feed. In addition to Cd, the criteria present minimum requirements for stabilization of sludge, time periods between the most recent sludge application and growth of vegetables, and maximum concentrations of polychlorinated biphenyls (10 mg/kg) allowed in sludges that are surface applied.

The criteria limit the Cd applied to soils used for growing food-chain crops on both an annual and cumulative basis. Maximum annual Cd additions will undergo a phased reduction from 2 kg/ha in 1980 to 0.5 kg/ha in 1985, except for leafy vegetables, root crops or tobacco, where the current and future maximum is 0.5 kg/ha. The soil-sewage sludge mixture must be adjusted to pH 6.5 at the time of sludge application to minimize Cd uptake by crops. The cumulative amount of Cd applied to soils is subdivided based on 1) native soil pH, and 2) soil cation exchange capacity. If a soil is naturally acid (pH <6.5) and the soil will not be maintained at pH 6.5 after termination of sludge applications, then the total amount of Cd that can be applied is 5 kg/ha. Alternatively, if a soil is naturally at pH $\geq$ 6.5 (e.g., calcareous) or will be maintained at pH $\geq$ 6.5 through liming, then the cumulative amount of Cd allowed is a function of soil CEC as follows: CEC < 5 meq/100g, 5 kg Cd/ha; CEC = 5-15 meq/100g, 10 kg Cd/ha; CEC > 15 meq/100g, 20 kg Cd/ha. Neither annual nor cumulative limits on Cd are imposed when all crops produced are used only for animal feed, and thus, increased Cd will not directly enter the human food-chain.

A soundly managed system for use of sewage sludge on cropland will consider parameters other than Cd. These additional parameters are the annual rate of N applied and the cumulative amounts of Pb, Ni, Zn, and Cu added to the soil [61, 62]. The application of sludges containing 10 to 20 mg Cd/kg at a 2 kg Cd/ha rate will add plant available N at a rate from 5 to 10 times that required for the most N demanding crops. The amount of N applied to soil in sewage sludge should be consistent with the N required by the crop in order to minimize the potential for NO_3^- leaching into groundwaters. The amount of plant available N in sewage sludge is equivalent to the NH_4^+ and NO_3^- content plus a fraction of the organic N that will be mineralized during decomposition of sludge in soils.

To enable the growth of any crop after sewage sludge applications cease, it is necessary to limit the cumulative amounts of Pb, Zn, Cu, and Ni applied to soil [61, 62]. The limits presented in Table XIV were developed through cooperative efforts of researchers in the northcentral and eastern regions of the U.S. In essence, metal limits are set for three categories of soil CEC. These limits assume that soil pH is maintained at 6.5 to prevent potential phytotoxicities from Zn, Cu, or Ni when metal sensitive crops are grown. As stated previously, the main reason for including Pb is to minimize problems resulting from direct ingestion of soil or dust because plant uptake of sludge-borne Pb is minimal. It should be noted that the use of CEC in Table XIV does not imply that sludge-borne metals are present as exchangeable cations in soils. Several studies have demonstrated that an insignificant fraction of the Pb, Zn, Cu, Ni, or Cd added to soils in sludge is held by the exchange complex of either the sludge or soil [e.g., 28, 97]. Soil CEC is an easily measured property and is positively correlated with soil organic matter and clay content, both of which tend to minimize, either directly or indirectly, the solubility of metals. Subsequent research [63, 64] has indicated that soil CEC does not influence to a great extent the plant availability of Cd in soils. Nevertheless, phytotoxicity problems could result if a cumulative metal limit was not employed for Zn, Cu, and Ni applications on agricultural soils. In all soils, the major factor influencing metal availability to plants is pH and its control within the 6.2 to 6.5 range will preclude excessive uptake of Cd and phytotoxicity from Zn, Cu, and Ni following addition of sludge to cropland. The design of systems for application of sludge on cropland has been described in detail using the above considerations for plant nutrients (N, P, K) and heavy metals (Cd, Pb, Zn, Cu, and Ni) [98, 99].

Table XIV. Maximum Amounts of Pb, Zn, Cu, and Ni That Can Be Applied to Agricultural Cropland [61, 62][a].

Metal	Soil Cation Exchange Capacity, meq/100g		
	<5	5-15	>15
	------- kg/ha ---------		
Pb	500	1000	2000
Zn	250	500	1000
Cu	125	250	500
Ni	125	250	500

[a] Assumes soil pH $\geq$ 6.5.

REFERENCES

1. "Application of Sewage Sludge to Cropland: Appraisal of Potential Hazards of the Heavy Metals to Plants and Animals," Council for Agricultural Science and Technology Report 64, Ames, IA (1976).

2. Sommers, L. E. "Chemical Composition of Sewage Sludges and Analysis of Their Potential Use As Fertilizers," *J. Environ. Qual.* 6:225-232 (1977).

3. Furr, A. K., A. W. Lawrence, S. S. C. Tong, M. C. Grandolfa, R. A. Hofsteader, C. A. Bache, W. H. Gutenmann, and D. J. Lisk. "Multielement and chlorinated hydrocarbon analysis of municipal sewages of American cities," *Environ. Sci. Technol.* 10:683-687 (1976).

4. Chaney, R. L., S. B. Hornick, and P. W. Simon. "Heavy Metal Relationships During Land Utilization of Sewage Sludge in the Northeast," in *Land As a Waste Management Alternative*, R. C. Loehr, Ed. (Ann Arbor, MI: Ann Arbor Science Publishers, 1977), pp. 283-314.

5. Tabatabai, M. A. and W. T. Frankenberger, Jr. "Chemical Composition of Sewage Sludges in Iowa," Research Bulletin 586, Agriculture and Home Economics Experiment Station, Iowa State Univ., Ames, IA (1979).

6. Doty, W. T., D. E. Baker, and R. F. Shipp. "Chemical Monitoring of Sewage Sludge in Pennsylvania," *J. Environ. Qual.* 6:421-426 (1977).

7. Page, A. L. "Fate and Effects of Trace Elements in Sewage Sludge When Applied to Agricultural Lands. A Literature Review Study," National Environmental Research Center, USEPA Report 670/2-74-005 (1974).

8. Blakeslee, P. A. "Monitoring Considerations for Municipal Wastewater Effluent and Sludge Application to the Land," in *Recycling Municipal Sludges and Effluents on Land* (Washington, D.C.: Nat. Assoc. State Univ. and Land-Grant Colleges, 1974), pp. 183-198.

9. Sommers, L. E., D. W. Nelson, J. E. Yahner, and J. V. Mannering. "Chemical Composition of Sewage Sludge From Selected Indiana Cities," *Indiana Acad. Sci.* 82:424-432 (1973).

10. Sommers, L. E., D. W. Nelson, K. J. Yost. "Variable Nature of Chemical Composition of Sewage Sludges," J. Environ. Qual. 5:303-306 (1976).

11. Berggren, B. and S. Oden. "Analyresultat Rorande Fung Metaller Och Klorerade Kolvaten. I Rötslam Fran Svenska Reningsverk, 1968-1971. Institution für Markvetsenskap Lantkrukshogskölan, 750 07, Uppsala, Sweden.

12. Berrow, J. L. and J. Webber. "Trace Elements in Sewage Sludges," J. Sci. Food Agric. 23:93-100 (1972).

13. Chaney, R. L. and P. M. Giordano. "Microelements as Related to Plant Deficiencies and Toxicities," in Soils for Management and Utilization of Organic Wastes and Waste Waters, L. F. Elliott and F. J. Stevenson, Eds. (Madison, WI: Soil Sci. Soc. Amer., 1977), pp. 234-279.

14. VanLoon, J. C. "Mercury Contamination of Vegetation Due to the Application of Sewage Sludge as Fertilizer," Environ. Letters 6:211-218 (1974).

15. Furr, A. K., W. C. Kelly, C. A. Bache, W. H. Gutenmann, and D. J. Lisk. "Multielement Absorption by Crops Grown in Pots on Municipal Sludge-Amended Soil," J. Agric. Food Chem. 24:889-892 (1976).

16. Baerug, R. and J. H. Martinsen. "The Influence of Sewage Sludge on the Content of Heavy Metals in Potatoes and on Tuber Yield," Plant and Soil 47:407-418.

17. Garcia, W. J., C. W. Blessin, G. E. Inglett, and R. O. Carlson. "Physical-Chemical Characteristics and Heavy Metal Content of Corn Grown on Sludge-Treated Strip-Mine Soil," J. Agric. Food Chem. 22:810-815 (1974).

18. Garcia, W. J., C. W. Blessin, H. W. Sandford, and G. E. Englett. "Translocation and Accumulation of Seven Heavy Metals in Tissues of Corn Plants Grown on Sludge-Treated Strip-Mined Soil," J. Agric. Food Chem. 27:1088-1094. (1979).

19. Elliott, L. F. and F. J. Stevenson, Eds. Soils for Management and Utilization of Organic Wastes and Waste Waters. (Madison, WI: Soil Science Society of America, 1977).

20. Allaway, W. N. "Agronomic Controls Over the Environmental Cycling of Trace Elements," Adv. Agron. 20:235-274 (1968).

21. Lisk, D. J. "Trace Elements in Soils, Plants, and Animals," Adv. Agron. 24:267-325 (1972).

22. Baker, D. E. and L. Chesnin. "Chemical Monitoring of Soils for Environmental Quality and Animal and Human Health," Adv. Agron. 27:305-374 (1976).

23. Loehr, R. C., Ed. Land as a Waste Management Alternative (Ann Arbor, MI: Ann Arbor Sci. Publishers, 1977).

24. Chaney, R. L. "Crop and Food Chain Effects of Toxic Elements in Sludges and Effluents," in Proceedings of Recycling Municipal Sludges and Effluents on Land (Washington, D.C.: The Nat. Assoc. State Univ. and Land-Grant Colleges, 1974), pp. 129-141.

25. Bloomfield, C. and G. Pruden. "The Effects of Aerobic and Anaerobic Incubation on the Extractabilities of Heavy Metals in Digested Sewage Sludge," Environ. Pollution 8:217-232 (1975).

26. Silviera, D. J. and L. E. Sommers. "Extractability of Copper, Zinc, Cadmium, and Lead in Soils Incubated With Sewage Sludge," J. Environ. Qual. 6:47-52 (1977).

27. Stover, R. C., L. E. Sommers, and D. J. Silviera. "Evaluation of Metals in Wastewater Sludge," J. Water Pollution Control Fed. 48:2165-2175 (1976).

28. Silviera, D. J., L. E. Sommers, and D. W. Nelson. "Evaluation of Crystalline Components in Sewage Sludge," Commun. Soil Sci. Plant Anal. 8:509-519 (1977).

29. Boyd, S. A., L. E. Sommers, and D. W. Nelson. "Infrared Spectra of Sewage Sludge Fractions: Evidence for an Amide Binding Site," Soil Sci. Soc. Amer. J. 43:893-899 (1979).

30. Tan, K. H., L. D. King, and H. D. Morris. "Complex Reactions of Zinc with Organic Matter Extracted from Sewage Sludge," Soil Sci. Soc. Amer. Proc. 35:748-752 (1971).

31. Sposito, G., K. M. Holtzclaw, and C. S. LeVesque-Madore. "Cupric Ion Complexation by Fulvic Acid Extracted from Sewage Sludge-Soil Mixtures," Soil Sci. Soc. Amer. J. 43:1148-1155 (1979).

32. Corey, R. B. and D. R. Keeney. Dept. of Soil Sci., Univ. of Wisconsin. Unpublished data.

33. Pietz, R. I., J. R. Peterson, C. Lue-Hing, and L. F. Welch. "Variability in the Concentration of Twelve Elements in Corn Grain," J. Environ. Qual. 7:106-110 (1978).

34. Bingham, F. T., A. L. Page, R. J. Mahler, and T. J. Ganje. "Growth and Cadmium Accumulation of Plants Grown on a Soil Treated with a Cadmium-Enriched Sewage Sludge," J. Environ. Qual. 4:207-211 (1975).

35. Bingham, F. T., A. L. Page, R. J. Mahler, and T. J. Ganje. "Yield and Cadmium Accumulation of Forage Species in Relation to Cadmium Content of Sludge-Amended Soil," J. Environ. Qual. 5:57-59 (1976).

36. Bingham, F. T. "Bioavailability of Cd to Food Crops in Relation to Heavy Metal Content of Sludge-Amended Soil," Environ. Health Perspectives 28:39-43 (1979).

37. Dowdy, R. H. and W. E. Larson. "The Availability of Sludge-borne Metals to Various Vegetable Crops," J. Environ. Qual. 4:278-282 (1975).

38. Kelling, K. A., D. R. Keeney, L. M. Walsh, and J. A. Ryan. "A Field Study of the Agricultural Use of Sewage Sludge: III. Effect on Uptake and Extractability of Sludge-borne Metals," J. Environ. Qual. 6:353-358 (1977).

39. Ham, G. E. and R. H. Dowdy. "Soybean Growth and Composition as Influenced by Soil Amendments of Sewage Sludge and Heavy Metals: Field Studies," Agron. J. 70:326-330 (1978).

40. Schauer, P. S., W. R. Wright, and J. Pelchat. "Sludge-borne Heavy Metal Availability and Uptake by Vegetable Crops Under Field Conditions," J. Environ. Qual. 9:69-73 (1980).

41. Hinesly, T. D., D. E. Alexander, E. L. Ziegler, and G. L. Barrett. "Zinc and Cadmium Accumulation by Corn Inbreds Grown on Sludge-Amended Soil," Agron. J. 70:425-428 (1978).

42. Boggess, S. F., S. Willavze, and D. E. Koeppe. "Differential Response of Soybean Varieties to Soil Cadmium," Agron. J. 70:756-760 (1978).

43. John, M. K. and C. J. VanLaerhoven. "Differential Effects of Cadmium on Lettuce Varieties," Environ. Pollution 10:163-173 (1976).

44. Giordano, P. M., D. A. Mays, and A. D. Behel, Jr. "Soil Temperature Effects on Uptake of Cadmium and Zinc by Vegetables Grown on Sludge-Amended Soil," *J. Environ. Qual.* 8:233-236 (1979).

45. Sommers, L. E. Dept. of Agronomy, Purdue University. Unpublished data.

46. Cunningham, J. D., D. R. Keeney, and J. A. Ryan. "Yield and Metal Composition of Corn and Rye Grown on Sewage Sludge-amended Soil," *J. Environ. Qual.* 4:448-454 (1975).

47. Lagerwerff, J. V., G. T. Biersdorf, R. P. Milberg, and D. L. Brower. "Effects of Incubation and Liming on Yield and Heavy Metal Uptake by Rye from Sewage-sludged Soil," *J. Environ. Qual.* 6:427-431 (1977).

48. John, M. K., and C. J. VanLaerhoven. "Effects of Sewage Sludge Composition, Application Rate, and Lime Regime on Plant Availability of Heavy Metals," *J. Environ. Qual.* 5:246-251 (1976).

49. Street, J. J., B. R. Sabey, and W. L. Lindsay, "Influence of pH, Phosphorus, Cadmium, Sewage Sludge, and Incubation Time on the Solubility and Plant Uptake of Cadmium," *J. Environ. Qual.* 7:286-290 (1978).

50. Bingham, F. T., A. L. Page, G. A. Mitchell, J. E. Strong. "Effects of Liming an Acid Soil Amended with Sewage Sludge Enriched with Cd, Cu, Ni, and Zn on Yield and Cd Content of Wheat Grain," *J. Environ. Qual.* 8:202-207 (1979).

51. Mahler, R. J., F. T. Bingham, and A. L. Page. "Cadmium-enriched Sewage Sludge Application to Acid and Calcareous Soils: Effect on Yield and Cadmium Uptake by Lettuce and Chard," *J. Environ. Qual.* 7:274-281 (1978).

52. Chaney, R. L., P. T. Hundemann, W. T. Palmer, R. J. Small, M. C. White, and A. M. Decker. "Plant Accumulation of Heavy Metals and Phytotoxicity Resulting from Utilization of Sewage Sludge and Sludge Composts on Cropland," in *Proceedings of National Conference on Composting of Municipal Residues and Sludges* (Rockville, MD: Information Transfer, Inc., 1978), pp. 86-97.

53. Chaney, R. L. and S. B. Hornick. "Accumulation and Effects of Cadmium on Crops," in *Proceedings First International Cadmium Conference* (London: Metal Bulletin Limited, 1978), pp. 125-140.

54. Street, J. J., W. L. Lindsay, and B. R. Sabey. "Solubility and Plant Uptake of Cadmium in Soils Amended with Cadmium and Sewage Sludge," *J. Environ. Qual.* 6:72-77 (1977).

55. Lindsay, W. L. Chemical Equilibria in Soils (New York: John Wiley & Sons, Inc., 1979), pp. 210-266, 315-342.

56. Lindsay, W. L. "Inorganic Phase Equilibria in Soils," in *Micronutrients in Agriculture*, J. J. Mortvedt, P. M. Giordano, and W. L. Lindsay, Eds. (Madison, WI: Soil Sci. Soc. Amer., 1972), pp. 41-47.

57. Santillan-Medrano, J. and J. J. Jurinak. "The Chemistry of Lead and Cadmium in Soil: Solid Phase Formation," *Soil Sci. Soc. Amer. Proc.* 39:851-856 (1975).

58. Jenne, E. A. and S. N. Luoma. "Forms of Trace Elements in Soils, Sediments, and Associated Waters: An Overview of Their Determination and Biological Availability," in *Biological Implications of Metals in the Environment*. H. Drucker and R. E. Wildung, Eds. (Oak Ridge, TN: Technical Information Center, ERDA, Conf-750929, 1977), pp. 110-143.

59. Cataldo, P. A. and R. E. Wildung. "Soil and Plant Factors Influencing the Accumulation of Heavy Metals by Plants," *Environ. Health Perspectives* 27:149-159 (1978).

60. "Criteria for the Classification of Solid Disposal Facilities and Practices: Final, Interim Final, and Proposed Regulations," *Federal Register* 44:53438-53468 (1979).

61. *Application of Sludges and Waste Waters on Agricultural Land; A Planning and Educational Guide*. B. D. Knezek and R. H. Miller, Eds., North Central Regional Publication No. 235 (Wooster, OH: Ohio Agricultural and Development Center, 1976) - Reprinted by U.S. EPA as MCD-35 (1978).

62. "Municipal Sludge Management: Environmental Factors," Office of Water Program Operations, U.S. EPA Report 430/9-77-004 (1977).

63. Haghiri, F. "Plant Uptake of Cadmium as Influenced by Cation Exchange Capacity, Organic Matter, Zinc, and Soil Temperature," *J. Environ. Qual.* 3:180-183 (1974).

64. Latterell, J. J., R. H. Dowdy, and G. E. Ham. "Sludge-borne Metal Uptake by Soybeans as a Function of Soil Cation Exchange Capacity," *Commun. Soil Sci. and Plant Anal.* 7:465-476 (1976).

65. Sheaffer, C. C., A. M. Decker, R. L. Chaney, and L. W. Douglass. "Soil Temperature and Sewage Sludge Effects on Corn Yield and Macronutrient Content," *J. Environ. Qual.* 8:450-454 (1979).

66. Sheaffer, C. C., A. M. Decker, R. L. Chaney, and L. W. Douglass. "Soil Temperature and Sewage Sludge Effects on Metals in Crop Tissue and Soils." *J. Environ. Qual.* 8:455-459 (1979).

67. Mitchell, G. A., F. T. Bingham and A. L. Page. "Yield and Metal Composition of Lettuce and Wheat Grown on Soils Amended with Sewage Sludge Enriched with Cadmium, Copper, Nickel, and Zinc. *J. Environ. Qual.* 7:165-171 (1978).

68. MacLean, A. J. and A. J. Dekker. "Availability of Zinc, Copper, and Nickel to Plants Grown in Sewage-treated Soils," *Can. J. Soil Sci.* 58:381-389 (1978).

69. Bingham, F. T., A. L. Page, G. A. Mitchell, and J. E. Strong. "Effects of Liming an Acid Soil Amended with Sewage Sludge Enriched with Cd, Cu, Ni, and Zn on Yield and Cd Content of Wheat Grain," *J. Environ. Qual.* 8:202-207 (1979).

70. Cunningham, J. D., J. A. Ryan, and D. R. Keeney. "Phytotoxicity in and Metal Uptake From Soil Treated with Metal-amended Sewage Sludge," *J. Environ. Qual.* 4:455-460 (1975).

71. Dowdy, R. H. and G. E. Ham. "Soybean Growth and Elemental Content as Influenced by Soil Amendments of Sewage Sludge and Heavy Metals: Seedling Studies," *Agron. J.* 69:300-303 (1977).

72. Chaney, R. L., M. C. White, and P. W. Simon. "Plant Uptake of Heavy Metals from Sewage Sludge Applied to Land," *Proc 2nd Nat'l Conf. Municipal Sludge Management* (Rockville, MD: Information Transfer, Inc., 1975) pp. 169-178.

73. Mortvedt, J. J. and P. M. Giordano. "Response of Corn to Zinc and Chromium in Municipal Wastes Applied to Soil," *J. Environ. Qual.* 4:170-174 (1975).

74. Kelling, K. A., A. E. Peterson, L. M. Walsh, J. A. Ryan, and D. R. Keeney. "A Field Study of the Agricultural Use of Sewage Sludge: I. Effect on Crop Yield and Uptake of N and P," *J. Environ. Qual.* 6:339-345 (1977).

75. Vlamis, J., D. E. Williams, K. Fong, and J. E. Cory. "Metal Uptake by Barley From Field Plots Fertilized with Sludge," Soil Sci. 126:49-55 (1978).

76. Kelling, K. A., L. M. Walsh, and A. E. Peterson. "Crop Response to Tank Truck Application of Liquid Sludge," J. Water Pollution Control Fed. 48:2190-2197.

77. Hinesly, T. D., E. L. Ziegler, and G. L. Barrett. "Residual Effects of Irrigating Corn With Digested Sewage Sludge," J. Environ. Qual. 8:35-38 (1979).

78. Hinesly, T. D., R. J. Jones, E. L. Ziegler, and J. J. Tyler. "Effects of Annual and Accumulative Applications of Sewage Sludge on Assimilation of Zinc and Cadmium by Corn (Zea mays L.)," Environ. Sci. Technol. 11:182-188 (1977).

79. Soon, Y. K., T. E. Bates, E. G. Beauchamp, and J. R. Mayer. "Land Application of Chemically-treated Sewage Sludge: I. Effects on Crop Yield and Nitrogen Availability," J. Environ. Qual. 7:264-269 (1978).

80. King, L. D., A. J. Leyshon and L. R. Webber. "Application of Municipal Refuse and Liquid Sewage Sludge to Agricultural Land: II. Lysimeter Study," J. Environ. Qual. 6:67-71 (1977).

81. Jones, R. L., T. D. Hinesly, E. L. Ziegler, and J. J. Tyler. "Cadmium and Zinc Contents of Corn Leaf and Grain Produced by Sludge-amended Soil," J. Environ. Qual. 4:509-514 (1975).

82. Lutrick, M. C., J. E. Bertrand, and H. L. Breland. "The Utilization of Liquid Digested Sludge on Agricultural Land," Soil and Crop Sci. Soc. Florida 35:101-106 (1976).

83. Hinesly, T. D., R. L. Jones, J. J. Tyler and E. L. Ziegler. "Soybean Yield Responses and Assimilation of Zn and Cd from Sewage Sludge-amended Soil," J. Water Pollution Control Fed. 48:2137-2152 (1976).

84. Sabey, B. R. and W. E. Hart. "Land Application of Sewage Sludge: I. Effect on Growth and Chemical Composition of Plants," J. Environ. Qual. 4:252-256 (1975).

85. Giordano, P. M. and D. A. Mays. "Yield and Heavy-Metal Content of Several Vegetable Species Grown in Soil Amended with Sewage Sludge," in Biological Implications of Metals in the Environment, H. Drucker and R. E. Wildung, Eds.

(Oak Ridge, TN: Technical Information Center, ERDA, Conf-750929, 1977), pp. 315-334.

86. King, L. D. and H. D. Morris. "Land Disposal of Liquid Sewage Sludge: II. The Effect on Soil pH, Manganese, Zinc, and Growth and Chemical Composition of Rye (Secale cereale L.). J. Environ. Qual. 1:425-429.

87. Sommers, L. E. and D. W. Nelson. Dept. of Agronomy, Purdue University. Unpublished data.

88. Keefer, R. F., R. N. Singh, D. J. Horvath, and A. R. Khawaja. "Heavy Metal Availability to Plants from Sludge Application," Compost Sci. 20:31-34 (1979).

89. Giordano, P. M., J. J. Mortvedt, and D. A. Mays. "Effect of Municipal Wastes on Crop Yields and Uptake of Heavy Metals," J. Environ. Qual. 4:394-399 (1975).

90. Ritter, W. F. and R. P. Eastburn. "The Uptake of Heavy Metals from Sewage Sludge Applied to Land by Corn and Soybeans," Commun. Soil Sci. Plant Anal. 9:799-811 (1978).

91. Webber, L. R. and E. G. Beauchamp. "Cadmium Concentration and Distribution in Corn (Zea mays L.) Grown on a Calcareous Soil for Three Years After Three Annual Sludge Applications," J. Environ. Sci. Health B14(5):459-474 (1979).

92. Kirkham, M. B. "Trace Elements in Corn Grown on Long-Term Sludge Disposal Site," Environ. Sci. Technol. 9:765-768 (1975).

93. Williams, C. H. and D. J. David. "The Effect of Superphosphate on the Cadmium Content of Soils and Plants," Aust. J. Soil Res. 11:43-56 (1973).

94. Williams, C. H. and D. J. David, "The Accumulation in Soil of Cadmium Residues from Phosphate Fertilizers and Their Effect on the Cadmium Content of Plants," Soil Sci. 121:86-93 (1976).

95. Mulla, D. J., A. L. Page, and T. J. Ganje. "Cadmium Accumulations and Bioavailbility in Soils from Long-Term Phosphorus Fertilization," J. Environ. Qual. (in press).

96. DeVries, M. P. C. and K. G. Tiller. "Sewage Sludge as a Soil Amendment, with Special Reference to Cd, Cu, Mn, Ni, Pb, and Zn--Comparison of Results from Experiments Conducted Inside and Outside a Glasshouse," Environ. Pollution 16:231-240 (1978).

97. Latterell, J. J., R. H. Dowdy, and W. E. Larson. "Correlation of Extractable Metals and Metal Uptake of Snap Beans Grown on Soil Amended with Sewage Sludge," J. Environ. Qual. 7:435-440 (1978).

98. "Principles and Design Criteria for Sewage Sludge Application on Land," in Sludge Treatment and Disposal Part 2, Sludge Disposal Environmental Research Information Center, U.S. EPA Report 625/4-78-012 (1978), pp. 57-112.

99. Sommers, L. E. and D. W. Nelson. "A Model for Application of Sewage Sludge on Cropland," in Proceedings First Annual Conference of Applied Research and Practice on Municipal and Industrial Waste (Madison, WI, 1978), pp. 307-326.

100. Webber, J. "Effects of Toxic Metals in Sewage on Crops," Water Pollution Control (London) 71:404-413 (1972).

101. Williams, J. H. "Use of Sewage Sludge on Agricultural Land and the Effects of Metals on Crops," Water Pollution Control (London) 74:635-644 (1975).

CHAPTER 6

AGRICULTURAL CROPS: PATHOGENS

Dean O. Cliver. Food Research Institute (Department of Food Microbiology and Toxicology), W.H.O. Collaborating Centre on Food Virology, and Department of Bacteriology, University of Wisconsin, Madison, Wisconsin.

INTRODUCTION

I shall limit this discussion to pathogens of concern to human health that may occur in sludge and thus contaminate crops grown in fields to which the sludge is applied. The pathogens to be considered here will include those that might cause infections as a result of being ingested with food.

The selection of agents to be discussed would have been much easier if there were a substantial record of foodborne disease in this country in which sludge had served as a medium of contamination. Unfortunately, I have not found any reported outbreak of this kind among outbreak reports to which I have access [1].

PATHOGENS

The agents to be considered here are those which can infect humans via the food vehicle. Infections, rather than intoxications, are the concern; for the agents of foodborne intoxications are so widely distributed in the environment that one need not consider sludge as their source.

Infectious agents capable of being transmitted through foods are, with few exceptions, produced in the human intestines. These agents are present in sewage primarily because of the use of water in disposing of human feces. Although the water-carriage toilet has come to be the virtual seat of modern Western culture, there must be environmentally, esthetically, and economically acceptable alternatives that would not entail the discharge of feces to the public sewers. The development of such alternatives does not seem to me to have attained its merited position among our nation's priorities.

Perhaps the only significant alternate source of agents in sludge that might infect man are wastes from the slaughter and processing of domestic animals and their products for human food. Sanitation methods for slaughter plants have long involved use of large volumes of water that ultimately contributed significantly to the waste treatment loads in the plants to which the resulting wastewaters were conducted. However, there seems now to be a trend to allow water-conserving sanitation procedures and to encourage recovery of some of the potentially useful substances that were lost previously with the plant wastewaters. This should have some effect in reducing the loads of animal-derived infectious agents in the wastewaters of communities in which slaughter plants are located.

The spectrum of significant infectious agents that might occur in sludge is not remarkable. Only a few bacterial genera, the enteric viruses, and some protozoan and metazoan parasites need be considered [2].

Bacteria

The bacterial agents of foodborne infections may be numerous, but only Salmonella, enteropathogenic Escherichia coli, Shigella, and perhaps Yersinia probably are prevalent enough in the US to warrant consideration here. Salmonella, and E. coli that is not necessarily enteropathogenic, are consistently present in sewage and sludge, whereas the other two may be more difficult to find.

It requires little imagination to envision sludge containing these bacteria being applied to food crops and causing outbreaks of foodborne disease in humans. Some imagination is necessary, however, for I have been quite unable to find a recorded outbreak of foodborne bacterial disease in the US in which the infectious agent was shown to have been derived from sludge. Such incidents probably do occur occasionally in the US, but they must be few and small to continue to escape notice as they have.

Animals of many species, four-legged, two-legged, winged, and cold-blooded, are known to be reservoirs of Salmonella, so members of this genus that occurred in sludge might sometimes be of nonhuman origin. Yersinia, too, has several alternate host species [3]. Enteropathogenic strains of E. coli, on the other hand, seem to show a considerable degree of host species specificity; and shigellae are, for practical purposes, exclusively of human origin.

Infections with these bacteria may be asymptomatic or produce illnesses as severe as typhoid fever. Gastroenteritis is the most common manifestation; the majority of cases are undoubtedly self-limiting without the aid of a physician and, for this and other reasons, unreported. Though transmission from person to person, without food or water serving as a vehicle, is common, foodborne outbreaks that have been recorded have usually been self-limiting in the sense that subsequent person-to-person transmission did not carry the agent through an entire community before subsiding.

Viruses

The viruses transmitted through foods in the US appear to be exclusively of human enteric origin [4]. There is no "normal flora" of viruses in the intestines, but enough people are infected at any time to ensure the presence of enteric viruses in community sewage and sludge.

The diseases that may be caused by enteric viruses include hepatitis, poliomyelitis, a variety of other illnesses ranging from ill-defined fevers to meningitis, and, perhaps most frequently, gastroenteritis. On the other hand, the majority of infections with any of these viruses are probably inapparent. Gastroenteritis is the most common foodborne disease of presumed viral etiology, but only recently has it been possible to confirm a virus cause for foodborne gastroenteritis by laboratory means. Viral hepatitis A is, by far, the most frequently reported foodborne viral disease: the extensive record of outbreaks shows that virtually any food can be involved, because a frequent cause of contamination is mishandling by an infected person. Poliomyelitis may once have been transmitted through foods with some frequency but is now virtually eradicated in the US through vaccination. The viruses of poliomyelitis and, apparently, of hepatitis A belong to the enterovirus group, so it is probably to be expected that others of the more than 60 members of this group would be transmitted through foods at least occasionally.

Among the food-associated outbreaks of virus disease of which I am aware, none seems actually to have involved sludge. There is no doubt that virus is often detectable in sludge, even after the sludge has been digested, but transmission of viral disease via food contaminated by sludge seems not to have been reported.

Parasites

The word "parasites" as used here refers only to protozoan and metazoan parasites, although it is clear that all of the viruses and perhaps some of the bacteria that have been discussed are parasites as well. Significant protozoan species that occur in human feces and sewage include Entamoeba histolytica and Giardia lamblia, which are stable outside the host in the cyst stage [5]. Metazoa include *Ascaris lumbricoides*, *Taenia saginata*, and *Taenia solium*; the eggs of these species (which may be shed in the proglottids or free in the case of the *Taenia* species) are fairly durable outside the host [5]. The eggs of *A. lumbricoides* and *T. solium* are infectious if ingested by humans, but the *Taenia* species are of concern principally because they may infest meat animals (cattle in the case of *T. saginata* and swine in the case of *T. solium*) and thus give rise indirectly to human disease.

SLUDGE

The term sludge as used here includes any of the sediment-rich fluids that are produced in the treatment of urban residential and, perhaps, industrial wastewater. Sludge that is handled as a liquid does not exceed 15% solids content, and even thickened sludges seldom exceed 5% solids.

Sources of pathogens

As was mentioned previously, most of the infectious agents that might occur in sludge and be transmitted through foods emanate originally from the human intestines. Slaughter plants and meat processing facilities have a limited potential for contributing agents infectious for man, but the majority of industrial wastewaters are unlikely to cause any problems of this kind.

The agents produced in the intestines are, of course, shed in feces. The degree to which the infectious agents sediment during wastewater treatment will depend first on the stability of their association with sedimentable fecal solids, second on their own size and density, and third on their affinity for other sediments that are present in the wastewater or are generated or added during treatment.

Sludge production

Any unit treatment process that involves sedimentation generates sludge that may harbor pathogens. Even "grit," which is often separated before primary treatment and is supposed to be primarily mineral matter, has been shown to be rich in virus [6].

Primary treatment entails settling, usually spontaneous, of fecal solids and much other organic material. Considerable sedimentation of parasites can be expected, whether or not they are intimately associated with fecal solids. One sees quite varied estimates as to the degree of virus removal accomplished in primary treatment. Disagreement seems to stem from the difficulty of doing valid estimates of quantities of virus associated with solids. I believe that the majority of influent virus is removed in primary treatment and is present in the primary sludge solids, which are a relatively hospitable environment for viruses. Enteric bacterial pathogens, as well as viruses and parasites, abound in primary sludge [7].

Secondary treatments generally entail the conversion of dissolved organic substances into sedimentable microbial cell mass. The organisms that grow on a trickling filter medium or in an aeration tank are not enteric in nature, nor have they any significant pathogenic potential. On the contrary, they seem to be capable of some inactivation of viruses [8] and probably have some effect against bacterial pathogens as well. This means that removal of pathogens in secondary treatment differs in that not all of the pathogens taken out of the wastewater are inevitably present in infectious form in the sludge (e.g., waste activated sludge). The difference is relative, however: solids produced in secondary wastewater treatment must be assumed to contain some pathogens and can only produce a partial dilution if blended with primary sludge.

Tertiary treatment is difficult to generalize. To the degree that the treatment may be a chemical reaction that yields new sedimentable solids, another, sometimes voluminous, sludge results. The pathogens that remain after secondary treatment are likely to sediment with these new solids, but the low level of pathogens and the large volume of sludge add up to a lower-hazard product than the primary and secondary sludges. Where excess lime treatment is used, the extreme alkalinity is likely to exert a significant antimicrobial effect [9].

Sludge treatment

Raw sludge, typically a dewatered or thickened blend of sludges from primary or secondary treatment, will ordinarily be treated in some way if it is going to be disposed to land. Treatments applied to sludge (anaerobic, mesophilic digestion is probably most common in the US at this time) are usually intended primarily to stabilize putrescible solids, to improve dewatering characteristics, and sometimes to produce combustible gas. Any destruction or inactivation of pathogens during these processes is generally fortuitous, but not necessarily insignificant. Salmonella kills as high as 99.7% were reported in a study of anaerobic, mesophilic sludge digestion, but growth of the organism was suspected in one of the digestors that was studied [10]. Aerobic digestion killed more salmonellae.

Anaerobic, mesophilic digestion (∼35°C, 20-day mean residence time) caused 80 to 90% virus inactivation in a recent study, whereas anaerobic, thermophilic digestion (∼49°C, 20-day mean residence time) inactivated 99% or more [11]. There is little doubt that the conditions of anaerobic, mesophilic digestion are significantly adverse to enteric viruses, and probably to protozoan cysts as well [1]. However, a problem inheres to the operation of completely mixed reactors: completely mixed operation with a mean residence time of 20 days means that one-twentieth of the volume in the reactor is removed each day, that one-twentieth of the reactor's capacity is added each day, and that one-twentieth of this newly added volume leaves the reactor by the following day. Unless conditions in the digestor are extremely lethal, this inherent one-in-four-hundred short circuiting may pass a significant amount of infectious pathogen when compared to what would have survived 20 days' residence. Prolonged subsequent holding (secondary digestion or lagooning), even at lower temperatures, will result in further reduction of pathogen levels.

The rate of pathogen destruction appears to vary strongly with temperature. Thermophilic anaerobic digestion is much more rapidly effective than mesophilic anaerobic digestion against microbes [11], and even small increases in mesophilic digestion temperatures can produce significant gains in inactivation rate [12]. Pasteurization (70°C for 30 to 60 min) and composting can yield essentially pathogen-free products, though the cost-effectiveness of these treatments has yet to be proven [13]. Ionizing radiation is a physical alternative to heat that also seems capable of killing pathogens in sludge [14,15]. Irradiation, composting, and pasteurization are pathogen-oriented processes, whereas digestion and lagooning, as was stated above, are not. Later, the question of whether pathogen-oriented sludge treatments are needed will be addressed.

Sludge application

Sludge may be applied to cropland by some type of spray apparatus, either truck-mounted or installed in the field [16]. If the sludge is applied to the soil surface, it may subsequently be tilled into the soil. Alternately, it may be drilled below the soil surface with an injection apparatus mounted on a truck or tractor.

The choice of an application method is likely to be based on the amount of land and type of soil available, as well as whether application is done seasonally or on a year-around basis. Subsurface application is preferable from the standpoint of containment of pathogens, but the energy cost of this mode of application is likely to be greater than for surface application. On the other hand, additional energy costs might be incurred by applying special treatments to kill pathogens just so as to make sludge safe for surface application.

CROPS

Where there is a concern for eventual groundwater contamination with nitrate, sludge should be applied where crops are being grown for harvest, and the rate of sludge application should be predicated on the nitrogen content of the portion of the crop that will be harvested and removed from the field [16]. Microbial action is required to convert the nitrogen from the many forms in which it occurs in sludge to nitrate so crops can use it. Because of the complex forms in which much nitrogen occurs in sludge, total mineralization may take several years.

Crops may be classified in several ways. One might first consider whether the crop is being grown for food, feed, or fiber. Fiber crops, such as cotton or flax, would merit little concern for consumer health. Feed for domestic, food-source animals might carry pathogens such as _Salmonella_ and parasites that would ultimately concern human health [5]; threats to animal health from sludge-borne pathogens are discussed in another paper at this symposium.

Assuming that a crop is a food that will be consumed by humans, one would further consider whether the edible portion grows above or below the soil surface and how the food is likely to be processed and prepared before consumption. It is probably reasonable to be more concerned about the use of sludge on "root crops" (true roots plus rhizomes, bulbs, and tubers) except where the sludge is sprayed directly on the crop standing in the field; however, expected processing has also to be considered. For example, one would be less concerned about application of sludge to sugar beets, which would undergo extensive processing, than to red beets, which might reach the consumer's home raw, or to carrots, which might be eaten completely raw.

"Root crops" may be subject to surface contamination by sludge-borne pathogens no matter how the sludge is applied, but plants whose edible portions grow above ground should become directly contaminated only if sludge is sprayed onto the standing crop. The translocation of human pathogens within the tissue of a plant seems to be a concern only in the case of viruses: two published studies provide evidence, but not absolutely convincing proof, that no practical risk attaches to this phenomenon [17,18]. That is, viruses to which the roots of a plant are exposed in sludge-amended soil are unlikely to be translocated through the plant to above-ground, edible portions.

Commercial processing is likely to afford more effective safeguards than whatever preparation takes place in the home, for consumers have often shown a very limited instinct for self-protection where food handling is concerned [19]. Even so, one major food processor is said to be curtailing use of vegetables grown on sludge-amended soils. The stated concern is for cadmium and other heavy metals, but a reported exemption for sludge that has been composted suggests that pathogens were the real focus of the action. This reported action probably ought not be taken too seriously as an indicator of health risk associated with vegetables from sludge-amended soils, first because the majority of food processors evidently view the matter differently, and second because decisions of this kind are frequently made by a processor's legal staff, rather than quality assurance personnel who are trained in microbiology. Whatever else, there should be quite adequate safety in the old rule of thumb that there are not likely to be pathogens in a food that has been thermally processed in a hermetically sealed container to the point of being "shelf-stable" at room temperature.

RISK ASSESSMENT

Two hypothetical extremes might be: (1) spraying raw primary sludge directly onto head lettuce just before it is harvested and (2) subsurface application of composted sludge on a field that is then planted with corn that will be fed to swine. No great insight is needed to assert that situation (1) is highly hazardous and that situation (2) is essentially risk-free. It is the intermediate levels of risk to which more considered judgment must be applied.

The components of risk assessment have been presented above. Among pathogens that might occur in sludge, viruses may be overrated and parasites underrated as hazards. Raw sludge is likely to be pathogen-rich; sludge digested by a mesophilic, anaerobic process should contain lower levels of most pathogens; and sludge that has received such antimicrobial treatments as irradiation, pasteurization, or composting should present only a minimal risk from infectious agents. Even so, there may be ample reason not to spray composted sludge on lettuce or apply it to fields in which radishes are grown, if only because consumers seldom cook either lettuce or radishes before eating them.

It would be easy enough to conclude that sludge should only be applied, with tillage during or after application, to land that is then planted with animal feed crops. All of the urban sludge produced in Wisconsin, if applied on the basis of nitrogen utilization by the crop, could be disposed on less than 1% of the state's corn land [16]. This would appear to offer some latitude in rotating application sites to minimize the buildup of heavy metals. The problem with this approach is that not every state has cornfields near enough to its urban centers to allow hauling the sludge at reasonable costs. Although sludge has some value as fertilizer and it must be disposed somehow in any case, costs of over-the-road transport can soon become prohibitive, as distances increase, at today's motor fuel prices. Barges or pipelines may offer alternatives but are not within the reach of all communities.

This may require that a different kind of bargain be struck. I shall not propose any high-risk options, such as application of raw sludge to cropland or application of almost any class of sludge to lettuce or radishes. Neither do I assert that all sludge must be disposed to land. However, it seems reasonable to apply a medium-risk sludge (e.g., anaerobically, mesophilically digested sludge) so that it is incorporated into soil on which feed grains will be grown. If the sludge is to be applied to soil that will grow crops nearer in the food chain to the human consumer, it may be necessary to intensify treatment (e.g., by adding a composting or pasteurization step) to mitigate risk of transmitting human disease. Pathogens that may be present in sludge do not seem likely to persist for more than a year, so use of the land during subsequent growing seasons (during which time sludge nitrogen continues to be mineralized and thus made available to plants) may not need to be constrained. Fecal coliforms and coliphages, for example, were reduced to negligible levels within 2 to 3 wk after sludge application to pasture grasses [20]. There is, it appears, some disagreement on this point: the US Food and Drug Administration recommends that food crops to be eaten raw by humans not be grown on sludge-amended soils until 3 yr after sludge application has ceased [21].

SUMMARY

The factors that govern the safety, from the standpoint of human pathogens potentially transmissible through foods, of applying sludges to agricultural crops have been reviewed. Enteric parasites, viruses, and bacteria may be present at high levels in raw sludge, which should probably never be applied to soil. Several treatments are available and used to reduce, incidentally or purposely, the levels of pathogens in sludge. Anaerobic, mesophilic digestion may be quite sufficient for sludge that is used to fertilize grain that will be consumed by meat animals. More rigorous treatment is appropriate where sludge is applied to soils growing crops to be consumed directly by humans, and foods that are likely to be eaten raw probably ought not be fertilized with sludge. On the other hand, sludge treatment and disposal are important components of the costs of treating urban wastewater, and the high costs of treating and transporting sludge can hardly be balanced by the limited value of sludge as fertilizer, so application sites should be chosen as near to a sewage plant as possible (to minimize hauling costs) without requiring inordinately expensive processing to destroy pathogens. Pathogens in sludge that is drilled or tilled beneath the surface of the soil are unlikely to contaminate above-ground parts of plants, and pathogens below the soil surface are likely to persist for less than a year under most circumstances.

REFERENCES

1. Bryan, F. L. "Diseases Transmitted by Foods Contaminated by Wastewater," J. Food Prot. 40(1): 45-56 (1977).
2. Bryan, F. L. "Diseases Transmitted by Foods (A Classification and Summary)," Center for Disease Control, DHEW Publication No. (CDC) 76-8237 (1976).
3. Bryan, F. L. "Infections and Intoxications Caused by Other Bacteria," in Food-borne Infections and Intoxications, 2nd ed., H. Riemann and F. L. Bryan, Eds. (New York: Academic Press, Inc., 1979), p. 250.
4. Cliver, D. O. "Viral Infections," in Food-borne Infections and Intoxications, 2nd ed., H. Riemann and F. L. Bryan, Eds. (New York: Academic Press, Inc., 1979), p. 299.
5. Healy, G. R., and D. Juranek. "Parasitic Infections," in Food-borne Infections and Intoxications, 2nd ed., H. Riemann and F. L. Bryan, Eds. (New York: Academic Press, Inc., 1979), p. 343.
6. Cliver, D. O. "Virus Association with Wastewater Solids," Envir. Letters 10(3):215-223 (1975).
7. Dudley, D. J., M. N. Guentzel, M. J. Ibarra, B. E. Moore and B. P. Sagik. "Enumeration of Potentially Pathogenic Bacteria from Sewage Sludges," Appl. Envir. Microbiol. 39(1):118-126 (1980).
8. Malina, J. F., Jr., K. R. Ranganathan, B. E. D. Moore and B. P. Sagik. "Poliovirus Inactivation by Activated Sludge," in Virus Survival in Water and Wastewater Systems, J. F. Malina, Jr., and B. P. Sagik, Eds. (Austin, TX: Center for Research in Water Resources, The University of Texas at Austin, 1974), p. 95.
9. Wolf, H. W., R. S. Safferman, A. R. Mixson and C. E. Stringer. "Virus Inactivation During Tertiary Treatment," in Virus Survival in Water and Waste-

water Systems, J. F. Malina, Jr., and B. P. Sagik, Eds. (Austin, TX: Center for Research in Water Resources, The University of Texas at Austin, 1974), p. 145.
10. Jones, P. W., L. M. Rennison, V. H. Lewin and D. L. Redhead. "The Occurrence and Significance to Animal Health of Salmonellas in Sewage and Sewage Sludges," J. Hyg., Camb. 84(1):47-62 (1980).
11. Berg, G., and D. Berman. "Destruction by Anaerobic Mesophilic and Thermophilic Digestion of Viruses and Indicator Bacteria Indigenous to Domestic Sludges," Appl. Envir. Microbiol. 39(2):361-368 (1980).
12. Palfi, A. "Survival of Enteroviruses During Anaerobic Sludge Digestion," in Proc. Sixth Int. Conf. Wat. Poll. Res. (Oxford, England: Pergamon Press, 1972), p. 99.
13. Sagik, B. P., and C. A. Sorber. "Assessing Risk for Effluent Land Application," Water Sew. Works 125(10):40-42 (1978).
14. Massachusetts Institute of Technology. "High Energy Electron Radiation of Wastewater Liquid Residuals," Report to the U.S. National Science Foundation, NSF/RA-770440 (December 31, 1977).
15. Melmed, L. N., and D. K. Comninos. "Disinfection of Sewage Sludge with Gamma Radiation," Water SA 5(4):153-159 (1979).
16. Keeney, D. R., K. W. Lee and L. M. Walsh. "Guidelines for the Application of Wastewater Sludge to Agricultural Land in Wisconsin," Wisconsin Department of Natural Resources Technical Bulletin No. 88 (1975).
17. Mazur, B., and W. Paciorkiewicz. "Dissemination of Enteroviruses in the Human Environment. The Presence of Poliovirus in Various Parts of Vegetable Plants Grown on Infected Soil," Med. Dosw. Mikrobiol. 25:93-98 (1973).
18. Murphey, W. H., O. R. Eylar, E. L. Schmidt and J. T. Syverton. "Absorption and Translocation of Mammalian Viruses by Plants," Virology 6:612-636 (1958).
19. Cliver, D. O. "Transmission of Viruses Through Foods," Crit. Rev. Envir. Contr. 1(4):551-579 (1971).
20. Brown, K. W., S. G. Jones, and K. C. Donnelly. "The Influence of Simulated Rainfall on Residual Bacteria and Virus on Grass Treated with Sewage Sludge," J. Envir. Qual., in press.
21. Larkin, E. P., J. T. Tierney, J. Lovett, D. Van Donsel and D. W. Francis. "Land Application of

Sewage Wastes: Potential for Contamination of Foodstuffs and Agricultural Soils by Viruses and Bacterial Pathogens," in *Risk Assessment and Health Effects of Land Application of Municipal Wastewater and Sludges*, B. P. Sagik and C. A. Sorber, Eds. (San Antonio, TX: Center for Applied Research and Technology, The University of Texas at San Antonio, 1978), p. 102.

CHAPTER 7

EFFECT OF TOXIC CHEMICALS PRESENT IN SEWAGE SLUDGE
ON ANIMAL HEALTH

Eldon W. Kienholz, Department of Animal Sciences,
Colorado State University, Fort Collins, Colorado 80523.

Currently at Biomass Research Center, University of
Arkansas, Fayetteville, Arkansas 72701.

INTRODUCTION

   Persons who manage municipal sludge disposal must take
into account the fears in the minds of community leaders.
Those fears center on the possibilities that sewage sludge
and wastewater will disseminate pathogenic organisms and
toxic minerals and organic chemicals into our environment.
There is no quick and complete answer to how human and animal
health would be affected by movement of such wastes onto
agricultural land. Thus, much needs to be known about the
specific details of each kind of potential contamination or
health risk associated with each kind of sewage waste utilization. The compiling of a list of organic and inorganic
materials in sewage wastes and their potential health effects
would be a never ending job; yet we need some effective
evaluation of health risks from application of conventionally
treated sewage wastes onto land.
   The Federal Water Pollution Control Act Ammendments of
1972 (P.L. 92-500) encouraged the recycling of municipal
wastewater treatment products onto land. However, there is
a lack of information on which to base a valid judgment about
the potential health hazards which would result from such a
country wide practice. There have been attempts to expand
the knowledge base; research projects supported by EPA, FDA,
USDA and other organizations. Up to the present time, the
chemicals and heavy metals of main interest in the contamination of foods included mercury, lead, cadmium, arsenic,
selenium, zinc, and organic chemicals, principally pesticides
and polychlorinated biphenyls (PCBs). A review of the
literature for effects of those materials in sewage sludge

upon health of animals is the purpose of this paper. In addition, the recent discovery at Colorado State University of vitamin A reductions in liver of animals eating digested municipal sewage sludge will also be reviewed and discussed.

Disposal of sewage sludge (SS) and wastewater onto land provides potential benefits, but also carries the risk of contaminating human and other animal food. Sufficient consideration should be given to the possible adverse effects of such application to land and food crops before this practice is widely instigated (1). Contamination of the food chain can occur unless measures are taken to establish tolerable levels of both organic and inorganic materials from the sewage (2). A recent statement of levels of solid waste toxicants allowed by law includes how to deal with sludges high in Cd, PCBs and other organic and inorganic contaminants (3). However, relatively little information is available on levels of refractory organic materials (ROR) in municipal sludges; and current chemicals of concern include aldrin, dieldrin, chlordane and DDT metabolites as well as various industrial chemicals such as PCBs (4). FDA officials have stated that sewage sludges should not contain more than 20 ppm Cd, 1000 ppm Pb and 10 ppm PCBs (5). Human food levels of Cd and Pb are close enough to maximum tolerable limits that further increases must be avoided, and FDA has committed itself to decreasing such contaminants in food to the lowest possible levels consistent with also insuring an adequate human food supply in the USA (6). There is certainly a need for more information about low level effects of all toxicants; particularly of Cd (7) and Pb (8). There is also need for more information about factors that affect release of toxicants from sewage sludge (9). EPA officials have expressed concern that there are insufficient guidelines for control of contaminants in sludges applied to crops that are ultimately consumed by humans and other animals (10). A recent review of the consequences of environmental hazards to human health is available; including some case studies, extrapolation of animal studies to man, and consideration of risk/benefit analyses (11).

Animals may get exposure to environmental hazards in a number of ways. Potential contaminants can be found in diets, water and in the bedding or litter. Recommendations for maximum content of PCBs, Cd and Pb in bedding are 0.050, 0.5 and 1.5 ppm, respectively (12). Animals also ingest raw soil when grazing. It has been estimated that each dairy cow ingests roughly 200 to 300 kg of soil per year; some individuals may consume from 100 kg to nearly 500 kg/year, depending upon grazing conditions (13). DDT residues have been found in sheep and passed on to lambs via ewe milk when sheep ingested surface soil containing DDT (14). Half the lindane applied to a pasture was recovered in the topsoil 10 months later (15).

Chemical contents of all kinds of sludges, residues and wastes need to be known if we are to successfully recycle those chemicals back into our environment. Levels of 30 inorganic elements have been reported for a few feedlot cattle diets, cattle manures, poultry manures and metropolitan sewage sludges (16). Toxic inorganic elements averaged 1700 times as much in sewage sludge as in cattle feedlot diets. Such findings lead to conclusions that the potential safety of recycled wastes should be thoroughly investigated. Detailed bi-weekly analyses of rural domestic digested sewage sludge in England revealed seasonal variation for the 26 inorganic elements that were assayed (17). More assays of this kind are needed.

EFFECTS OF FEEDING SEWAGE SLUDGES (SS) DIRECTLY TO ANIMALS
(See Table I)

Table I. Health Defects Reported When Animals Have Eaten Various Dietary Levels of Digested Sewage Sludge

Investigator	Dietary Level (%, dry)	Health Defect Noted (animal type)
Hartenstein et al.	100	none (earthworm)
Kinzell et al.	50	enlarged livers (rats)
Edds et al.	50	cadmium toxicity (swine)
Cheek and Meyer	30	none (rats and quail)
Beaudouin et al.	20	none (swine)
Kienholz et al.	20	reduced liver vitamin A (chicks)
	12	reduced liver vitamin A (cattle)
Johnson et al.	17.5	none (cattle)
Firth and Johnson	10	none (chicks)
	5	none (swine)
Bertrand et al.	6	none (cattle)
Univ. of Maryland and USDA	6?	iron toxicity (cattle)
Fitzgerald	5?	none (cattle)
Sedgwick and Arthur	?	reduced reproduction of river fauna

Cattle

When anaerobically digested Chicago sewage sludge has been sprayed onto forage that animals grazed (the diet was probably less than 5% sewage sludge), no histological or pathological defects, no differences in weight gains of calves, no effect on reproduction, and no increased infections were noted (18). It was concluded that there was little risk to man or animals associated with land application of that sludge. The author further stated that land utilization of sludges should present no greater disease risk to the general public than the risk of the same disease in a community where sludge is not utilized (19).

When a high iron digested sludge was sprayed onto forage just prior to cattle grazing, sludge was 5 to 6% of the total dry weight diet; thus the overall diet was close to 2000 ppm Fe on a dry weight basis. After 4 months of grazing such pasture, livers were 17,000 ppm Fe compared to 216 ppm in control livers (20). In a follow up of the first study, there was much less iron in the sludge and in the diet. There were also no health defects noted. It was concluded that some high-iron sludges may cause iron toxicity in animals if such sludges are directly consumed by those animals (21).

When Denver sewage sludge was fed to constitute 12% of dry weight diet for steers, there were no health defect symptoms during a 94-day feeding study (22). When Chicago sewage sludge was consumed at 6% of the dry weight diet for 141 days, there were no differences in growth, feed efficiency or carcass quality measurements (23). Nor were there any health defects noted. However, livers and kidneys accumulated several heavy metals, causing the authors to caution about Cd uptake from sludge into animals.

Swine

Many years ago it was reported that there were no adverse effects when dried activated sewage sludge was fed as 5% of diet to baby pigs and as 10% of diet to baby chicks for periods of about a month each (24). But, when swine received Chicago sludge as 50% of dry weight of diet for 42 days, cadmium toxicity was found (25, 26). Liver function was reduced as measured by activities of several liver enzymes. It was concluded that Cd was interfering with iron utilization, causing anemia; also that Cd was blocking microsomal enzyme systems in liver. When female swine were fed 20% anaerobically digested sewage sludge for nearly 2 months, reproductive performance was not affected; weaning weights were slightly depressed, probably due to reduced energy in SS-diet (27). Sows had increasing concentrations of Cd, Ni,

Cu, Zn, Cr, Mn and Al in selected tissues; offspring had some higher metal levels in livers in the first, but not the second, litters. That caused the authors to wonder if sows were adapting to that diet and protecting their offspring. The SS used had few, if any, nutrients for those swine, but no health problems were noted.

Rats

Activated sewage sludge was fed as 30% of diet for 2 weeks to rats and Japanese Quail; no health effects were noted (28). When levels up to 50% of diet were fed to male rats for 111 days, growth was reduced; livers were enlarged and heavy metals accumulated (29). Reduced sodium pentobarbital sleeping time was the assay for suggesting that liver drug metabolism had been stimulated by eating of SS. It was concluded that there were toxic organic materials in that sewage sludge.

Other Animals

Baby chicks readily consumed SS as 10% of the diet for about a month; there was no evidence of adverse effects upon those animals (24). When activated sludge was discharged into the Thames River from London sewage treatment plants during a union strike of treatment plants, shrimps, sprats and soles were much reduced in the exposed river area for the next 3 to 4 months (30). When earthworms consumed activated sludge from San Jose, California, no health problems were seen, but each of the problematic heavy metals accumulated and concentrated in the earthworm (31). When diets contained 20% air dried anaerobically digested Fort Collins sewage sludge (FCSS), birds gained less body weight and the liver vitamin A levels were greatly reduced when birds consumed sludge diets (this paper). It was concluded that ingestion of FCSS impaired the metabolism of vitamin A in those chicks.

EFFECTS OF FEEDING UNDIGESTED SEWAGE SLUDGED DIRECTLY TO ANIMALS

It has been stressed that sludge utilization will be increasingly important as we learn how to recycle wastes for our common benefit; especially energy materials must be conserved and not discarded (32). One research group has been studying thermoradiated raw undigested sewage sludge (TRUSS) in diets for rats, sheep and cattle (33-36). The thermoradiation destroyed pathogens, then the TRUSS was dried and fed as part of the diet, utilizing the nutrients, including the energy. TRUSS often had an energy value of 4.7 kcal/g, compared to corn at 4.4 kcal/g of dry product. The energy was 40 to 70% digestible in all tests. There were no adverse health effects during growth, maintenance and reproduction tests. Whereas there are very little nutritional benefits of <u>digested</u> sewage sludges in diets, the TRUSS product was calculated to be worth half the cost of cottonseed meal to the animals tested. Thus, the authors point out that raw denatured SS could be considered a valuable resource for feeding domestic animals rather than digested in modern municipal waste treatment facilities at a staggering cost.

EFFECTS OF FEEDING SLUDGE FERTILIZED CROPS TO ANIMALS
(See Table II)

Table II. Health Defects Reported When Animals Have Eaten Crops Grown on Sludge-Amended Soils

Crop	Cd Level (ppm)	Animal	Health Defects	Reference
Corn	0.7	Chicken	None	39
Soybeans	2.4	Chicken	None	39
Corn	0.7	Chicken	None	40
Corn	0.6	Swine	None	37
Corn	0.6	Pheasants	None	38
Pasture?	?	Goats	None	46
Corn	2	Meadow Voles	None	41
Sorghum Forage	5	Meadow Voles	None	41
Corn	0.03	Steer	None	23
Silage Corn	1.7	Sheep	Liver Damage from Cd & PCBs	42
Cabbage	5.3 (0.76 PCBs)	Sheep	Liver & Thyroid Damage	43
Lettuce	26	Mice	Liver Damage	44
Swiss Chard	2.7	Guinea Pig	None	45

Chicago sewage sludge fertilized corn (SS-corn) when fed to pigs gave a modest growth increase; probably due to higher protein content of that corn compared to control corn (37). Higher serum glucose levels were noted and were suggested as being caused by higher Cd and Ni in SS-corn. No health lesions were found. Similar SS-corn was also fed to pheasants, and no adverse health effects were noted (38). Nor were there any adverse health or performance effects when hens or chicks received SS-corn or SS-soybeans (39). Uptake of Cd from sludge fertilized corn and soy was similar to uptake of Cd from an equal level of $CdCl_2$ added to the diet. Even when SS-corn was fed for 80 continuous weeks to female chickens during the growth and laying cycle, there were no adverse health effects noted (40). There was no movement of Cd from diet into hen eggs. When SS-corn and SS-sorghum forage (2 and 5 ppm Cd, respectively) were fed to meadow voles, no effect on animal performance was noted during 40 days of post-weaning study (41). Cd increased in liver and kidney, not muscle. Uptake of Cd was the same as from inorganic Cd salts added to diets. It was concluded that it was important to keep Cd out of soils. When SS-corn was used as 77% of dry weight of steer diets for 141 days, there were no changes in heavy metal content of tissues (23). Corn metal contents had been only slightly affected by moderate sludge application to land.

When SS-corn silage was derived from very heavily sludge amended soil, sheep health was impaired (42). At the end of the 274-day feeding study, Cd was elevated in many tissues. There were no histological differences as measured by traditional light microscopy. However, electron microscopy revealed enlarged and swollen mitochondria, necrosis in hepatocytes of each sheep fed the SS-corn silage. The authors pointed out that similar findings have been observed in rats exposed to cadmium, thus the 1.7 ppm Cd in SS-corn silage may have been the cause of health defects found. When SS-cabbage was fed as 30% of the dry weight diet for 77 days, sheep had 3 fold increases in Cd in some tissues, while PCBs were increased about 2 fold in liver (to about 1 ppm PCBs) (43). Degenerative changes were seen in liver and thyroid that suggested a direct effect of the PCBs that had been picked up from sludge by cabbage. Enhanced hepatic microsomal mixed function oxidase activity and intestinal aryl hydrocarbon hydroxylase activity were also found in those sheep livers. These changes were indicative of ingestion of foreign organic compounds such as PCBs. When romaine lettuce was grown on sludge amended soil and fed as 45% of dry weight diet to mice, Cd was elevated in kidneys and hepatic microsomal enzyme activities were lowered (44). When sewage fertilized swiss Chard was fed at 20 to 28% of dry weight diet to guinea pigs for 80 days, higher Br, Ca, Co, Eu, Fe, Ni and Sr levels were found in animal tissues (45). Composting the sludge prior to amending the soil appeared to render Cd, Cu, Ni and Zn less available to the growing chard. No adverse health effects were noted in the guinea pigs. Finally, when sewage sludge was applied to land that was then grazed by goats, milk and goat tissues were extensively analyzed for toxicant uptake and there was none (46). Animals were judged healthy and unaffected by SS fertilization of the land.

EFFECTS OF FEEDING PROCESSED MUNICIPAL SOLID WASTE (MSW) TO ANIMALS

Municipal solid waste (MSW) feeding is somewhat similar in some respects to sewage waste feeding; lead and PCBs have increased in animals receiving such a foodstuff (47). When such material was ensiled first, then fed as 18.5% of diet for 91 days to dairy heifers, dry matter digestibility was very good, but dry matter intake was depressed 12% below control (48). However, when fed as unensiled diet, intake was depressed by 29 to 42% compared to controls. When aerobic digested MSW was fed to cattle for 91 days, as 17.5% of diet, lead content of calf livers increased about 3 fold; also PCBs in body fat were increased about 25 fold (49). Thus, there may be problems from PCBs and lead if digested MSW is fed to cattle. Low levels of lead have been reported to decrease bone formation (50). An extensive evaluation of cattle feed components from solid waste has been published (51). That material was thoroughly reviewed and compared to many other data showing that waste paper (garbage) ordinarily depresses food intake of cattle (52). The authors concluded that the depressed food intake was apparently due to increased water holding capacity of paper fibers resulting in increased rumen fill and long water retention in the rumen. Even though accumulation of organic and inorganic compounds have been reported, adverse effects of MSW upon animal health have not been reported.

HEALTH DEFECTS FROM REFRACTORY ORGANIC RESIDUES (ROR) IN SEWAGE SLUDGES

Polychlorinated biphenyls (PCBs) in cabbage grown on digested sewage sludge have probably been the cause of degenerative changes seen in liver and thyroid of sheep (43). The way liver metabolizes PCBs and other related ROR has been reviewed (53-56). Serious injurious effects in infant primates have resulted when nursing mothers consumed 2.5 ppm of one PCB during gestation and lactation (57). Dairy cows have a PCB half time retention in their bodies of 40 to 70 days (58) and milk fat contains a constant ratio of PCBs and other related compounds compared to body fat burdens. ROR molecules differ in retention after ingestion with the diet. At the end of a 94-day feeding trial, retention of 7 different ROR varied from less than 1% to about 75% of amount ingested as part of the sewage sludge (table III). It is not known if the low retained molecules were absorbed or if they were rapidly eliminated from the body after absorption. Organic-chlorine residues have been described as causing reproductive failure in herring gulls (59).

Table III. Retention of Refractory Organic Residues (ROR) in Beef Animals After Consuming Metropolitan Denver Sewage Sludge (MDSS) as 12% of Finishing Diet for 94 Days

ppDDE	HCB	Dieldrin	PCBs	Oxychlordane	Chlordane	ppDDD
Total amount ingested in 94 days from MDSS (mg)							
5	0.7	10	263	10	138	2
Retention at end of 94 days (% of amount ingested)							
75	55	33	25	2	< 1	< 1

NOTE: Fifteen other specific ROR were sought, but not found in MDSS.

PBBs, a recent environmental pollutant, appears even more toxic than PCBs (60-62), but has not yet been described as causing concern via digested sewage sludge. A novel idea for reducing animal burdens of ROR has been recently described (63). That method involves low level feeding of mineral oil, and such treatment has increased HCB removal from rats (64) and Rhesus Monkeys (65) by 3 fold and 10 fold, respectively, when compared to controls. Cattle and sheep also experienced a 3 fold increase in HCB removal when 5% mineral oil was included in their diet (66).

VITAMIN A METABOLISM WHILE ANIMALS INGEST DIGESTED SEWAGE SLUDGE

It was recently discovered that ingestion of a digested sewage sludge impaired vitamin A metabolism in two species of animals. When a diet containing 20% of air dried anaerobically digested Fort Collins, Colorado sewage sludge (FCSS) was fed to day-old broiler chicks for 4 weeks, liver vitamin A levels were reduced to only 8% as much as vitamin A in control chick livers (Table IV). That same sludge reduced steer liver vitamin A levels to 40% of control liver values when FCSS was fed in steer finishing diets for 15 weeks (Table IV).

The reason for such vitamin A reduction is unknown at the present time. PCBs have reduced liver vitamin A levels, but quail and rats had only a 50% liver vitamin A reduction after consuming 100 ppm PCB in diets for 2 months (67). That is much more dietary PCB than the approximately 1 ppm PCB in FCSS chick diets.

Table IV. Effect of Dried Anaerobically Digested Fort Collins, Colorado, Sewage Sludge in Animal Diets upon Vitamin A Content of Livers

Dietary Treatment	Vitamin A content (IU/g) of fresh wt. liver	
	Steers	Chickens
Controls	1173	960
Sludge-fed	471**[1]	77**[2]

** different (P<.01) from controls.

[1] fed 12% (dry basis) sewage sludge (6 animals/treatment) for 15 weeks

[2] fed 20% sewage sludge (30 chicks/treatment) for 4 weeks

Cadmium has caused vitamin A metabolism difficulties when fed to animals. Dietary Cd lowered serum vitamin A concentrations, and it appeared that the mobilization of vitamin A from liver was impaired (68). Retinol-binding protein appears to be the predominant low molecular weight protein excreted in the proteinuria occurring in cadmium-exposed workers and in Cd-treated rats (69). Cd has long been known to interract with zinc (70) and there is a known zinc-vitamin A interraction.

Sewage sludge contains a lipid biopolymer, cutin, which is essentially insoluble (71). Cutin accounted for nearly 25% of total anaerobic sludge organic matter. A few organisms and enzymes have been found which degrade cutin, but it is probably not degraded in the intestine of animals. That lipid biopolymer could be absorbing or adsorbing lipid soluble materials such as vitamin A and preventing vitamin A movement from lumen of the gut across gut mucosa and into the blood stream; effectively slowing or stopping vitamin A absorption. However, at the present time the reason for reduced liver vitamin A levels of animals fed digested sewage sludge is not known. When a raw sewage sludge was fed at high levels for long periods, there was no reduction in liver vitamin A levels (36). Thus, this particular health defect is probably restricted to digested sludge.

ANIMAL FOOD CONCENTRATES DERIVED FROM SEWAGE SLUDGE

An organic feed concentrate (OFC) has been prepared from activated sewage sludge and included in diets for rats, rabbits and chickens at levels up to 20% of the diet (72). Reduced fertility, viability and lactation were found with 10% or higher OFC in the diet. Chicken egg laying was suppressed with 5% OFC and hens entirely ceased egg laying when fed 15 or 20% levels of OFC. High phosphate levels and high levels of heavy metals were thought to account for the animal health problems. Potentially toxic substances were said to be removable, but at present OFC was recommended at no higher than 5% of diets for animals. In the future there may be more of these kinds of animal food materials extracted from sewage waste, and effects upon animals must be determined before such foodstuffs are distributed for dietary use.

CONCLUSIONS

There is ample reason for caution and concern as sewage sludge is amended to soils or used as fertilizer in the growing of crops. Adverse animal health effects have resulted from such practices. Adequate laws, regulations, guidelines and monitoring are essential for adequately protecting animal food chains from contamination from toxic materials found in sewage sludges.

Research projects supported by governmental agencies have been very important in building the knowledge base we now have, but much more needs to be known about the various sludges and the effects of those sludges upon crops and animals. Ways also need to be found to minimize the known problems as well as to tap the potential benefits from recycling sewage wastes.

REFERENCES

1. Braude, G. L., R. B. Read, Jr. and C. F. Jelinek. "Comparison of Health Considerations for Land Treatment of Wastewater," in *State of Knowledge in Land Treatment of Wastewater, Vol. 1* (Hanover, NH: U.S. Army Corps of Engineers, Cold Regions Research and Engineering Lab., 1978) pp. 59-65.

2. Braude, G., B. P. Sagik and C. A. Sorber. "Human Health Risk: Using Sludge for Crops," *Water and Sewage Works* (December 1978) pp. 62-64.

3. "Criteria for Classification of Solid Waste Disposal Facilities and Practices: Final, Interim Final and Proposed Regulations," Federal Register, Part IX Environmental Protection Agency (September 13, 1979).

4. Jelinek, C. F. and G. L. Braude. "Management of Sludge Use on Land," J. Food Protection 41:476-480 (1978).

5. Jelinek, C. F. and G. L. Braude. "Management of Sludge Use on Land: FDA Considerations," in *Proceedings of the 3rd National Conference on Sludge Management Disposal and Utilization* (Rockville, MD: Information Transfer Inc., 1977), pp. 35-38.

6. Wagstaff, D. J., J. F. Brown and J. R. McDowell. "Metals and Other Feed Contaminants: Regulatory Concerns," *FDA ByLines* 9(6):271-286 (1979).

7. Doyle, J. J. "Effects of Low Levels of Dietary Cadmium in Animals: A Review", *J. Environmental Quality* 6(2):111-116 (1977).

8. Underwood, E. J. *Trace Elements in Human and Animal Nutrition.* Academic Press, NY, NY pp. 420.

9. Lagerwerff, J. V., G. T. Biersdorf and D. L. Brower. "Retention of Metals in Sewage Sludge I: Constituent Heavy Metals," *J. Environmental Quality* 5(1):19-23 (1976).

10. Bastian, R. K. and W. A. Whittington. "EPA Guidance on Disposal of Municipal Sewage Sludge Onto Land," in *Proceedings of the 3rd National Conference on Sludge Management Disposal and Utilization* (Rockville, MD: Information Transfer Inc., 1977), pp. 32-34.

11. Hammond, E. C. and I. J. Selikoff. Public Control of Environmental Health Hazards, Vol. 329 (New York: New York Academy of Sciences, 1979), pp. 1-405.

12. Weisbroth, S. H. "Chemical Contamination of Lab Animal Beedings: Problems and Recommendations," Lab Animal 8(7): 24-34 (1979)

13. Healy, W. B. "Ingestion of Soil by Dairy Cows." New Zealand J. Agr. Res. 11:487-499 (1968).

14. Harrison, D. L., J. C. M. Mol and W. B. Healy. "DDT Residues in Sheep from Ingestion of Soil", New Zealand J. Agr. Res. 13:664-672 (1970).

15. Collett, J. N. and D. L. Harrison. "Lindane Residues of Pasture and in the Fat of Sheep Grazing Pasture Treated with Lindane Prills," New Zealand J. Agr. Res. 11:589-600 (1968).

16. Capar, S. G., J.T. Tanner, M. H. Friedman and K. W. Boyer. "Multielement Analysis of Animal Feed, Animal Wastes and Sewage Sludge", Environmental Science and Technology 12(7): 785-790 (1978)

17. Tacon, A. G. J. and P. N. Ferns. "Activated Sewage Sludge, a Potential Animal Foodstuff. I. Proximate and Mineral Content, Seasonal Variation", Agric. and Environment 4:247-269 (1979)

18. Fitzgerald, P. R. "Toxicology of Heavy Metals in Sludges Applied to the Land", in Proceedings of the 5th National Conference on Acceptable Sludge Disposal Techniques: Cost, Benefit, Risk, Health and Public Acceptance (Rockville, MD: Information Transfer Inc., 1978) pp. 106-116.

19. Fitzgerald, P. R. "Potential Impact on the Public Health Due to Parasites in Soil/Sludge Systems," in Proceedings of the 8th National Conference on Municipal Sludge Management: Impact of Industrial Toxic Materials on POTW Sludge (Rockville, MD: Information Transfer Inc., 1979), pp. 214-221.

20. University of Maryland and USDA. "Feasibility of Using Sewer Sludge for Plant and Animal Production," Final Report, Contract # 2JAACA-B/2JAAQA/28-XXCA, Cooperative Research Project, funded by Washington Suburban Sanitary Commission (1976 and 1977).

21. University of Maryland and USDA. "Feasibility of Using Sewer Sludge for Plant and Animal Production," Final Report, Contract # 2JAACA-B/2JAAQA/28-XXCA, Cooperative Research Project, funded by Washington Suburban Sanitary Commission (1978 and 1978).

22. Kienholz, E. W., G. M. Ward, D. E. Johnson, J. Baxter, G. Braude and G. Stern. "Metropolitan Denver Sewage Sludge Fed to Feedlot Steers," J. Animal Science 48(4): 734-741 (1979).

23. Bertrand, J. E., M. C. Lutrick, H. L. Breland and R.L. West. "Effects of Dried Digested Sludge and Corn Grown on Soil Treated with Liquid Digested Sludge on Performance, Carcass Quality and Tissue Residues in Beef Steers," J. Animal Science 50(1): 35-40 (1980).

24. Firth, J. A. and B. C. Johnson. "Sewage Sludge as a Feed Ingredient for Swine and Poultry," Agr. and Food Chemistry 3(9):795-796 (1955).

25. Edds, G. T., K. E. Ferslew and R. A. Bortell. "Feeding of Urban Sewage Sludge to Swine," in Proceedings of the 82nd Annual Meeting of U.S. Animal Health Association, (Hilton Hotel, Buffalo, N.Y. 1978), pp. 207-220.

26. Osuna, O. "Toxicology of Aflotoxin B-1, Warfarin, and Cadmium in Young Pigs," Ph.D. Thesis, University of Florida, Gainesville, FL (1980).

27. Beaudouin, J., R.L. Shirley and D. L. Hammell. "Effect of Sewage Sludge Diets Fed Swine or Nutrient Digestibility, Reproduction, Growth and Minerals in Tissues," J. Animal Science (in press).

28. Cheeke, P. R. and R. O. Meyer. "Evaluation of the Nutritive Value of Activated Sewage Sludge with Rats and Japanese Quail," Nutrition Reports International 8 (6):385-392 (1973).

29. Kinzell, J. H., P. R. Cheeke and R. W. Chen. "Tissue Heavy Metal Accumulation Sleeping Times and Muligeneration Reproductive Performance of Rats Fed Activated Sewer Sludge," Nutrition Reports International 15(6): 645-650 (1977).

30. Sedgwick, R. and D. R. Arthur. "A Natural Pollution Experiment: The Effects of a Sewage Strike on the Fauna of the Thames Estuary," Environ. Pollut. 11:137-160 (1976).

31. Hartenstein, R., E. F. Neuhauser and J. Collier. "Accumulation of Heavy Metals in the Earthworm Eisenia foetida," J. Environmental Quality 9:23-26 (1980)

32. Meyers, S. "EPA's Sludge Management Program: New Responsibilities and Challenges," in Proceedings of the 3rd National Conference on Sludge Management Disposal and Utilization, (Rockville, MD: Information Transfer Inc., 1977), pp. 1-2.

33. Smith, G. S., H. E. Kiesling, J. M. Cadle, C. Staples, L. B. Bruce and H. D. Sivinski. "Recycling Sewage Solids as Feedstuffs for Livestock," in Proceedings of the 3rd National Conference on Sludge Management Disposal and Utilization, (Rockville, MD: Information Transfer Inc., 1979), pp. 119-127.

34. Smith, G. S., H. W. Kiesling and E. E. Ray. "Prospective Usage of Sewage Solids as Feed for Cattle," in Proceedings of 8th National Conference on Municipal Sludge Management: Impact of Industrial Toxic Materials on POTW Sludge (Silver Springs, MD: Information Transfer Inc., 1979) pp. 190-200.

35. Smith, G. S., H. E. Kiesling, C. H. Herbel, E. E. Ray, P. Truijillo, R. Orcasberro and J. S. Sivinski. "Sewage Solids as Supplemental Feed for Ruminants Grazing Rangeland Forage," Transactions American Nuclear Society (in press).

36. Smith, G. S. Personal Communication (1980).

37. Hansen, L. G., J. L. Dorner, C. S. Byerly, R. P. Tarara and T. D. Hinesly. "Effects of Sewage Sludge-Fertilized Corn Fed to Growing Swine," Am. J. Vet. Res. 37:711-720 (1976).

38. Hinesley, T. D., E. L. Ziegler and J. J. Tyler. "Selected Chemical Elements in Tissues of Pheasants Fed Corn Grain from Sewage Sludge Amended Soil", Agro-Ecosystems 3:11-26 (1976).

39. Buck, J. S., T. D. Hinesley, R. D. Rowland and D. J. Bray. "Cadmium Retention by Chickens Fed Crops From Soils Fertilized with Sewage Sludge," Poultry Science 58:1039-1040 (Abstract) (1979).

40. Bray, D. J., E. L. Ziegler, K. E. Redborg, T. D. Hinesly and R. D. Rowland. "Effect of Lifetime Feeding of Crops from Sludge-Amended Soils to Pullets," J. Poultry Science (in press)

41. Williams, P. H., J. S. Shenk and D. E. Baker. "Cadmium Accumulation by Meadow Voles (Microtus Pennsylavanicus) from Crops Grown on Sludge Treated Soil," J. Environmental Quality 7(3):450-454 (1978).

42. Heffron, C. L., J. T. Reid, D. C. Elfving, G. S. Stoewsand, W. M. Haschek, J. N. Telford, A. K. Furr, T. F. Parkinson, C. A. Bache, W. H. Gutenmann, P. C. Wszolek and D. J. Lisk. "Cadmium and Zinc in Growing Sheep Fed Silage Corn Grown on Municipal Sludge Amended Soil," J. Agric. Food Chemistry 28:58-61 (1980).

43. Haschek, W. M., A. K. Furr, T. F. Parkinson, C. L. Heffron, J. T. Reid, C.A. Bache, P. C. Wszolek, W. H. Gutenmann and D. J. Lisk. "Elements and Polychlorinated Biphenyl Deposition and Effects in Sheep Fed Cabbage Grown on Municipal Sewage Sludge," Cornell Vet. 69:302-314 (1979).

44. Chaney, R. L., G. S. Stoewsand, C. A. Bache and D. J. Lisk. "Cadmium Deposition and Hepatic Microsomal Induction in Mice Fed Lettuce Grown on Municipal Sludge Amended Soil," J. Agric. Food Chem. 26(4): 992-994 (1978)

45. Chaney, R. L., G. S. Stoewsand, A. K. Furr, C. A. Bache and D. J. Lisk. "Elemental Content of Tissues of Guinea Pigs Fed Swiss Chard Grown on Municipal Sewage Sludge Amended Soil," J. Agric. Food Chem. 26(4): 994-997 (1978).

46. Dowdy, R. H. Personal Communication (1980).

47. Utley, P. R., O. H. Jones, Jr. and W. C. McCormick. "Processed Municipal Solid Waste as a Roughage and Supplemental Protein Source in Beef Cattle Finishing Diets," J. Animal Science 35(1): 139-143 (1972).

48. Johnson, J. C., Jr. and W. C. McCormick. "Ensiled Diet Containing Processed Municipal Garbage and Sorghum Forage for Heifers," J. Dairy Science 58(11):1672-1676 (1975).

49. Johnson, J. C., Jr., P. R. Utley, R. L. Jones and W. C. McCormick. "Aerobic Digested Municipal Garbage as a Feedstuff for Cattle," J. Animal Science 41(5):1487-1495 (1975).

50. Anderson, C. and K. D. Danylchuk. "Retardation of Bone Formation in Dogs Exposed to Lead," in Animals as Monitors of Enviromental Pollutants, (Washington, DC: National Academy of Sciences, 1979), pp. 377.

51. McIlroy, W. and F. A. Martz. "Development and Evaluation of Cattle Feed Components Prepared from Solid Waste Materials," NSF-RANN contract #AEN-75-14652A01, (Springfield, VA: National Technical Information Service, U.S. Department of Commerce, 1976),pp. 1-15.

52. Gillies, M. T. Cattle Feed from Solid Waste Components: Animal Feeds from Waste Materials, (Park Ridge, NJ: Noyes Data Corporation, 1978), pp. 263-303.

53. Kappas, A. and A. P. Alvares. "How the Liver Metabolizes Foreign Substances," Scientific American 232(6): 22-31 (1975).

54. Davis, G. K. "Chlorinated Hydrocarbons in Nutrition and Metabolism," Federation Proceedings 32(9):1993-2016 (1972).

55. McNally, J. M. "Polychlorinated Biphenyls: Environmental Contamination of Food," Office of Technology Assessment, Congress of the United States, Washington, DC, (1978).

56. Collins, W. T., Jr., L. Kasza and C. C. Capen. "Ultrastructural and Biochemical Effects of Polychlorinated Byphenyls on the Thyroid Gland of Osborne Mendel Rats," in Animals as Monitors of Environmental Pollutants, (Washington, DC: National Academy of Sciences, 1979), pp. 327-337.

57. Carstens, L. A., D.A. Barsotti and J. R. Allen. "Exposure of Infant Rhesus Macaques to Polychlorinated Biphenyl," in Animals as Monitors of Environmental Pollutants, (Washington, DC: National Academy of Sciences, 1979), pp. 339-344

58. Fries, G. F. "The Kinetics of Halogenated Hydrocarbon Retention and Elimination in Dairy Cattle," in *Fate of Pesticides in Large Animals,* (New York: Academic Press, 1977) pp. 159-174.

59. Hallett, D. J., R. J. Norstrom, F. I. Onuska and M. Comba. "Mirex, Chlordane, Dieldrin, DDT, and PCBs: Metabolites and Photoisomers in Lake Ontario Herring Gulls," in *Fate of Pesticides in Large Animals,* (New York: Academic Press, 1977), pp. 183-192.

60. Sleight, S. D. "Polybrominated Biphenyls" a Recent Environmental Pollutants," in Animals as *Monitors of Environmental Pollutants,* (Washington, DC: National Academy of Sciences, 1979), pp. 366-372.

61. Fries, G. F., G. S. Marrow and R. M. Cook. "Distribution and Kinetics of PBB Residues in Cattle," *Environmental Health Perspectives* 23:43-50 (1978).

62. McNally, M. M. "Polybrominated Biphenyls: Environmental Contamination of Food." *Office of Technology Assessment.* (Congress of the United States, Washington D.C.) pp. 1-30. (1978)

63. "Oiling the Way to a Cleaner Body," *Science News* 117:188 (1980).

64. Rozman, T., K. Rozman, J. Williams and H. Grein. "Intestinal Excretion of Hexachlorobenzene in the Rat," *Am. Soc. Toxicology,* (in press).

65. Rozman, T., J. Williams, K. Rozman and H. Greim. "Quantitative Determination of Intestinal Excretion of Hexachlorobenzene in Mineral Oil Treated and Untreated Rhesus Monkeys with Complete Biliary Bypasses," *Am. Soc. Toxicology* (in press).

66. Smith, G. S., I. Neiman and K. Rozman. "Prospective Usage of Dietary Mineral Oil to Enhance Pesticide Decontamination of Livestock," in *Proceedings of 1980 Cattle Growers Short Course,* (Las Cruces, NM: New Mexico State University, 1980), pp. 1-3.

67. Cecil, H. C., S. J. Harris, J. Bitman and G. F. Fries. "Polychlorinated Biphenyl-Induced Decrease in Liver Vitamin A in Japanese Quail and Rats," *Bull. Environ. Contam. and Toxicology* 9(3):179-185 (1973)

68. Sugawara, C. and N. Sugawara. "The effect of Cadmium on Vitamin A Metabolism," Toxicol. Appl. Pharmacol. 46:19-27 (1978).

69. Bernard, A., A. Goret, H. Goels, J. P. Buchet and R. Lauwerys. "Experimental Confirmation in Rats of the Mixed Type Proteinuria Observed in Workers Exposed to Cadmium," Toxicology 10:369-375 (1978).

70. Frieberg, L., M. Piscator and G. Nordberg. Cadmium in the Environment (Cleveland: Chemical Rubber Publishing Company, 1971), pp. 1-115.

71. Kolattukudy, P. E. and R. E. Purdy. "Identification of Cutin, A Lipid Biopolymer, as Significant Component of Sewages Sludge," Environmental Science and Technology 7(7):619-622 (1973).

72. Weinberg, M. S., H. K. Weiss, A. L. Palanker, and A. L. Sheffner. "Sludge Conditioning Using SO_2 and Low Pressure for Production of Organic Feed Concentrate," EPA Project #14-12-813, (Springfield, VA: National Technical Information Service, U.S. Department of Commerce, 1973), pp. 1-35.

CHAPTER 8

RISK TO ANIMAL HEALTH FROM PATHOGENS IN MUNICIPAL SLUDGE

J. Gary Yeager. The BDM Corporation, Albuquerque, New Mexico.

INTRODUCTION

Public and legislative concerns directed toward resource and materials recycling have stimulated widespread interest in the use of sewage sludge to improve the fertility and water-holding capacity of soil. The use of sludge on land to grow crops for human or animal consumption has raised concerns over the health hazards from the sludge pathogens. This concern has been chiefly directed toward human health risks. Relatively little attention has been focused on the risks to the health of animals that may graze on sludge-amended pastures or consume feedstuffs grown on these lands. Concern about the animal health risks is justified because economic losses from animal disease that may be associated with the use of sewage sludge could be quite large [1,2]. In fact, these losses may exceed potential economic losses from human disease associated with sludge use.

Most relevant literature on the subject is concerned with zoonoses, animal diseases that may be secondarily transmitted to man. However, when considering the risks to animal health from municipal sewage sludge, it seems appropriate that consideration should also be given to human diseases that may be transmitted to animals. This review emphasizes the risk to animal health from zoonotic and human pathogens in sludge and from specific animal pathogens that may be found in sludge.

The pathogens shown in Table I will be discussed in this review. These pathogens were selected because of their veterinary importance and/or their demonstrated presence in sewage and sludge. The list is not intended to be complete, but rather to serve as a point of departure for thought and discussion on the subject.

Table I. Some Pathogens Affecting Animal Health That May be Found in Municipal Sludge

Pathogen	Disease	Hosts of Major Concern
Salmonella sp.	Salmonellosis	Domestic and Wild Animals, Man
Mycobacterium sp.	Tuberculosis	Domestic Animals, Man
Brucella sp.	Brucellosis	Domestic Animals, Man
Escherichia Coli	Diarrheal Disease	Domestic Animals, Man
Bacillus Anthracis	Anthrax	Domestic Animals, Man
Leptospira sp.	Leptospirosis	Domestic and Wild Animals, Man
Reovirus	Possible Respiratory/Enteric Disease	Domestic and Wild Animals, Man
Rotavirus	Diarrheal Disease	Domestic and Wild Animals, Man
Taenia sp.	Taeniasis	Man (Definitive Host) Cattle, Swine, Man, (Intermediate Hosts)

Sources of Pathogens in Sludge

Pathogens that pose potential hazards to animal health enter municipal wastewater from a variety of sources. Humans infected with enteric diseases are a major source of bacterial, viral, and parasitic pathogens found in sewage and sludges. To pose an animal health risk, these pathogens must also be capable of producing disease in animals after surviving sewage treatment and exposure to environmental stresses. Other pathogen sources that may sometimes be of greater importance to animal health are effluents (from abattoirs, rendering plants, and dairies) that may be discharged into municipal wastewater treatment systems. In some local areas, the contribution to the sludge pathogen load from these sources may be significant. In a study on parasites in sludges in the southern United States, Riemers et al. [3] noted a significant increase in Ascaris sp. ova in the sludge from a plant receiving wastes from a swine slaughterhouse. Other pathogen sources may also be important, especially in specific local areas, but the ones mentioned above are the main sources of animal pathogens that might be in sludge.

Animal Exposure to Sludge-Borne Pathogens

Because other portions of this symposium address the animal health hazard associated with the consumption of sludge-based or processed grain feeds that may be contaminated with sludge pathogens, this review will emphasize the health risks to animals grazing on sludge-treated pastureland. These animals would be exposed to pathogens by ingesting forage contaminated with sludge and by ingesting sludge-containing soil along with the forage. (A recent study showed that up to 6% of an animal's diet could be composed of sludge from these sources [4].) An additional route of exposure would be the inhalation of pathogens associated with airborne particulates.

Regulation of Sludge Application to Pastureland

Concern over the health risks associated with sludge application to land has resulted in the establishment of regulations by the Environmental Protection Agency. Any discussion of the health risk to animals grazing sludge-treated pasture must be made in the context of these regulations as contained in 40 CFR part 257 [5]. Concerning animals, these regulations state:

> Sewage sludge applied to the land surface or incorporated into the soil must be treated by a Process to Significantly Reduce Pathogens. Aerobic digestion, air drying, anaerobic digestion, composting, lime stabilization, or other similar techniques will satisfy this requirement. In addition ... grazing by animals whose products are consumed by humans must be prevented for at least one month.

Although not explicit in the regulations, it can be implied that treatment of sludge with a Process to Further Reduce Pathogens eliminates the one-month waiting period. These processes include gamma ray irradiation, beta ray irradiation, pasteurization, or equivalent methods performed after a Process to Significantly Reduce Pathogens. Sludges treated by high-temperature composting, heat drying, heat treatment, and thermophilic digestion require no prior treatment.

BACTERIAL PATHOGENS OF POTENTIAL RISK TO ANIMALS

The following discussion of bacterial pathogens of veterinary importance includes those known to be found in municipal wastewater sludge.

Salmonella

Members of the genus Salmonella are pathogens of man and animals that are commonly isolated from municipal sewage and sludge. Salmonellae commonly occur in a wide range of hosts. They are pathogenic for both man and a variety of animals [6,7]. The genus can be divided into three general groups based upon their host range and specificity [6,7].
1. Salmonellae specifically pathogenic for man - This group contains the etiologic agents of human typhoid and paratyphoid fevers. Members of this group are not generally pathogenic for animals.
2. Salmonellae specifically pathogenic for animals - Members of this group cause such problems as enteritis, septicemia, and abortion in their specific hosts. Effluents from abattoirs and rendering plants are potential sources of these organisms in sludge. These organisms do not usually cause human disease.
3. Non-host-adapted salmonellae - This group contains the majority of Salmonella serotypes, the reservoirs of which are wild and domestic animals. Members of the group produce nonspecific or subclinical infections in animals and salmonellosis in man.

The latter category, the non-host-adapted salmonellae, is of greatest concern when considering the risks to human and animal health from an application of sludge. These serotypes produce the typical zoonoses while the human and animal specific salmonellae play an insignificant role in the epidimiology of salmonellosis shared by man and animals [6]. Salmonella typhimurium is usually the most commonly isolated member of the non-host-adapted group [6,7,8].

Salmonellosis in Animals

Animal infection with non-host-adapted salmonellae may be responsible for explosive outbreaks of animal disease with high mortality, especially in young animals [6,9,10]. However, the most common manifestation of these salmonelloses is an unspectacular generalized infection followed by a long symptomless period characterized by infection of mesenteric lymph nodes and chronic shedding of salmonellae in the feces [6,7,10]. Excretion of pathogens increases when the infected animal is subjected to environmental stresses or the trauma and crowding associated with transportation and slaughter [11]. This intermittent shedding of pathogens is largely responsible for the preservation and expansion of the reservoir of infection in domestic animals and the contamination of meat upon slaughtering [6].

Fate of Salmonellae During Sewage Treatment

Salmonellae enter municipal wastewater with the feces of infected individuals and with the effluents of abattoirs, rendering plants, and dairies. An estimated annual human morbidity of over 2,000,000 in the United States provides a constant source of pathogens to wastewater [7]. Salmonellae are also isolated from the effluents and waste from slaughterhouses [12,13].

While the level of salmonellae depends upon local morbidity, 5000 organisms per liter of raw sewage is probably a reasonable average figure [14]. Primary sedimentation will remove about 80 to 90% of the organisms to sludge [15]. Dudley et al. [16] recently detected 2400 salmonellae per kilogram of sludge. Depending on parameters such as digestion time and temperature, salmonellae are generally reduced about 90 to 99% by anaerobic digestion [15,17]. If the sludge is not dried or dewatered, no further treatment reductions will occur prior to land application. The dewatering of sludge by evaporation in drying beds will reduce the levels of salmonellae. Recent research at Sandia National Laboratories into the effects of sludge moisture content on the inactivation of $\underline{S}$. typhimurium in sludge showed a reduction of about 90% as the sludge was dried to 95% solids at room temperature. The persistence of salmonellae following drying was a function of moisture content. At a solids content of 96% virtually no change in levels of salmonellae occurred over a 4-month period, while one order of magnitude was lost each 18 to 20 days at a solids content of 70 to 80%. Similar findings have also been observed in foods [6]. Heat or radiation based Processes to Further Reduce Pathogens are generally effective in reducing or eliminating salmonellae from sludge. Our recent studies at Sandia National Laboratories have shown D-values (the dose of radiation required to reduce a population of organisms by 90%) for gamma radiation of $\underline{S}$. typhimurium ranging from less than 54 krad with liquid sludge (5% solids) to 120 krad with sludge dried to 96% solids. Sustained temperatures in excess of 45°C quickly eliminate salmonellae in sludge [18].

Survival of Salmonellae After Land Application

The persistence of salmonellae in the environment following dissemination depends upon a variety of factors including temperature, moisture availability, pH, and sunlight [12]. Once applied to pasture, salmonellae have been shown to persist for extended periods in the soil environment and in association with pasture grasses [7,9,19,20, 21,22]. Bicknell [23] reported an outbreak of $\underline{S}$. aberdeen in cattle associated with overflow of raw sewage from a sewer system, and other outbreaks in animals associated with sewage irrigation or accidental contamination have also been reported [10,24,25,26].

Animal Salmonellosis and Human Health

Contaminated meat and meat products are the most common vehicles of transmission for human salmonellosis [27]. Generally, the meat products are secondarily contaminated by intestinal contents, mesenteric lymph nodes and contaminated equipment in abattoirs rather than by primary contamination of the tissues due to animal infection [6,27]. Upon grinding or inadequate cooking of meat, the salmonellae may multiply to the levels required to initiate salmonellosis [27]. In one study, 35% of the carcasses in a slaughterhouse were contaminated with salmonellae [27]. Milk-borne salmonellosis is sometimes reported, but the organisms are almost always present because of fecal contamination of milk rather than as a result of infection [6,25,28].

Significance of Sludge-borne Salmonellosis

The interdependence of human and animal salmonellosis is well documented. However, human salmonellosis is undoubtedly more dependent on animal reservoirs than the reverse. The infection of animals by sludge-borne salmonellae would contribute to the perpetuation of the natural reservoir of infection, but the magnitude of the reservoir is already so great that the significance of this occurrence is probably small.

Mycobacterium

Mycobacteria are acid-fast rods that are responsible for tuberculosis and leprosy in humans and tuberculosis in animals. The concern over human tuberculosis resulting from exposure to tubercular animals is longstanding and has been responsible for religious proscription against the consumption of certain meats [29].

In the 19th century United States, bovine tuberculosis caused by Mycobacterium bovis was responsible for large losses in cattle herds [30]. However, aggressive campaigns involving tuberculin testing and slaughter of reactors has greatly reduced the incidence of this disease [1]. It is estimated that the current U.S. incidence is 0.1%, occurring most commonly in dairy herds [1]. In 1978, 92 infected cattle herds were reported [31].

Disease in Animals

Tuberculosis in cattle may be pulmonary or visceral. Most infections of cattle with M. bovis are chronic, but the virulent organisms may be excreted in milk or in feces [30]. In acute tuberculosis mastitis as many as 1.5×10^8 organisms per day may be excreted into milk [32]. While tuberculosis in cattle may be inapparent in nature, M. bovis causes disease in man that is clinically indistinguishable from tuberculosis caused by M. tuberculosis. There is, however, little evidence to suggest that animal tuberculosis caused by M. tuberculosis is a significant problem [29,32]. Several reports have indicated that cattle have become tuberculine reactive as a result of drinking water or grazing on pasture contaminated with effluent from sanitoria [12,32,33]. Opinions as to the significance of bovine infection with M. tuberculosis range from the idea that the infection would merely serve to immunize against subsequent infection with M. bovis to concern over the possibility that subclinical infection could lead to contamination of meat and milk with pathogens [30].

Mycobacteria in Sewage, Sludge, and the Environment

Viable mycobacteria can enter sewage and sludge from several sources. Early studies on the effluents from sanitoria demonstrated that virtually all tubercular patients excrete M. tuberculosis in their feces [34,35]. Levels of about 1×10^6 per liter have been found in sewage and up to 1×10^8 per liter were recovered from sludge [32]. Abattoirs may also contribute M. bovis to the load of mycobacteria in sewage and sludge. Jensen [35] isolated mycobacteria in the effluents of four out of seven abattoirs surveyed. Today, with tuberculosis patients no longer housed in sanitoria, any organisms they excrete are discharged into private or municipal wastewater treatment facilities where the fastidious nature and slow growth of these pathogens make the organisms difficult to isolate.

Primary sedimentation removes about 50% of the mycobacteria in sewage to sludge [36,37]. Anaerobic digestion inactivates about 90% of the mycobacterium population in sludge [32,36]. Mycobacteria are relatively stable in drying sludge. Jensen [35] recovered virulent organisms after 11½ to 15 months in drying sludge. Processes with extended elevated temperatures or irradiation treatment would effectively reduce or eliminate the levels of mycobacteria in sludge.

Once released to the environment, the organisms are quite stable and would likely persist for long periods [38]. Survival of M. bovis for up to 5 months in soil has been observed [39, 40]. Recent studies reported by Larkin [41] showed a 29-day survival of M. bovis (BCG) in soil. Viable mycobacteria persisted about 2 weeks on pasture grass in one study and up to 49 days in another [39,42].

Before the widespread adoption of pasteurization and tuberculin testing, numerous cases of human tuberculosis were attributed to M. bovis contaminated milk [12,30]. This risk is now virtually nonexistent. While there is the possibility that mycobacteria are present in the tissues of slaughtered, asymptomatic animals, an analysis of the incidence of M. avian-intracellulare reactivity in pork eaters showed no differences between a control group and pork eaters even though an estimated 3×10^6 swine with tubercular lesions have passed inspection since 1972 [43]. The findings of this study are supported by the informed opinions of others [29,32,38].

Even though cases of animal tuberculosis caused by M. tuberculosis have been documented, most authors agree that the risk to animal health from mycobacteria in sewage and sludge is probably insignificant [29,32,38]. Atypical mycobacteria have been isolated in high numbers from sewage sludge, but their significance to animal or human health is not clear [16].

Brucella

The brucellae are small Gram-negative bacteria that are capable of causing disease in man and a variety of animal hosts [44]. Infections in animals of economic concern are caused by B. abortus (cattle), B. suis (swine and cattle), and B. melitensis (sheep and goats). In man, each of these organisms is also capable of causing a febrile disease characterized by acute or relapsing fever.

Animal infection with brucellae is usually localized to the genital tissues, especially in the gravid uterus where infection often results in abortion of the fetus. The disease may also localize in the mammary glands and organisms may be excreted in milk for long periods. Brucellosis in animals, and the resulting economic losses, have decreased because of testing and immunization programs.

Because Brucella sp. infection in man does not lead to shedding in the feces or urine, any organisms present in sewage and sludge would originate in the wastes from dairies or slaughterhouses [45]. Brucellae have been isolated from dairy wastes [12]; however, in a recent study, no brucellae were isolated from the effluents of seven abattoirs or from the sludge in treatment plants receiving the effluents [45].

Even though the presence of brucellae in sludges is unlikely, numerous studies have been conducted on the survival of the organism in sludge and in the environment. Because the brucellae are slow growing and have complex nutritional requirements, studies indicate that they would not compete well with less fastidious organisms either during sewage treatment or in the environment [44]. Bell, et al. [45] found that B. abortus survived less than 1 day in raw primary sludge, and a maximum survival of 26 days was observed in neutralized raw sludge. In another study, B. abortus was reduced to undetectable levels after 11 weeks in cattle slurry [38]. Persistence for up to 100 days in feces, 20 days in a soil-manure mixture, and up to 66 days in wet soil at room temperature has been reported, but 4½ hours in direct sunlight may eliminate B. abortus [12,33].

Reported U.S. human brucellosis morbidity was 172 cases in 1978 [46]. As in previous years, the majority of human infections occurred in individuals working in the animal processing industry. The transmission of the disease by milk and milk products has been almost eliminated by the widespread practice of pasteurization.

The risk to animal health from brucellae in sewage sludge is probably insignificant. Wray [12] states, "with regard to Brucella the greatest risk of animal infection arises from foetal products of abortion; environmental contamination from other sources are not considered to be important pathways of infection."

Esherichia

In recent years, enteropathogenic strains of E. coli have been identified as the etiologic agents of diarrheal disease in newborn humans and domestic animals [47]. Diarrheal disease caused by enteropathogenic E. coli is responsible for considerable losses of newborn animals in parts of the United States and in other countries [48]. Enteropathogenic strains of E. coli produce two enterotoxins which cause watery, profuse diarrhea accompanied by dehydration and often death in infected animals. It is not clear that enteropathogenic E. coli of human origin that might be found in sludge would be capable of initiating disease in young animals. The strains causing animal diarrheal disease are generally host specific, and human strains may demonstrate the same specificity [3]. In humans the organism causes disease in infants and adults that may range from mild diarrhea to cholera-like illness [47]. In developing countries, E. coli and other diarrheal diseases are a leading cause of death in children less than 5 years of age [10].

It is impossible to differentiate enteropathogenic E. coli from normal intestinal flora by routine cultural procedures. Consequently, its abundance in municipal sewage has not been determined. The concentration in municipal sewage and sludge would depend largely on local morbidity, but, because of the low morbidity in the United States, the numbers are probably not large. Enteropathogenic E. coli would probably behave similarly to other Enterobacteriacae like salmonellae during sewage treatment and in the environment. In one study, fecal coliforms were inactivated on grass by 50 hours of sunlight and could not be recovered from alfalfa after 10 hours [49].

The significance to animal health of these organisms in municipal sludge used on pasture is not clear, but further study is warranted.

Anthrax

Anthrax caused by Bacillus anthracis is a severe, usually fatal, disease of herbivorous animals that is found in isolated spots in many parts of the world. Humans can also contract the disease by contact with infected animals or animal products [50]. In man the disease may be cutaneous without complications or it may be a fatal pulmonary infection known as "Wool Sorters Disease".

In animals anthrax is generally contracted by the consumption of forage contaminated with B. anthracis. The course of the disease is usually rapid, resulting in death in a few days. In the terminal stages of the illness, the organism is found in excretions and bloody discharges from the infected animal. These discharges and spores from the decomposing carcass contaminate the area surrounding the animal. The contaminated carcass and ground serve as the sources of infection for subsequent cases of anthrax. The vegetative form of B. anthracis is not resistant to environmental stresses, but the spores have been shown to persist in the soil environment for extended periods [33]. An outbreak of anthrax in England was tentatively associated with the carcass of an animal buried 30 years previously [51].

With proper disposal of the carcasses of infected animals, it is unlikely that B. anthracis would ever enter the municipal waste stream. Additionally, the course of the disease is so short and the illness is so severe that infected animals are unlikely to be slaughtered in abattoirs. Spores of B. anthracis are often found in the hair and hides of animals from enzootic areas [50]. Effluent from tanneries where these hides are processed could be a source of organisms for waste treatment facilities, and any spores in the sludge would likely be unaltered by treatment processes or land application. Untreated effluent from tanneries has been associated with outbreaks of the disease, but no recorded cases have been attributed to the use of contaminated sewage sludge or pastures [29].

The lack of evidence linking sludge use to anthrax and the improbability that the organism would ever be introduced into municipal sludge suggest that the risk to animal health from B. anthracis associated with land application of sludge is probably insignificant.

Leptospirosis

Leptospirosis is a zoonotic disease caused by spirochetes in the genus Leptospira. The principal reservoirs of the infection are domestic animals and wild rodents [52]. All leptospires pathogenic for animals are pathogenic for man, who usually contracts the disease by exposure to the urine of infected animals [52]. Human leptospirosis ranges from a subclinical infection to acute, fatal illness. Clinical features are generally nonspecific, but occasionally, the jaundice associated with classic Weil's disease will occur. In the common animal hosts, the disease is generally very nonspecific in nature. Complications include abortion, decreased milk production, and poor weight gain that may have significant economic impact [1]. The disease often becomes chronic and subclinical with leptospires being shed in the urine for over 6 months [52].

Leptospires originating from infected humans, pet feces, dairy wastes, abattoirs, and rat infestation of sewers may enter municipal wastewater treatment systems. In studies cited by Wray [12], Varfolomeeva isolated the organism from 6.3% of the untreated sewage samples tested. The presence of leptospires in sewage is confirmed by documented cases of sewage-acquired leptospirosis, mainly caused by L. icterohaemorrhagiae and L. canicola [52].

Because of their complex nutritional requirements, strict aerobic nature, and sensitivity to environmental stress, leptospires do not generally grow in the environment and would not be expected to fare well during the sewage treatment. Chang [53] reported 12- to 14-hours survival of L. icterohaemorrhagiae in anaerobic sewage and 2- to 3-day survival under aerobic conditions. Leptospires have been shown to persist 183 days in saturated soil, but soil drying eliminated viable organisms in 2½ hours [54].

Man is an accidental host for leptospires. The route of infection is most commonly through exposure to animal urine, primarily from the pet dog, and human transmission is rare [52]. While the organisms have been isolated from the milk and tissues of infected animals, no human infections have been reported from these sources. Farmers, veterinarians, abattoir workers, and other animal handlers comprise the group at greatest risk.

The animal health risk from human or animal leptospires in municipal sewage and sludges is not clear. However, given the relative fragility of the organism, the risk is probably small.

VIRAL PATHOGENS

Domestic animals, like humans, are subject to infection by a large number of virus types. Animal viruses are generally characterized by a high degree of host specificity; thus, it is likely that few of the large number of known human enteric viruses in sewage and sludge would pose any health risk to animals. Animal viruses have been detected in the effluents of abattoirs and could find their way into municipal waste streams [55]. The numbers of these viruses in municipal sewage and their significance to animal and human health are not known.

Two enteric viruses have host-range characteristics that allow speculation as to their potential health significance in sludges applied to land.

Reovirus

Reoviruses are enteric viruses with an extremely broad host range. Identical serotypes have been recovered from man and a variety of animals including cattle and swine, and cross infectivity of human and animal isolates has been demonstrated [56]. Reoviruses are often associated with respiratory and enteric disease in man and animals, but proof of their contribution to the etiology of the diseases is only circumstantial. With the exception of studies by Ward and Ashley [57,58] in liquid and dried sludge, the fate of reoviruses in sludge has not been widely investigated.

Rotavirus

Rotaviruses, like reoviruses, have a double-stranded, segmented RNA genome surrounded by a double protein capsid. These viruses have been shown to be a common cause of infantile diarrheal disease, a leading cause of infant mortality in developing countries [59]. Other rotaviruses under a variety of names are responsible for diarrheal disease in domestic and wild animals. Human and animal rotaviruses share common group antigens, and experimental infection of gnotobitic calves with human rotavirus has been demonstrated [59]. Other studies have shown that immunity induced by bovine rotavirus protected calves against challenge with human rotavirus [60].

The survival of rotaviruses during sewage treatment and in the environment has been the subject of little research. Ward and Ashley [61] studied the effects of detergents in sludge on rotavirus infectivity, and Farrah et al. [62] have studied adsorption on activated sludge and aluminum hydroxide flocs.

Fate of Enteric Viruses in Sludge Treatment

From the preceding discussion of reovirus and rotavirus, it can be seen that little is known about their survival during sludge treatment and in the environment. However, considerable research has been done on the survival of enteric picornaviruses during sludge treatment. The results of these studies may provide insights into the fate of other animal pathogenic viruses during common sludge treatment processes.

Anaerobic digestion generally reduces enteric virus populations by 2 to 3 orders of magnitude depending upon digestion time and other process parameters [63,64]. Drying of sludge to a solids content of about 80% or greater has been shown to greatly reduce virus numbers [57]. Viruses in sludge are more resistant to irradiation than bacterial or parasitic pathogens. Ward and Ashley [65] showed that doses of about 300 krads or greater may be required for a 90% reduction in virus infectivity. Sustained high-temperature treatments such as pasteurization and composting are effective in reducing or eliminating viruses in sludge [66,67].

More research is required before the health significance of animal pathogenic viruses in sludge is known.

PARASITES

The ova and resistant stages of parasites found in sewage generally have a specific gravity greater than one and will settle into sludge during primary sedimentation [68, 69].

Nematodes

The morphologically identical ova of the human parasite Ascaris lumbricoides and A. suum from swine are commonly isolated from municipal sewage and sludge in the United States and elsewhere [3,69,70]. These ova should pose little health hazard to animals grazing on sludge-treated fields, because their host range is relatively narrow. The ova of these parasites may hatch and larval migration may occur in animals other than the appropriate host, but mature adults will not generally develop [71]. Swine coming into contact with sludge containing A. suum from slaughterhouses could become infected, but intraspecific transfer of ova is probably the dominant route of infection. The ascarid parasites Neoascaris vitulorum of cattle and Parascaris equorum of horses may possibly be found in sewage and sludge as the result of effluent discharge from abattoirs, but no reports of their occurrence in sludge or resulting disease problems have been made.

Cestodes

Of the potentially wastewater-borne pathogens discussed so far, none have an absolute dependence on sewage for their transmission and perpetuation. Taenia saginata, the beef tapeworm, and T. solium, the tapeworm of swine, on the other hand, require man and human feces for the completion of their life cycles. As the definitive host for these parasites, man houses the adult form in his small intestine [72]. The adult cestodes produce large numbers of eggs in gravid segments that are passed in human feces. The ingestion of these ova by the intermediate host (cattle or swine) results in hatching of the eggs in the intestine and migration of the larvae to the skeletal muscles of the host where they produce cysticerci called measles. The ingestion of inadequately cooked "measly" meat by man completes the life cycle of the parasite [73]. The disease is not generally serious in man but is aesthetically unacceptable [74].

There are no accurate figures on taeniasis morbidity, but incidence in cattle and humans in the United States is estimated at less than 1% [74,75]. Infected individuals may excrete up to 1×10^6 ova per day, but actual numbers in municipal wastewater and sewage are variable and depend upon local morbidity [69,74]. In one Russian study reported by Greenberg [74], 1935 ova per liter were recovered during one period while only 19 ova per liter were isolated in later samples. During sedimentation, about 90% of the ova will be partitioned into the settled sludge [74,76]. Even though incidence of taeniasis is low in the United States, several investigators have recovered ova from municipal sewage sludges. [3,70,77].

Anaerobic digestion is relatively ineffective in removing tapeworm ova from sludge. Studies by Newton et al. [76] showed that 10 to 15% of inoculated T. saginata ova were morphologically normal after 201 days of mesophilic, anaerobic digestion. Greenberg [74] cites experiments by Jepsen and Roth in which ova survived 16 days in sewage and 71 days in manure. Taenia sp. ova are more sensitive to desiccation than ascarid ova [69]. The ova survive no longer than 14 days in the absence of surface moisture [75]. Penfold et al. [78] found no viable ova in sludge that had been air-dried for 2 years, but recovered viable ova after 13½ weeks in dry soil.

Extended survival of ova on grass or pasture has been observed. Cattle grazing on pasture irrigated with sewage 2 months earlier were infected with T. saginata and artificially contaminated grass yielded viable ova 159 days after inoculation [74].

It is generally felt that conventional sewage treatment processes are inadequate in eliminating the animal health hazard from tapeworm ova in sewage and sludge [69,74,76]. The documented outbreaks of cysticercosis in cattle associated with irrigation of pastures with sewage and the demonstrated ability of the ova to survive treatment processes, suggest that this may be the pathogen of greatest concern when considering the use of sludge on pasture [79,80]. Treating contaminated sludge with a Process to Further Reduce Pathogens prior to land application should eliminate this hazard.

QUANTIFICATION OF THE RISK TO ANIMAL HEALTH

The risk to animal health posed by sludge-borne pathogens in the environment depends upon the exposure of a susceptible animal to a dose of virulent pathogens capable of initiating infection and disease. The presence of sufficient numbers of pathogens at the point of animal exposure depends upon the ability of the pathogens to survive sludge treatment and any subsequent stresses associated with the environment or further processing. From the preceding discussion of pathogens, it is clear that pathogen survival in sludge during treatment and in the environment is highly variable and depends upon a complex interaction of environmental stresses.

The complex nature and interaction of the factors that cause the pathogen risk to animal health are only beginning to be understood by researchers. Most individuals who are required to make and implement sludge management decisions are not likely to be prepared to adequately address this subject. To the author's knowledge, no comprehensive evaluations of animal health risks have been made prior to the use of sewage or sludge on pastures. The potential health and economic impact of widespread animal disease resulting from the use of municipal sludge indicate that an informed risk evaluation would be desirable.

A computer model is a tool that would allow the quick and easy evaluation of animal health risks associated with a variety of sludge treatment and use alternatives. A computer model can rapidly deal with the complex, simultaneous interaction of risk factors and can provide meaningful, understandable data for consideration in a decision making process.

The Beneficial Uses Program at Sandia National Laboratories in Albuquerque, New Mexico has been actively conducting research on the use of cesium-137 in conjunction with conventional sewage treatment to reduce or eliminate the

pathogens in municipal sewage sludge. Recognizing the need for a comprehensive assessment of the pathogen risks associated with the beneficial use of sludge, Sandia National Laboratories (with funding from the Environmental Protection Agency's Health Effects and Municipal Environmental Research Laboratories and the Department of Energy) contracted with The BDM Corporation to develop a risk assessment computer model. The model provides an evaluation of the risks associated with the following sludge use alternatives:
1. Dried raw or digested sludge used as an agricultural fertilizer.
2. Liquid sludge used as an agricultural fertilizer.
3. Dried raw sludge used as an animal feed supplement.
4. Composted sludge used as a residential soil amendment.

A comprehensive literature review was used to define 13 pathways describing the fate of three representative pathogens (_Salmonella_ sp., _Ascaris_, and poliovirus) during sewage treatment and after release to the environment. Information from the literature review was also used to develop mathematical descriptions of the changes in pathogen populations during sludge treatment processes and at locations in the environment where human exposure is possible. A separate portion of the model calculates information about the chances of human infection resulting from an exposure to pathogens at the environmental locations described by the model. The example pathways from the model, shown in Figures 1 and 2, demonstrate that this approach would also be useful in assessing the risk to animals from pathogens in municipal sludge.

The pathway diagram shown in Figure 1 describes the movement of pathogens through a liquid sludge treatment process. Associated with each compartment or box in the diagram is a mathematical expression (process function) that calculates the change in pathogen population that occurs in the compartment. The dashed line around the "Anaerobic Digestion" compartment indicates that the process can be included or eliminated at the discretion of the model user. Transfer functions (denoted by the symbol τ connecting compartments) mathematically describe the transfer of pathogens between pathway compartments.

The model user supplies both the number of pathogens entering the pathway with raw sludge and certain treatment and environmental parameters such as ambient temperature, radiation dose, and storage time. Default values based on information from the literature review are provided in the event that the user cannot or does not wish to supply the information. Upon execution of the computer code associated

PATHOGENS AND ANIMAL HEALTH   189

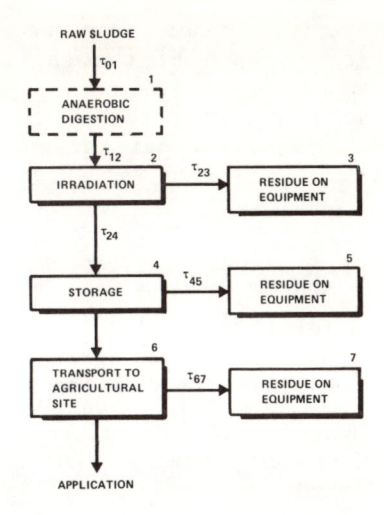

Figure 1. Model pathway for treatment of liquid sludge

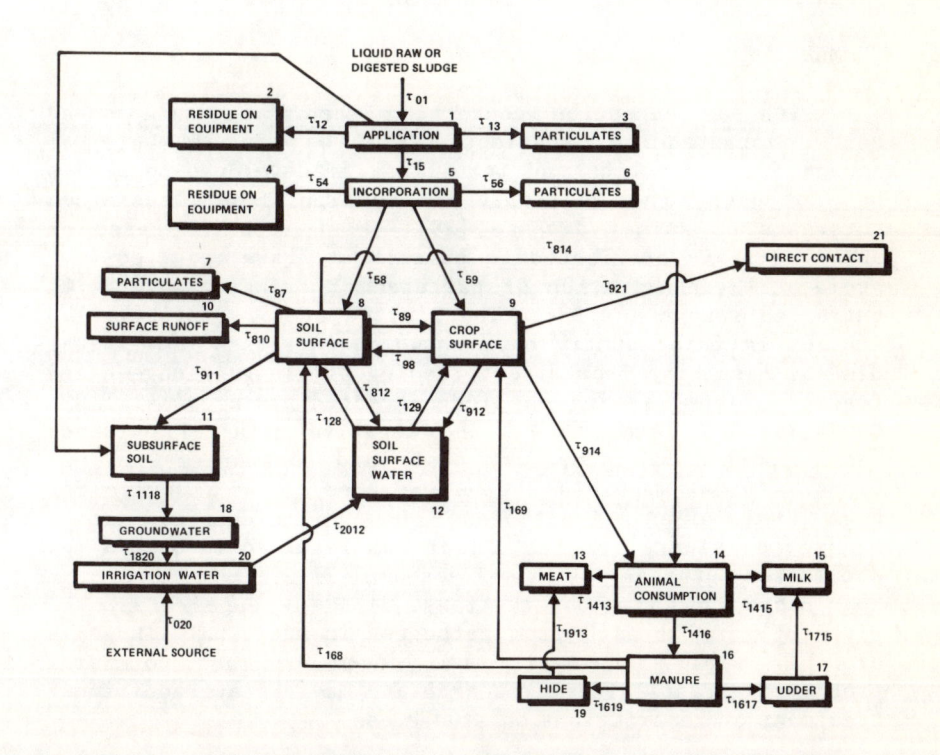

Figure 2. Model pathway for application of liquid sludge to pasture

with this pathway, the model provides estimates of the pathogen population in each of the environmental compartments. In an evaluation of the risk associated with this liquid sludge, the number of pathogens in the "Transport to Agricultural Site" compartment at the end of the model run would be present if the sludge was subsequently applied to land.

The more complicated model pathway shown in Figure 2 describes the fate of sludge pathogens after liquid sludge is applied to a pasture. Most of the compartments in this diagram represent potential points of human exposure. However, if the risk to animal health was being evaluated, the "Soil Surface" and "Crop Surface" compartments would contain estimates of the pathogens available for animal consumption at any point in time after sludge application. The "Meat" and "Milk" compartments contain estimates of the pathogens transferred to these locations as a result of infection or fecal contamination. These compartments represent potential sources of exposure to man.

With modifications to account for any unique characteristics of selected animal pathogens, such a computer model could be an effective tool for evaluating the animal health risks associated with certain uses of municipal sludge.

SUMMARY AND CONCLUSIONS

With an exception concerning Taenia saginata noted below, longstanding experience with the use of municipal sewage for irrigation and fertilizer in developed countries has not been shown to result in any significant, sustained outbreaks of animal disease [29]. Most reported cases of animal disease attributed to human waste have resulted from accidental contamination of pastures or drinking water with untreated sewage [10,23]. The limited number of controlled studies in this country have shown no increases in clinical disease resulting from direct feeding of sludge, sludge-based feeds, or from grazing on sludge-treated pastures [4,81,82, 83,84,85]. In some cases, the negative clinical evidence was supported by laboratory tests and postmortem examinations. Animal manure might be expected to be a source for high levels of animal pathogens. However, widespread use of manure and slurry for fertilizer has resulted in only a few documented outbreaks of animal disease [21,29].

Based on these observations, it would appear that the risk to animal health from pathogens in municipal sludge is probably small. However, some documented and potential problems warrant further research and may dictate additional treatment or restrictions on sludge use.

Estimates of continued low levels of taeniasis in this country and apparent increases in the disease in England demonstrate that the human feces/animal infection link so critical to the perpetuation of this parasite is still in existence [2,72].

In recent studies, tapeworm ova have been infrequently isolated from U.S. sludges [3,70]. However, the resistance of the ova to sewage treatment and environmental stresses may require that sludges originating in any high morbidity areas receive treatment by a Process to Further Reduce Pathogens prior to land application.

Current human tuberculosis therapy does not include confinement of patients in sanitoria. Consequently, any discharge of mycobacteria to municipal sewage is likely to be diffuse rather than concentrated. The hardiness of pathogenic mycobacteria during sewage treatment and in the environment may lead to a potential health hazard in areas where local human or animal morbidity has resulted in elevated numbers of the organisms in sludge. Additional treatment or restrictions on the landspreading of sludge would be required in such areas.

The ability of salmonellae and enteropathogenic E. coli to regrow in the environment may represent a risk to animal health when local conditions provide the moisture and nutrients required for regrowth. These pathogens and any resulting hazard would be eliminated by subjecting sludge to additional treatment.

The risk from viral pathogens in sludge is not well defined. The general resistance of enteric viruses to some sewage treatment processes and their ability to persist for long periods under certain environmental conditions may lead to future animal health risks. The numbers of pathogenic viruses in sludge can be considerably reduced by drying of the sludge and completely eliminated with high temperature treatment or large doses of radiation.

As previously mentioned, the wastes from abattoirs have been shown to contain bacterial, viral, and parasitic animal pathogens [3,13,55]. Routine meat inspection processes will generally lead to the rejection of heavily affected carcasses from diseased animals, but the slaughter of clinically normal animals may lead to the introduction of pathogens into the waste stream. The waste from abattoirs and the rendering plants that process diseased animals are potential sources of high concentrations of specific animal pathogens in sludge. Because relatively little is known about the actual numbers and risks due to pathogens from this source, it may be advisable to consider abattoirs, dairies and rendering plants as point sources of pathogen pollution with their wastes requiring pretreatment prior to discharge into municipal treatment plants. Mandatory additional treatment of sludges from plants receiving significant abattoir wastes would also reduce the magnitude of this risk.

Other potential problems of a speculative nature are interesting to consider. The broad host range of reoviruses and rotaviruses has been mentioned, but little is known about the ability of human enteric pathogens to cause disease in animals. Also, the long-term evolution of animal pathogens from human pathogens should be considered. Host-specific animal pathogens could result from gradual adaptation of human pathogens or from a transfer of genetic material in the environment.

Likewise, there is little information on the fate of human pathogens as they pass through the digestive systems of non-host animals. Passive concentration of hepatitis A virus in shellfish and human disease resulting from shellfish consumption are well documented [85]. The author is not aware of any similar occurrences in wild or domestic animals. Recent studies by O'Brien [87] with a variety of enteric viruses have demonstrated that rumen fluid is not virucidal and is, in fact, sometimes protective of viruses at normal rumen temperatures.

It is possible that future increases in the beneficial use of sewage sludge may result in the evolution of unforeseen animal health risks. However, potential and real animal health risks justify continued research and implementation of risk assessment and epidemiologic surveillance programs in areas where animals may be exposed to municipal sludge. Sludge management practices and regulations should remain flexible and responsive enough to permit rapid imposition of corrective action when warranted by epidemiological evidence.

ACKNOWLEDGEMENT

The development of the computer model described in this review was supported by a contract with Sandia National Laboratories Division 4535 under an interagency agreement with the U.S. Environmental Protection Agency and the U.S. Department of Energy.

NOTICE

This report was prepared as an account of work sponsored by the U.S. Government. Neither the United States nor the Department of Energy, nor any of their employees, nor any of their contractors, subcontractors, or their employees, makes any warranty, expressed or implied, or assumes any legal liability or responsibility for the accuracy, completeness or usefullness of any information, apparatus, product or process disclosed or represents that its use would not infringe privately owned rights.

REFERENCES

1. Jensen, R. and D. R. Mackey. Diseases of Feedlot Cattle (Philadelphia: Lea and Febiger, 1979).
2. Crewe, B., and R. Owen. "750,000 Eggs a Day - £750,000 a Year," New Scientist 80: 344-346 (1978).
3. Reimers, R. S., M. D. Little, D. B. Leftwich, D. D. Bowman, A. J. Englande, and R. F. Wilkinson. "Investigation of Parasites in Southern Sludges and Disinfection by Standard Sludge Treatment Processes," EPA Draft Final Report (1980).
4. "Feasibility of Using Sewage Sludge for Plant and Animal Production, Final Report 1978-1979," University of Maryland, U.S. Department of Agriculture Research Project, Beltsville, Maryland (1980).
5. "Criteria for Classification of Solid Waste Disposal Facilities and Practices," Federal Register 44: 53438-53468 (September 13, 1979).
6. Prost, E., and H. Riemann. "Food-borne Salmonellosis," Ann. Rev. Microbiol. 21:495-528 (1967).
7. "An Evaluation of the Salmonella Problem," National Academy of Sciences, Washington (1969).
8. Joint Working Party of the Veterinary Laboratory Services of the Ministry of Agriculture, Fisheries, and Food, and the Public Health Laboratory Service. "Salmonellae in Cattle and their Feedingstuffs and the Relation to Human Infection," J. Hyg. 63:223-241 (1965).
9. Moore, G. R., H. Rothenbacker, M. V. Bennett, and R. D. Borner. "Bovine Salmonellosis," J. Amer. Vet. Med. Assn. 141:841-844 (1962).
10. Williams, L. P., and B. C. Hobbs. "Enterobacteriaceae Infections." in Diseases Transmitted from Animals to Man, W. T. Hubbert, W. F. McCulloch, and P. R. Schnurrenberger, Eds. (Springfield, IL: Charles C. Thomas Publisher, 1975), pp. 33-109.
11. Taylor, J., and J. H. McCoy. "Salmonella and Arizona Infections," in Food-borne Infections and Intoxications, H. Riemann, Ed. (New York: Academic Press, 1969), p. 3-73.
12. Wray, C., "Survival and Spread of Pathogenic Bacteria of Veterinary Importance Within the Environment," Vet. Bull. 45:543-550 (1975).

13. Galton, M. M., W. V. Smith, H. B. McElroth, and A. V. Hardy. "Salmonella in Swine, Cattle, and the Environment of Abattoirs," J. Infect. Dis. 95:236-245 (1954).
14. Akin, E. W., H. P. Pahren, W. Jakubowski, and J. B. Lucas. "Health Hazards Associated with Wastewater Effluents and Sludges: Microbial Considerations," in Risk Assessment and the Health Effects of Land Application of Municipal Wastewater and Sludges, B. P. Sagik and C. A. Sorber, Eds. (University of Texas at San Antonio, 1978), pp. 9-27.
15. Mom, C. P., and C. O. Schaeffer. "Typhoid Bacteria in Sewage and in Sludge - An Investigation into the Hygienic Significance of Sewage Purification in the Tropics with Regard to Typhoid Fever," Sewage Works J. 12:715:737 (1940).
16. Dudley, D. J., M. N. Guentzel, M. J. Ibarra, B. E. Moore, and B. P. Sagik. "Enumeration of Potentially Pathogenic Bacteria from Sewage Sludges," Appl. Environ. Microbiol. 39:118-126 (1980).
17. Spray Waste Inc. "Microbiological Considerations in the Use of Sewage Sludge for Agricultural Purposes, Annual Report of the East Bay Municipal Utility District Study for 1976," (Davis, CA: 1977).
18. Brandon, J. R., and K. S. Neuhauser. "Moisture Effects on Inactivation and Growth of Bacteria and Fungi in Sludges," Sandia Laboratories Report SAND 77-1970 (1978).
19. Josland, S. W. "Survival of Salmonella typhimurium in Various Substances under Natural Conditions," Aust. Vet. J. 27:264-266 (1951).
20. Findlay, C. R. "The Persistence of Salmonella dublin in Slurry Tanks and on Pasture," Vet. Rec. 91-233-235 (1972).
21. Jack, E. J., and P. T. Hepper. "An Outbreak of Salmonella typhimurium Infection in Cattle Associated with the Spreading of Slurry," Vet. Rec. 84:196-199 (1969).
22. Burge, W. D., and P. B. Marsh. "Infectious Disease Hazards of Landspreading Sewage Wastes," J. Environ. Qual. 7:1-9 (1978).
23. Bicknell, S. R. "Salmonella aberdeen Infection in Cattle Associated with Human Sewage," J. Hyg. 70:121-126 (1972).
24. Baker, J. R. "An Outbreak of Salmonellosis in Sheep," Vet. Rec. 88:270-276 (1971).
25. Gibson, E. A. "Reviews on the Progress of Dairy Science. Section E. Diseases of Dairy Cattle. Salmonella Infection in Cattle," J. Dairy Res. 32:97-134 (1965).

26. George, J. T. A., J. G. Wallace, H. R. Morrison, and J. F. Harbourne. "Paratyphoid in Man and Cattle," Br. Med. J. 3:208-211 (1973).
27. Weissman, M. A., and J. A. Carpenter. "Incidence of Salmonellae in Meat and Meat Products," Appl. Microbiol. 17:899-902 (1969).
28. Hobbs, B. C. "Salmonella in Foods," in Proceedings of a National Conference on Salmonellosis, (U.S. Dept. of Health, Education and Welfare, 1964), pp. 84-94.
29. Williams, B. M. "The Animal Health Risks from the Use of Sewage Sludge on Pastures," in Utilisation of Sewage Sludge on Land (England: Water Research Centre Laboratories, 1979), pp. 1-14.
30. Kleeburg, H. H. "Tuberculosis and Other Mycobacterioses," in Diseases Transmitted from Animals to Man, W. T. Hubbert, W. F. McCulloch, and P. R. Schnurrenberger, Eds. (Springfield, IL: Charles C. Thomas Publisher, 1975), pp. 303-360.
31. "Animal Morbidity Report, Calendar Year 1978," U.S.D.A. Animal Plant Health Inspection Service (1979).
32. Greenberg, A. E., and E. Kupka. "Tuberculosis Transmission by Wastewater - A Review," Sewage Ind. Wastes 5:524-537 (1957).
33. Morrison, S. M., and K. L. Martin. "Pathogen Survival in Soils Receiving Waste," in Land as a Waste Management Alternative, R. C. Loehr, Ed. (Ann Arbor, MI: Ann Arbor Science Publishers, Inc., 1977), pp. 371-389.
34. Heukelekian, H., and M. Albanese. "Enumeration and Survival of Human Tubercle Bacilli in Polluted Waters II. Effect of Sewage Treatment and Natural Purification," Sewage Ind. Wastes 28:1094-1102 (1956).
35. Jensen, K. E. "Presence and Destruction of Tubercle Bacilli in Sewage," Bull. World Health Org. 10:171-179 (1954).
36. Sproul, O. J. "The Efficiency of Wastewater Unit Processes in Risk Reduction," in Risk Assessment and Health Effects of Land Application of Municipal Wastewater and Sludges, B. P. Sagik and C. A. Sorber, Eds. (University of Texas at San Antonio, 1978), pp. 282-296.
37. Bryan, F. L. "Disease Transmitted by Foods Contaminated by Wastewater," Wastewater Use in the Production of Food and Fiber, National Environmental Research Center, USEPA Report-660/2-704-041 (1974), pp. 16-45.
38. Rankin, J. D., and R. J. Taylor. "A Study of Some Disease Hazards Which Could be Associated With the System of Applying Cattle Slurry to Pasture," Vet. Rec. 85:578-581 (1969).

39. Maddock, E. C. G. "Studies on the Survival Time of the Bovine Tubercle Bacillus in Soil, Soil and Dung, in Dung, and on Grass, with Experiments on the Preliminary Treatment of Infected Organic Matter, and the Cultivation of the Organism," J. Hyg. 38:103-117 (1933).
40. Williams, R. S., and W. A. Hoy. "Viability of B. tuberculosis (bovinus) on Pasture Land, in Stored Feces, and in Liquid Manure," J. Hyg. 30:413-419 (1930).
41. Larkin, E. P., J. T. Tierney, J. Lovett, D. van Donsel, and D. W. Francis. "Land Application of Sewage Wastes Potential for Contamination of Foodstuffs and Agricultural Soils by Viruses and Bacterial Pathogens," in Risk Assessment and Health Effects of Land Application of Municipal Wastewater and Sludges, B. P. Sagik and C. A. Sorber, Eds. (University of Texas at San Antonio, 1978), pp. 102-115.
42. Sepp, E. "The Use of Sewage for Irrigation, a Literature Review," Bureau of Sanitary Engineering, California State Dept. of Public Health (1971).
43. Brown, J., and J. W. Tollison. "Influence of Pork Consumption on Human Infection with Mycobacterium avian-intracellulare," Appl. Environ. Microbiol. 38:1144-1146 (1979).
44. Schwartz, M. N. "The Brucellae," in Microbiology, B. D. Davis, R. Dulbecco, H. N. Eisen, H. S. Ginsberg, and W. B. Wood, Jr., Eds. (Hagerstown, MD: Harper and Row Publishers, 1973), pp. 812-817.
45. Bell, J. C., V. A. Argent, and D. Edgar. "The Survival of Brucella abortis in Sewage Sludge," in Utilisation of Sewage Sludge on Land (England: Water Research Centre Laboratories, 1979), pp. 475-483.
46. Center for Disease Control. "Brucellosis U.S-1978," Morbidity and Mortality Weekly Report 28:437-438 (1979).
47. Sack, R. B. "Human Diarrheal Disease Caused by Enteropathogenic Escherichia coli," Ann. Rev. Microbiol. 29:333-353 (1975).
48. Myers, L. L. "Enteric Colibacillosis in Calves: Immunogenicity and Antigenicity of Escherichia coli Antigens," Am. J. Vet. Res. 39:761-765 (1978).
49. Bell, R. G., and J. B. Bole. "Elimination of Fecal Coliform Bacteria from Dried Canary-Grass Irrigated with Municipal Sewage Lagoon Effluent," J. Env. Qual. 5:417-418 (1976).
50. Wright, G. C. "Anthrax," in Diseases Transmitted from Animals to Man, W. J. Hubbert, W. F. McCulloch, and P. R. Schnurrenberger, Eds. (Springfield, IL: Charles C. Thomas Publisher, 1975), pp. 237-250.

51. Hugh-Jones, M. E., and S. N. Hussaini. "An Anthrax Outbreak in Berkshire," Vet. Rec. 94:228-232 (1974).
52. Diesch, S. L., and H. C. Erlinghausen. "Leptospiroses," in Diseases Transmitted from Animals to Man, W. T. Hubbert, W. F. McCulloch, and P. R. Schnurrenberger, Eds. (Springfield, IL: Charles C. Thomas Publisher, 1975), pp. 436-462.
53. Chang, S. L., M. Buckingham, and M. P. Taylor. "Studies on Leptospira icterohaemorrhagiae IV. Survival in Water and Sewage: Destruction in Water by Halogen Compounds, Synthetic Detergents, and Heat," J. Infect. Dis. 82:256-266 (1948).
54. Okazaki, W., and L. M. Ringen. "Some Effects of Various Environmental Conditions on the Survival of Leptospira pomona," Am. J. Vet. Res. 18:219-223 (1957).
55. Malherbe, H. H., M. Strickland-Cholmley, and S. M. Geyer. "Viruses in Abattoir Effluents," in Transmission of Viruses by the Water Route, G. Berg, Ed. (New York: Interscience Publishers, 1967), pp. 347-354.
56. Rosen, L. "Reoviruses," in Diseases Transmitted from Animals to Man, W. T. Hubbert, W. F. McCulloch, and P. R. Schnurrenberger, Eds. (Springfield, IL: Charles C. Thomas, Publisher, 1975), pp. 919-921.
57. Ward, R. L., and C. S. Ashley. "Inactivation of Enteric Viruses in Wastewater Sludge Through Dewatering by Evaporation," Appl. Environ. Microbiol. 34:564-570 (1977).
58. Ward, R. L., and C. S. Ashley. "Identification of Detergents as Components of Wastewater Sludge that Modify the Thermal Stability of Reovirus and Enteroviruses," Appl. Environ. Microbiol. 36:889-897 (1978).
59. McNulty, M. S. "Rotaviruses-Review Article," J. Gen. Virol. 40:1-18 (1978).
60. Wyatt, R. G., R. H. Yolken, A. R. Kalica, H. D. James, Jr., A. Z. Kapikian, and R. M. Chanock. "Rotaviral Immunity in Gnotobiotic Calves: Heterologous Resistance to Human Rotavirus Induced by Bovine Virus," Science 203:548-550 (1979).
61. Ward, R. L., and C. S. Ashley. "Effects of Wastewater Sludge and its Detergents on the Stability of Rotavirus," Appl. Environ. Microbiol. (in press).
62. Farrah, S. R., S. M. Goyal, C. P. Gerba, R. H. Conklin, and E. M. Smith. "Comparison Between Adsorption of Poliovirus and Rotavirus by Aluminum Hydroxide and Activated Sludge Flocs," Appl. Environ. Microbiol. 35:360-363 (1978).
63. Bertucci, J. J., C. Lue-Hing, D. R. Zenz, and S. J. Sedita. "Inactivation of Viruses During Anaerobic Sludge Digestion," J. Water Pollut. Cont. Fed. 49:1642-1651 (1977).

64. Moore, B. E., B. P. Sagik, and C. A. Sorber. "Land Application of Sludges: Minimizing the Impact of Viruses on Groundwater," in <u>Risk Assessment and Health Effects of Land Application of Municipal Wastewater and Sludges</u>, B. P. Sagik and C. A. Sorber, Eds. (University of Texas at San Antonio, 1978), pp. 154-167.
65. Ward, R. L., and C. S. Ashley. "Inactivation of Poliovirus in Wastewater Sludge with Radiation and Thermoradiation," <u>Appl. Environ. Microbiol.</u> 33:1218-1219 (1977).
66. Wiley, B. B., and S. C. Westerberg. "Survival of Human Pathogens in Composted Sewage," <u>Appl. Microbiol.</u> 18:994-1001 (1969).
67. Ward, R. L., C. S. Ashley, and R. H. Moseley. "Heat Inactivation of Poliovirus in Wastewater Sludge," <u>Appl. Environ. Microbiol.</u> 32:339-346 (1976).
68. Shepard, M. R. N. "Part B - Helminthological Aspects of Sewage Treatment in Hot Climates," in <u>Water Wastes and Health in Hot Climates</u>, R. Feachern, M. McGarry, and D. Mara, Eds. (New York: John Wiley and Sons, 1977), pp. 299-309.
69. Hays, B. D. "Potential for Parasitic Disease Transmission with Land Application of Sewage Plant Effluents and Sludges," <u>Water Res.</u> 11:583-595 (1977).
70. Theis, J. H., V. Bolton, and D. R. Storm. "Helminth Ova in Soil and Sludge from Twelve U.S. Urban Areas," <u>J. Water Pollut. Cont. Fed.</u> 50:2485-2493 (1978).
71. Levine, N. A. <u>Nematode Parasites of Domestic Animals and Man</u> (Minneapolis: Burgess Publishing Co., 1968).
72. Rausch, R. L. "Taeniidae," in <u>Diseases Transmitted from Animals to Man</u>, W. T. Hubbert, W. F. McCulloch, and P. R. Schnurrenberger, Eds. (Springfield, IL: Charles C. Thomas Publisher, 1975), pp. 678-707.
73. Gemmell, M. A., and P. D. Johnstone. "Experimental Epidemiology of Hydatidosis and Cysticerocis," <u>Adv. Parisitol.</u> 15:310-369 (1977).
74. Greenberg, A., and B. H. Dean. "The Beef Tapeworm, Measly Beef, and Sewage-A Review," <u>Sewage Ind. Wastes</u> 30:262-269 (1958).
75. Pawlowski, Z., and M. G. Schultz. "Taeniasis and Cysticercosis, (<u>Taenia saginata</u>)," <u>Adv. Parasitol.</u> 10:269-343 (1972).
76. Newton, W. L., H. J. Bennett, and W. B. Figgat. "Observations on the Effects of Various Sewage Treatment Processes upon Eggs of <u>Taenia saginata</u>," <u>Am. J. Hyg.</u> 49:166-175 (1949).
77. Wang, W. L., and S. G. Dunlop. "Animal Parasites in Sewage and Irrigation Water," <u>Sewage Ind. Wastes</u> 26:1020-1032 (1954).

78. Penfold, W. J., H. B. Penfold, and M. Phillips. "A Survey of the Incidence of Taenia saginata Infestation in the Population of the State of Victoria from January 1934 to July 1935," Med. J. Aust. 1:283-285 (1936).
79. Roberts, F. C. "Experiments with Sewage Farming in Southwest U.S.," Am. J. Public Health 25:122-125 (1935).
80. Rickard, M. D., and A. J. Adolph. "The Prevalence of Cysticerci of Taenia saginata in Cattle Reared in Sewage Irrigated Pasture," Med. J. Aust. 1:525-527 (1977).
81. Smith, G. S., H. E. Kiesling, J. M. Cadle, C. Staples, L. Bruce, and H. D Sivinski. "Recycling Sewage Solids as Feedstuffs for Livestock," in Proceedings of the Third National Conference on Sludge Management, Disposal, and Utilization. (Rockville, MD: Information Transfer Inc., 1977), pp. 119-127.
82. Crawford, A. B., and A. H. Frank. "Effect on Animal Health of Feeding Sewage," Civil Eng. 10:495-496 (1940).
83. Weaver, R. W., N. O. Dronen, B. G. Foster, F. C. Heck, and R. C. Fehrmann. "Sewage Disposal on Agricultural Soils, Chemical and Microbiological Implications. Vol. II: Microbiological Implications," USEPA Report 600/2-78-131b (1978).
84. Kienholz, E., G. M. Ward, D. E. Johnson, and J. C. Baxter. "Health Considerations Relating to Ingestion of Sludge by Farm Animals," in Proceedings of the Third National Conference on Sludge Management, Disposal, and Utilization, (Rockville, MD: Information Transfer Inc., 1977), pp. 128-134.
85. Jolley, W. R., and P. R. Fitzgerald. "Sludge in the Pasture?," Illinois Res. 18:10-11 (1976).
86. Cliver, D. O. "Viral Infections," in Food-Borne Infections and Intoxications, H. Riemann, Ed. (New York: Academic Press, 1969), pp. 73-113.
87. O'Brien, R. T., Personal Communication, New Mexico State University, Las Cruces, NM (1980).

CHAPTER 9

DISEASE TRANSMISSION BY WILD ANIMALS FROM SLUDGE-AMENDED LAND

Annie K. Prestwood. Department of Parasitology, College of Veterinary Medicine, University of Georgia, Athens, Georgia.

INTRODUCTION

In any discussion of health risks associated with land application of municipal sludge, the role of wild animals in disease transmission must be considered. Free-living wild animals are ubiquitous, and almost all sites chosen for land application of sludge will harbor multiple populations of vertebrates. The presence of some of these animals is obvious, e.g., the white-tailed deer browsing in a rapidly growing forest. Other animals, however, are secretive and their presence is inapparent, e.g., the numerous small rodents and synanthropic lizards which feed among the flowers and shrubs on a surburban lawn. Each of these animals occupies an integral portion of our total environment and must be regarded as potential vectors of disease organisms when they feed upon sludge-amended land. This paper will consider some aspects of potential disease transmission by wild animals from sludge-amended land to man, domestic animals, and to other populations of wild animals.

Upon examination of the literature dealing with sludge and land application thereof, references were not encountered which dealt with disease transmission by wild animals. Information was available on specific organisms or groups of organisms and chemical compounds found in sludge. From this, extrapolations were made upon which this discussion is based. Only terrestrial vertebrates will be considered.

Free-living wild animals potentially may be involved in disease transmission in several different ways. Wild animals may serve as mechanical vectors of microorganisms, simply carrying a bacterium or a parasite egg to another site where a susceptible animal can become infected. Wild animals may become actively infected, amplify the organism, and subsequently shed organisms into the environment, or they may store the organism without amplification or concentrat a chemical and become involved in disease transmission only upon entry into the food chain. Parasites, bacteria, and various chemicals were considered most likely to be transmitted by wildlife from sludge-amended land.

PARASITES

Protozoa

Cysts of several protozoan parasites have been found in municipal sludge and include amoebae, Giardia, and various coccidian oocysts (1-3). Of these, Giardia spp. probably are most likely to become established in wild fauna and subsequently transmitted to man or domestic animals.

Recently, outbreaks of waterborne giardiasis have been reported in communities in California, Colorado, Oregon, and Pennsylvania (4). These outbreaks have several features in common, namely, they occurred in communities in which surface water was used wholly or in part as the municipal water supply; water was disinfected by chlorination; and the water was not filtered. The source of infection for these communities was thought to be beavers, since in three of the four communities involved, watersheds contained beavers which were eliminating cysts of Giardia lamblia in the feces. The source of infection for beavers is unknown but probably is man since other reservoirs are unknown.

Oocysts of isosporid coccidia have been found in municipal wastes from the Chicago area (1). Coccidia of the Isospora type (oocysts having two sporocysts each containing four sporozoites) include genera which are pathogens for man and animals, e.g., Toxoplasma gondii (5). This organism is unique in that it is capable of infecting fish, amphibians, reptiles, birds, and mammals. All life cycle stages of T. gondii are infective, i.e., oocysts containing sporozoites, clones containing rapidly multiplying tachyzoites, and resting cysts containing bradyzoites. Domestic cats and other members of the Felidae are the only known definitive hosts. Refuse from cat litter boxes commonly is flushed for convenient disposal, and indeed, this is the recommended method for disposition (5). Considering the likelihood that Toxoplasma-infected cat feces often gain entry into the municipal sewage system, considerable research on survival of oocysts in sludge, including animal inoculation studies, should be conducted. Other isosporid coccidian oocysts (Besnoitia, Hammondia, Isospora) also are highly resistant to environmental factors and are capable of using wild or domestic animals as intermediate or paratenic hosts. Studies on survival of oocysts of these genera in municipal sludge similarly should be conducted, since these organisms produce disease in domestic animals or wildlife.

Cestodes

Eggs of tapeworms frequently are found in municipal sludge, and those of the Taeniidae and Hymenolepididae have been reported most often (1-3,6-7). Cestodes of the genus Taenia commonly parasitize domestic cats and dogs and occasionally are found in man. Species of Taenia are relatively host specific with regard to both definitive and intermediate hosts. Taenia taeniaeformis, for example, utilizes cats and other Felidae as definitive hosts and a variety of rats as intermediate hosts. Taenia pisiformis of dogs utilizes rabbits as intermediate hosts, while T. saginata and T. solium of man utilize cattle and swine, respectively, as intermediate hosts. Survival of eggs of Taenia spp. in municipal sludge was evidenced recently by an outbreak of Cysticercus bovis (Taenia saginata) among steers pastured on land fertilized with municipal sludge (8). This outbreak of cysticercosis elicited much concern by veterinarians and others concerned with public health, since nearly 30% of these cattle would have passed normal food inspection procedures (9). The source of this infection, land application of sludge, represented a deviation from the usual mode of infection for Cysticercus bovis and also demonstrated the ease with which organisms can be spread into new locales.

When considering the prevalence of dog and cat helminth ova in sludge (6), it is likely that the taeniid tapeworms infecting these animals also are perpetuated in part by land application of sludge containing viable eggs. Because of the nature of this wild-domestic animal cycle and its transmission at a usual low level, it is unlikely that sludge transmission would be suspected. Only when infections attained localized epizootic proportions might sludge transmission be incriminated.

Eggs of various species of Taenia cannot be differentiated morphologically nor can Taenia eggs be distinguished from those of the closely related genus Echinococcus. Adults of this latter genus parasitize dogs, cats, and related wild carnivores. Intermediate stages are found in domestic or wild ruminants (E. granulosus) or in microtine rodents (E. multilocularis). Unfortunately intermediate stages of both species can occur in man. Echinococcus infections may be highly pathogenic in the mammalian intermediate host. It is likely, when land application occurs, that infections with this helminth can be spread directly to man by contact with viable eggs or indirectly by infection of intermediate hosts with subsequent infection of domestic pets. Man is particularly vulnerable to infection by E. multilocularis since it is unlikely that intermediate stages in rodents would be discovered prior to infection of dogs or cats. In locales where Echinococcus is enzootic, heating of sewage to destroy taeniid eggs should be mandatory.

Hymenolepid tapeworms commonly are parasites of rodents and occasionally man. Eggs of these tapeworms probably gain entry into municipal sewers via the feces of rats dwelling within these systems. Eggs, which survive in municipal sludge upon land application, become available to a variety of beetles which serve as intermediate hosts. Wild rodents, in turn, become infected when feeding upon beetles; thus upon entry into a hymenolepid-free environment, the necessary prerequisites may be present for the organism to be cycled repeatedly without additional application of sludge.

Nematodes

Roundworm eggs frequently are found in municipal sludge (1-3,6-7), and their resistance to environmental and other factors is well established (10-11). Several genera of ascarids have been identified from sludge and include Ascaris, Toxocara, and Toxascaris. The former is a parasite of man, while the latter two helminths are found in dogs and cats. Although these organisms have direct life cycles, they readily utilize paratenic or transport hosts. Upon ingestion by an unusual host, frequently a small mammal, second stage larvae invade somatic tissues where they usually become dormant and await ingestion by a suitable definitive host. This route of infection commonly occurs with some ascarids, e.g., Toxocara cati, and is considered by some to constitute the primary route of infection of this particular species. Most ascarid larvae do little harm to the paratenic host; however, larvae of Ascaris spp. cause pneumonitis in abnormal hosts, and larvae of Toxocara canis produce the well-known syndrome of visceral larva migrans upon ingestion by man. Similar migrations by T. canis also occur in other animals. Ascarids of the genus Baylisascaris are neurotropic and may produce severe neurologic disturbances in paratenic hosts. Baylisascaris has been reported from domestic skunks sold in pet shops, with severe infections resulting in death (12). Since wastes from pet shops enter municipal sewage, eggs of these helminths ultimately gain access to the environment via land application of sludge. Subsequent ingestion of Baylisascaris eggs by herbivorous small mammals (rats, cottontail rabbits, woodchucks) results in neurologic disorders, occasional deaths, and probably most often, an increased susceptibility to predation.

Other nematode ova occurring in sludge are those of whipworms (Trichuris spp.), pinworms (Enterobius sp.), and hookworms (Ancylostoma, Uncinaria, Necatur). There is little likelihood of whipworms or pinworms being transmitted by wildlife from sludge-amended land by other than mechanical means. It is conceivable that hookworms could use wild animals as paratenic hosts; however, this is not considered a likely occurrence.

Nematode parasites of importance to domestic livestock may gain entry to municipal sludge via waste water from abbatoirs, veterinary hospitals, diagnostic laboratories, etc. It is unlikely that free-living wildlife would serve other than mechanical carriers of these parasites since most are relatively host specific.

Considerable variation exists in parasitic fauna found in sludge from different locales (6). The richest fauna, i.e., that with the largest number of parasitic genera, was found in large metropolitan cities which have a great diversity of peoples, e.g., Oakland, California. Exotic (foreign) parasites occasionally are present in municipal sludge samples, and these organisms potentially could become established if the necessary intermediate hosts are present. Fortunately, snail vectors needed for completion of life cycles of various schistosome flukes affecting man and domestic livestock are not present in the continental United States.

BACTERIA

Salmonella

Of the bacterial flora occurring in sewage sludge, the Salmonella probably have received the most attention. These organisms are capable pathogens and produce disease in both man and animals (13). Pathogenicity of different serotypes differs widely; some produce acute disease which usually is manifest as an intestinal infection resulting in enteritis and diarrhea. Occasionally these infections may become septicemic and cause death. Other serotypes, however, may exist in the host as apparently harmless commensals. Individual animals or man, upon infection with Salmonella, may become inapparent carriers and eliminate organisms in the feces for long periods of time.

Salmonella species often are found in animals, man, food, water, and sewage, and an analysis of 1.5 million strains isolated in 109 countries between 1935 and 1975 has been presented (14). Salmonellae have been isolated frequently from wild animals, and one author stated, "It would seem that salmonella have been isolated from many living creatures, in fact so many have been recorded that it seems reasonable to suppose that the absence of a record means that the particular creature may not have been studied" (15).

Salmonellae commonly occur in wild birds, and outbreaks of salmonellosis in free-flying birds occasionally have been recorded (15-16). Gulls, greenfinchs, and sparrows were involved in several different outbreaks of salmonellosis. The source of infection for free-flying birds often is difficult to determine. Some outbreaks have been associated with bird feeding stations. In England, S. enteritidis was isolated from 23 of 32 mute swans feeding in a sewage-polluted river. In another study, S. anatum was isolated from 1 of 587 birds feeding at a sewage treatment plant in southeast England (16). These authors concluded that there was little or no association between Salmonella infection and sewage at this particular plant. In contrast, more than 1000 fecal droppings from gulls were collected from the Hamburg, Germany, sewage disposal works, from the port, and from the city streets. Salmonellae were recovered from 78% of the samples from the sewage disposal works, 66% from the port, and 28% from the city streets. Samples taken from sewage-free areas were consistently negative (17). Based on a review of the available data, it was concluded, "Salmonella infections in wild birds are acquired primarily from their environment..." (13).

There is increasing evidence that Salmonella infections in other wild animals are associated with increased contact with man. In Florida, 14 serotypes of Salmonella were isolated from 28 of 168 raccoons (18). Infection rate for raccoons from state parks in Flordia was 22% as opposed to 13% for animals from surburban and rural areas. Despite regulations to the contrary, visitors to parks routinely feed the raccoons. Also in Florida, 10 of 124 free-ranging synanthropic lizards harbored 6 serotypes of Salmonella (19). Infected lizards were captured mainly in campsites in a county park. Lizards and other reptiles are thought to become infected by ingesting or imbibing contaminated substances. Lizards often are subject to predation and therefore may serve to spread these organisms to other wildlife. Salmonella also may be spread to children while playing with these reptiles.

Coliforms

Escherichia coli is a usual species of bacteria occurring in sludge. Although this organism is a normal gut inhabitant, certain strains are pathogenic. In a study of sewage sludge applied to a forest clearcut in Washington state, it was found that fecal coliform bacteria present in sludge at the time of application remained viable for many months, although bacterial populations decrease in time in sludge applied in both summer and winter (20). Survival was longest in sludge applied during the summer. There was little movement of sludge or coliforms into the groundwater system, and most sludge and organisms remained at the surface level where they presented a health hazard only through direct contact.

CHEMICALS

Perhaps the greatest threat to health from land application of sewage sludge is the accumulation in the food chain of organic chemical compounds or toxic elements. Polychlorinated biphenyls (PCBs) and chlorinated pesticides often are found in sludge (21). These organic compounds accumulate in the bodies of animals feeding on forage fertilized with sludge and may be eliminated in milk of lactating animals. In this manner these compounds can enter the food chain of other animals or man via consumption of contaminated meat or milk.

Cadmium and lead also commonly are found in sludge (21-22). Lead levels as high as 5000 ppm have been reported, and cadmium levels of about 1000 ppm of dried sludge have been detected. Cadmium is particularly important since many plants, particularly grains, legumes, and leafy vegetables, take up cadmium directly from the soil. Lead is less readily incorporated into vegetation.

Much concern has been elicited because of the accumulation of cadmium and lead in animals (21). This accumulation occurs when animals graze or browse on sludge-amended land and ingest sludge with soil, sludge clinging to vegetation, or vegetation which had incorporated chemicals from sludge-amended soil. The Food and Drug Administration has estimated the average daily intake of cadmium and lead in a teen-age boy's diet. The teen-age boy is the heartiest eater in the United States. The average intake for lead is 254 µg daily and for cadmium is 72 µg daily. The Food and Agriculture Organization and the World Health Organization propose a 420 µg daily intake of lead as maximum and a 72 µg daily intake of cadmium as maximum for adults. Thus, it is apparent that the maximum allowable dietary intake of cadmium already has been attained.

Cadmium accumulates in the kidney and liver of man and animals; however, ingestion of cadmium over a prolonged period of time is required to produce kidney damage (23). Clinical signs of primary cadmium toxicity in animals include anemia, retarded development of the gonads or degeneration of the gonads, enlarged joints, scaly skin, liver and kidney damage, reduced growth, and mortality rates that directly correlate with the amount of cadmium in the diet. These signs and changes are similar to those associated with relative or absolute zinc deficiency in the diet (24). Cadmium interacts with zinc in the diet and effectively may tie-up this mineral.

Mature female goats were fed 75 ppm cadmium in their diet, and all died within 19 months (24). Cadmium-treated goats experienced reproductive abnormalities including delayed conception, abortions, and stillbirths. Kids born to cadmium-treated dams had marked reduction in liver zinc and copper, and it was suggested that kids died of a combined zinc and copper deficiency following treatment of their dams with cadmium.

Adult cattle and sheep were fed a cadmium fungicide in the diet at levels of 50, 100, 200, 300, and 500 ppm cadmium for 49 and 41 weeks, respectively (25). Cattle and sheep fed cadmium at levels greater than 200 ppm were anemic as evidenced by decreased erythrocyte counts, packed cell volumes, and hemoglobin. Reproductive abnormalities were observed at all levels of cadmium fed and included infertility, stillbirths, abortions, and damage to the reproductive organs of the dam.

The use of sewage sludge as fertilizer in forest ecosystems and in reclamation of land has been advocated. This method of disposition has certain advantages in that it avoids placing potentially hazardous wastes on cropland or pastureland. The utilization of sludge on forestland or land undergoing reclamation for fertilization purposes is not without its hazards. Most game animals, for example, prefer an ecosystem that is in the early stages of succession, and it is well known that there is increased utilization of early successional stages when the vegetation is fertilized. Sludge application by mechanical spreading or by spraying undoubtedly will cause increased utilization of these plots. By feeding on sludge-treated vegetation, toxic chemicals as well as disease organisms can enter the food chain. It has been estimated that white-tailed deer and cottontail rabbits contribute over 135 million pounds of edible meat annually to the diet of the population of the United States (26). Since wild animals are not supplementally fed, it can be hypothesized that these animals will take in more toxic wastes from sludge-amended land than would domestic livestock. Thus considerable amounts of these potential toxins may enter the human food chain. In certain portions of the country, e.g., the Appalachian region which traditionally has had a depressed economy, wild game constitutes a large portion of the diet of individuals living within this area. The loss of this valuable source of protein because of the presence of toxic chemicals would be untenable.

In addition to the potentially harmful effects of chemicals to the human food chain via wildlife species, the effect of these agents on wildlife per se are unknown. The long-term effects of cadmium ingestion on white-tailed deer, for example, may be reduced productivity due to infertility or other reproductive abnormalities. Similarly, prolonged ingestion of PCBs may result in population reduction because of accumulation of PCBs in the milk and its subsequent transfer to the offspring. With continued ingestion of PCBs on vegetation, sufficient accumulation may occur to cause death or reproductive impairment of second generation animals. Furthermore, the effects of combinations of these chemicals are not known.

SUMMARY AND CONCLUSIONS

Parasites, bacteria, and toxic chemicals are present in sludge and may be transmitted to man, domestic animals, or wildlife through land application. Disease organisms potentially transmitted from sludge-amended land by wildlife include Giardia spp., Toxoplasma gondii, Taenia spp., Echinococcus spp., Hymenolepis spp., Ascaris spp., Toxocara spp., Baylisascaris spp., Salmonella spp., Escherichia coli, and toxic chemicals. Wildlife species may serve as mechanical carriers of disease organisms; they may serve as biological vectors by amplifying the organism; or they may store or concentrate the organism or chemical and in effect transmit the agent upon entering the food chain.

There is little information available on actual transmission of disease agents from sludge-amended land by wildlife. Because wildlife species prefer early successional stages which are rapidly growing and are influenced positively by sludge amendment, disease agents may thus be concentrated relatively rapidly by these animals. Since numerous game animals are used as human food, disease agents and toxins may enter the human food supply. Disease agents and chemicals may cause overt and covert disease in wild populations and thus contribute to population reduction (extinction?) of some species. The effects of combinations of toxic elements and/or disease organisms on the health of wild animals are unknown. Prior to widespread utilization of municipal sewage sludge on forestland or land reclamation projects, the long-term effects of disease agents and chemicals should be studied carefully.

REFERENCES

1. Fox, J.C. and P.R. Fitzgerald. "Parasite Content of Municipal Wastes from the Chicago Area," in Program and Abstracts of 52nd Annual Meeting of the American Society of Parasitologists (Las Vegas, NV: 1977), pp. 68-69.

2. Burge, W.D. and P.B. Marsh. "Infectious Disease Hazards of Landspreading Sewage Wastes," J. Environ. Qual. 7:1-8 (1978).

3. Hays, B.D. "Is There a Potential for Parasitic Disease Transmission from Land Application of Sewage Effluents and Sludges?", J. Environ. Hlth. 39:424-426 (1977).

4. "Waterborne Giardiasis--California, Colorado, Oregon, Pennsylvania," Center for Disease Control, Morbidity and Mortality Report (1980), 29):121-123.

5. Dubey, J.P. "Toxoplasma, Hammondia, Besnoitia, Sarcocysti and Other Tissue Cyst-forming Coccidia of Man and Animals," in Parasitic Protozoa, J.P. Kreier, Ed. (N.Y. Academic Press: 1977), pp. 101-237.

6. Theis, J.H. and D.R. Storm. "Helminth Ova in Soil and Sludge from Twelve U.S. Urban Areas," J. Water Pollut. Control Fed. 49:2485-2493 (1978).

7. Shephard, M.R.N. "The Role of Sewage Treatment in the Control of Human Helminthiasis," Helminthol. Abstracts, Series A., Part 1 (1971), pp. 1-16.

8. Hammerberg, B., G.A. MacInnis and T. Hyler. "Taenia saginata Cysticerci in Grazing Steers in Virginia," J.A.V.M.A. 173:1462-1464 (1978).

9. Herd, R. "Animal Health and Public Health Aspects of Bovine Parasitism," J.A.V.M.A. 176:737-743 (1980).

10. Fitzgerald, P.R. and R.F. Ashley. "Differential Survival of Ascaris Ova in Wastewater Sludge," J. Water Pollut. Control Fed. 48:1722-1724 (1977).

11. Arther, R.G. and P.R. Fitzgerald. "Viability of Nematode Ova Isolated from Anaerobically Digested Sewage Sludge," in Program and Abstracts of 52nd Annual Meeting of the American Society of Parasitologists (Las Vegas, NV: 1977), p. 69.

12. Nettles, V.F., W.R. Davidson and G.L. Doster, "Peritonitis Due to Intestinal Perforation by Ascarids in a Skunk." J.A.V.M.A. 173:1227-1228 (1978).

13. Steele, J.H. and M.M. Galton. "Salmonellosis," in Infections and Parasitic Diseases of Wild Birds, J.W. Davis, et al., Eds. (Iowa State Univ. Press, Ames: 1971), pp. 51-58.

14. Kelterborn, E. "On the Frequency of Occurrence of Salmonella Species. An Analysis of 1.5 Million Strains of Salmonellae Isolated in 109 Countries During the Period 1934-1975," Zbl. Bakt. Hyg., I. Abt. Orig. A. 243:289-307 (1979).

15. Taylor, J. "Salmonella in Wild Animals," in Diseases in Free-living Wild Animals, A. McDiarmid, Ed. (Symposium 24 Zoological Society of London, Academic Press: 1969), pp. 53-73.

16. Plant, C.W. "Salmonellosis in Wild Birds Feeding at Sewage Treatment Works," J. Hyg., Camb. 81:43-48 (1978).

17. Muller, G. "Salmonella in Bird Feces," Nature 207:1315 (1965).

18. Bigler, W.J., G.L. Hoff, A.M. Jasmin and F.H. White. "Salmonella Infections in Florida Raccoons, Procyon lotor," Arch. Environ. Hlth. 28:261-262 (1974).

19. Hoff, G.L. and F.H. White. "Salmonella in Reptiles: Isolation from Free-ranging Lizards (Reptilia: Lacertilia) in Florida," J. Herpetol. 11:123-129 (1977).

20. Edmonds, R.L. "Survival of Coliform Bacteria in Sewage Sludge Applied to a Forest Clearcut and Potential Movement into Groundwater," Appl. Environ. Microbiol. 32:537-546 (1976).

21. Jelinek, C.F. and G.L. Braude. "Management of Sludge Use on Land," J. Food Protection 41:476-480 (1978).

22. Capar, S.G., J.T. Tanner, M.H. Friedmann and K.W. Boyer. "Multielement Analysis of Animal Feed, Animal Wastes, and Sewage Sludge," Environ. Sci. & Tech. 12:785-790 (1978).

23. Willoughby, R.A. "A Review of Cadmium Toxicity in Domesticated Animals," in Cadmium 77, edited proceedings, First International Cadmium Conference (San Francisco, CA: Metal Bulletin Ltd., London, 1978), pp. 100-105.

24. Anke, M., A. Hennig, H.J. Schneider, H. Ludke, W. von Gagern and H. Schlegel. "The Interrelations between Cadmium, Zinc, Copper and Iron in Metabolism of Hens, Ruminants and Man," in Trace Element Metabolism in Animals, C.F. Mills, Ed. (Edinburgh, E. and S. Livingston: 1970), pp. 317-320.

25. Wright, F.C., J.S. Palmer, J.C. River, M. Haufler, J.A. Miller and C.A. McBeth. "Effects of Dietary Feeding of Organocadmium to Cattle and Sheep," J. Agric. Food Chem. 25:293-297 (1977).

26. Williamson, L.L. "A Study of Potential Game Animal Utilization as an Emergency Food Supply During Nuclear Disaster Recovery," M.S. Thesis, Univ. of Georgia, Athens (1969).

SECTION III

OCCUPATIONAL HAZARDS ASSOCIATED WITH
MUNICIPAL SLUDGE

CHAPTER 10

OCCUPATIONAL HAZARDS ASSOCIATED WITH SLUDGE HANDLING

C. S. Clark. Department of Environmental Health, University of Cincinnati College of Medicine, Cincinnati, Ohio.

H. S. Bjornson and J. W. Holland. Department of Surgery, University of Cincinnati College of Medicine, Cincinnati, Ohio.

T. L. Huge, V. A. Majeti and P. S. Gartside. Department of Environmental Health, University of Cincinnati College of Medicine, Cincinnati, Ohio.

Workers engaged in the handling of municipal wastewater treatment plant sludge are exposed to a variety of biological, physical and chemical factors which may present potential health risks. These risks depend upon the nature of the area served by the treatment plant and the type and condition of sludge treatment process(es) in use. Previous papers at this symposium have discussed some of the biological and chemical agents in municipal sludge which pose a potential health risk under certain conditions. At an earlier U.S. EPA-sponsored symposium [1] the results of several studies of the health risks of occupational exposure to wastewater treatment plant personnel, including those engaged in sludge treatment, were presented. Two of the occupational studies reported exposures to organic chemicals, one from a sludge discharge to the sewerage system in Louisville, Kentucky [2] and the other from an apparently continuous but variable entry of toxic compounds into a treatment plant in Memphis, Tennessee. Each of these studies involved exposure to similar groups of compounds [3]. In both of these studies the acute effects in most cases subsided after the exposure had ended; long term effects of these exposures are not known, however. A study of Manitoba, Canada, wastewater treatment plant workers [4] revealed an increased incidence of sinusitis and nasal disorders among the workers. The symptoms of sinusitis were observed upon exposure to the work environment and diminished after leaving work suggesting an allergen may be involved.

A three-city study of wastewater workers by Clark et al. [5] revealed that occupational exposure to wastewater posed a minimal risk of bacterial or viral infection. The only group with higher rates of gastrointestinal illnesses were workers newly employed in wastewater treatment plants. This study in general did not involve "worst case" exposure situations since the volunteers were working at relatively modern activated sludge wastewater treatment plants.

This paper will focus on potential health effects of exposure to microbial agents associated with the composting of municipal wastewater treatment plant sludge. Following a brief review of the literature, preliminary results will be presented of an ongoing health study of workers engaged in municipal sludge composting at several sites in the Eastern U.S.

BACKGROUND

Sewage sludge composting by the windrow method has been practiced for a number of years by the Los Angeles County Sanitation District. More recently the development of the aerated pile composting method by the U.S. Department of Agriculture, Research Station at Beltsville, Maryland, has resulted in considerable expansion of the use of sludge composting. Currently, Philadelphia, Pennsylvania; Camden, New Jersey; Washington, D.C.; Windsor, Ontario; Bangor and Portland, Maine; as well as other smaller cities are currently engaged in municipal sludge composting by the aerated pile method. Many other cities, including New York City are actively considering this method of sludge treatment. A number of factors are responsible for the expansion in the use of sludge composting. Among them are the prohibition of ocean dumping of sludge after 1981, the increase in the amount of sludge being produced nationally, and the growing emphasis on the land application of wastewater and wastewater sludges.

Municipal wastewaters may contain a variety of potentially pathogenic microorganisms. Conventional wastewater treatment does not completely destroy these microorganisms, many of which are concentrated in the sludge. Therefore, sludge should be stabilized by a method such as composting prior to its application on land. Composting is a thermophilic aerobic decomposition process. The heat generated during the compostin process has been shown to effectively reduce the numbers of viable microorganisms present. Two types of composting processes are currently in use -- windrow and forced aeration pile systems [6,7]. The windrow system consists of long, low piles which are aerated by periodic turning. The forced aeration pile system consists of a stationary compost pile constructed over a network of porous pipe attached to a blower which draws air through the pile. Temperatures in the range of 55°-65°C are usually attained during the composting process [6] provided the mixing or aeration is efficient. One of the most important objectives of composting is to obtain these high uniform temperatures throughout the system for sufficient duration so as to penetrate the entire mass. When this objective is fulfilled the composting process will inactivate most microorganisms including viruses.

The heat generated during composting results in temperatures ideal for the proliferation of many thermophilic microorganisms such as actinomycetes, murcorales and in particular, Aspergillus fumigatus. Therefore, compost workers potentially may be exposed not only to the enteric pathogenic microorganisms present in raw sludge (also referred to as primary pathogens since they are capable of initiating an infection in an apparently healthy individual) but also to the thermophilic fungi and actinomycetes that proliferate during composting. Some of the thermophilic fungi are capable of infecting individuals whose defenses have been compromised (these fungi are referred to as secondary pathogens). The dust at composting sites may also contain significant quantities of lipopolysaccharide (LPS or endotoxin) derived from viable and nonviable gram-negative microorganisms which are present in sludge. In addition, the fungus Aspergillus flavus which produces aflatoxin, one of the most potent known human carcinogens, may increase in numbers during composting and the aflatoxin produced by this fungus may present an additional hazard for compost workers. Detroy et al. [8], have shown that the optimal conditions for aflatoxin production (i.e., moisture content, humidity, temperature, incubation time, aeration and nitrogen and carbohydrate content) are similar to conditions present in portions of the aerated compost pile [9]. It has also been shown that aflatoxin is not destroyed by temperatures of 60-80°C and therefore would not be detoxified by the temperatures generated during composting [8]. The long term health effects of chronic exposure to the combination of microbial pathogens, microbial toxins and LPS which may be present in the work environment at composting sites have not been investigated.

Rylander et al., studied various parameters of acute and chronic inflammation in workers exposed to dust arising from heat treated sludge at a sewage treatment plant in Gothenburg, Sweden [10-12]. These investigators observed a significant elevation in immunoglobulins (IgG, IgM, and IgA), leukocytes and platelets in workers at the sewage treatment plant as compared to a group of age matched control workers at a neighboring oil refinery. In addition, elevated levels of C-reactive protein and fibrin degradation products were observed in significantly greater numbers of the sewage treatment plant workers as compared to the control population. Workers at the sewage treatment plant were also reported to experience the following clinical symptoms, apparently related to heavy dust exposure: a) fever, b) purulent discharge from the eyes, c) diarrhea, and d) fatigue. The investigators postulated that the serologic changes and clinical symptoms observed in the sewage treatment plant workers may be related to endotoxin which is present in the dust arising from the heat treated sludge. They expressed concern regarding the potential health risks associated with the chronic exposure to endotoxin containing dusts in the work environment.

Rylander et al. also studied the airway immune response to inhaled endotoxin in laboratory animals [13]. Rats were exposed to an aerosolized endotoxin solution (Escherichia coli O25:B6) daily for 10 days. The dose of LPS deposited in the lungs of each rat was estimated to be 0.3 µg per day. After the 10-day exposure, elevated levels of IgG, IgM and IgA antibodies specific for E. coli O25:B6 endotoxin were demonstrated in the sera of the exposed rats. Bronchial washings from the exposed rats contained IgG and IgA antibodies directed against the E. coli O25:B6 endotoxin. In addition, the bronchial lavage fluid was also observed to contain an increased number of polymorphonuclear leukocytes (PMN) following endotoxin challenge. At cessation of exposure, the number of PMN's in the bronchial lavage fluid decreased and reached values comparable to those observed in control rats within three days. These observations suggested that inhalation of endotoxin is capable of producing an acute inflammatory response in the lungs of laboratory animals and that long term exposure to dusts containing endotoxins may lead to pathologic changes in the pulmonary parenchyma caused by the persistent inflammation. McGuire et al. [14] recently reported acute pulmonary inflammation in rhesus monkeys induced by intravenous infusion of purified bacterial LPS. These investigators have demonstrated the presence of an enzyme which cleared components of the contact and complement systems in the lungs of the LPS challenged animals.

Rylander and his colleagues have extended their studies of the immune response to inhaled endotoxin to include workers exposed to cotton dust in cotton mills in England and Sweden [12,13]. Workers with a history of exposure to cotton dust and subjective pulmonary symptoms of chest tightness were shown to have, in their nasal secretions, elevated levels of

IgG and IgA antibodies specific for antigens prepared from a gram-negative bacteria isolated from cotton plants being processed in the mills. No elevation of IgG or IgA antibody levels with specificity for the antigen preparations employed were observed in nasal secretions obtained from a control population of workers with no history of exposure to cotton dust. Rylander postulated, based on his human and animal data, that chronic inhalation of endotoxin with the accompanying inflammatory response in the lungs, may play a critical role in the development of Byssinosis.

Dutkiewicz [15,16] conducted an immunologic survey of grain handlers in Poland who were exposed to high concentrations of the gram-negative bacterium Erwinia herbicola (synonym Enterobacter agglomerans) in their work environment. Workers exposed to dusts containing large numbers of E. herbicola (5×10^4 to 1×10^5 cfu/m^3) had a significantly higher frequency of serum precipitating antibody directed against E. herbicola antigen preparations than unexposed individuals. Furthermore, intradermal skin testing with antigens prepared from E. herbicola revealed a higher incidence of positive skin reactions in workers exposed to E. herbicola than in unexposed individuals. Grain handlers exposed to dust containing E. herbicola and who complained of respiratory symptoms such as cough, dyspnea, and wheezing, were shown to have a higher incidence of positive skin and precipitin reactions to E. herbicola antigens than asymptomatic exposed workers. This study provides further evidence to suggest that exposure to gram-negative LPS by inhalation is capable of inciting a systemic immune response and that it may play a primary role in the development of chronic pulmonary disease in exposed individuals.

Lundholm and Rylander [17] recently reported on a study of 11 workers at a plant where municipal sewage sludge and household garbage were crushed, milled, and allowed to compost for six months. Gram-negative bacteria were reported to be present in large numbers where garbage was loaded onto the conveyor belts (28,000 cfu/m^3) and even higher near the mill outlet (25,000-500,000 cfu/m^3). Elsewhere at the plant they ranged from an average of 330 to 11,000 cfu/m^3. At the central water treatment plant levels averaged 20 cfu/m^3. Six of the 11 compost plant workers reported symptoms of nausea, headache, fever or diarrhea compared to only two of the 41 water treatment plant employees who served as controls.

Environmental Monitoring

Millner, Bassett and Marsh [18] recently reported on their monitoring of airborne spores of A. fumigatus at two compost sites at Beltsville, Maryland and at the Blue Plains Wastewater Treatment Plant in Washington, D.C. By volumetric sampling they were able to determine that the dispersal of the spores from composting sewage sludge and wood chip mixtures behaved as a Gaussian plume. A major source of airborne spores was found to be the front-end loader which was estimated to aerosolize 4.6×10^6 A. fumigatus spores/second when moving and dropping compost. The sampling instrument used was the six-stage Andersen sampler with the sampling orifice directed into the wind and with the top cone removed to collect more non-respirable particles [19]. Other investigators [20] including us, have neither removed the top cone nor tilted the sampler in their sampling procedure. At the Beltsville compost site airborne concentrations of A. fumigatus between three and 30 meters downwind of a compost pile being agitated by a front-end loader were reported to range from 1400 to 3100 cfu/m^3. Fifteen minutes after agitation, levels were not above 39 cfu/m^3.

Potential Health Effects of Aspergillus fumigatus

Aspergillus species are ubiquitous in the environment of most countries of the world. The fungus grows well on a variety of substrates, including stored hay or grain, decaying vegetation, soil, and dung. A. fumigatus grows well at 45°C or even higher, making it one of the most common microorganisms found in compost sites [21]. A. fumigatus, and others of the A. species, have been shown to be capable of causing disease in both normal and compromised individuals. Therefore, workers involved in the composting process may be exposed to potential health risks due to their exposure to A. fumigatus.

The term "Aspergillosis" has been used to describe illness attributed to antigenic stimulation, colonization, or tissue invasion by A. species. Aspergillosis is usually acquired by inhalation of air-borne spores. These spores (conidia) are small enough (2.5-3.0µm for A. fumigatus) to reach alveoli or to gain entrance to paranasal sinuses. The disease varies in severity from an incidental, saprophytic relationship with the host to a fulminating, fatal infection.

The dose (number of cfu), portal of entry, and the immune status of the host are thought to be major determinants in the course of the infection caused by A. species. Immunosuppressed patients are at a greater risk to infection by A. fumigatus [19,21]. Serious and often fatal invasive infections caused by A. fumigatus have been reported in immunosuppressed patients following kidney transplanation and in patients with leukemia or lymphoma receiving chemotherapy.

Individuals receiving antibodies or adrenal cortical hormones have also been shown to have a higher incidence of mycotic infections. Additional predisposing factors to aspergillosis appear to be the presence of malignant or other debilitating disease, leukopenia or granulocytopenia, other infections, pneumonitis and underlying pulmonary disease [22].

Exposure of atopic individuals, who have a history of asthma, to spores of A. fumigatus may result in a disease known as allergic bronchopulmonary aspergillosis. Colonization of the bronchi in these patients with Aspergillus results in episodic bronchial plugging which appears to lead to areas of sacular bronchiectasis. These patients may produce sputum plugs which often reveal fungal mycelia on microscopic examination. Laboratory abnormalities in these patients include significant eosinophilia of blood and sputum, marked elevation of total serum IgE, and serum precipitating antibody to Aspergillus antigens. These patients usually have an immediate type skin response to Aspergillus antigen. Persistent bronchopulmonary aspergillosis may result in irreversible complications such as bronchiectasis and pulmonary fibrosis.

The pathophysiology of allergic bronchopulmonary aspergillosis is thought to be mediated by a combination of both Type I and Type III immunologic reactions. IgE-sensitized mast cells in the bronchi react to antigens from Aspergillus colonies growing in the bronchi. These mast cells release histamine, slow-reacting substance of anaphylaxis and eosinophilic chemotactic factor leading to bronchospasm, increased permeability of the bronchial mucosa, absorption of Aspergillus antigen into the circulation, and pulmonary and peripheral blood eosinophilia. The absorbed Aspergillus antigen can react with IgG, resulting in the formation of antigen-antibody complexes and complement fixation. This in turn may lead to chronic inflammation in the bronchi and peribronchial tissues with eventual bronchial destruction, bronchiectasis, and pulmonary fibrosis.

Due to the wide distribution of A. species in the environment, exposure to Aspergillus must be nearly universal, but disease, either invasive or allergic, is uncommon. As indicated above, a complex interaction between host factors and the challenge dose of A. species spores appears to determine the course of events which follow exposure to Aspergillus.

To date, no known adverse health effects have been detected among compost workers due to exposure to thermophilic fungi and/or actinomycetes. The Los Angeles County Sanitation District has been composting sewage sludge for many years using the windrow method. The compost produced is

marketed by Kellogg Supply Co., Inc., for use in a wide
variety of applications. In the several decades of the
operation of this system, there have been no reports of adverse health impacts [7]. However, due to the apparent transient nature of the work force the efficiency of the reporting system is probably low.

Composting by the forced aeration method has been carried
out in Beltsville since March 1973, and for shorter periods of
time in other locations. In the accumulated time of system
operation and worker and user exposure, there have thus far
been no reported cases of disease resulting from exposure to
compost.

ELEMENTS OF COMPOST WORKERS STUDY

This investigation was designed to evaluate the health
effects related to exposure to nonviable substances and viable
microorganisms present in dust arising from the composting of
sewage sludge. The clinical and serologic evaluation of
workers exposed to dust at the composting site included: 1)
comprehensive history and physical examination; 2) health
questionnaires; 3) illness monitoring; 4) liver and kidney
function profiles; 5) anterior nares and oropharyngeal swab
cultures; 6) chest X-ray; 7) complete blood count with differential; 8) determination of antibodies directed against
A. fumigatus; 9) determination of antibodies directed against
lipopolysaccharide (LPS) present in compost samples; 10) skin
testing with A. fumigatus antigen; 11) pulmonary function
tests; 12) quantitation of total IgG, IgM, IgA, and IgE; and
13) determination of C-reactive protein, levels of C3 and
CH50. Health questionnaires, anterior nares and oropharyngeal
swab cultures and sera for each of the serologic tests were
obtained five times during 1979, the first year of the investigation. Environmental monitoring for viable particles
was also conducted several times during the first year. The
scheduling of the above elements of the study protocol is
outlined in Table I.

Population Groups Selected

Workers at sludge composting facilities in Camden, N.J.,
Beltsville, MD and Washington, D.C. were selected because of
their geographic closeness and the willingness of their organizations to cooperate with the study. Control workers
were recruited either from sewage treatment plants producing
the sludge, if the job locations were judged to be sufficiently
distant from the compost operations, or from other nearby
sewage treatment plants. Compost treatment facilities in
Camden, N.J. and Washington, D.C. are operated by the Camden

Table I. Scheduling of Elements of Compost Workers Study

Study Element	Month 1	5	7	9	12
Comprehensive history	X				
Physical examination				X	
Illness monitoring	X	X	X	X	X
Anterior nares and oropharyngeal swabs for A. fumigatus	X	X	X	X	X
Chest X-ray			(ONCE)		
Complete blood count with differential	X	X	X	X	X
Fungal serology (A fumigatus, A. carneus, A. niger and A. flavus)*	X	X	X	X	X
Determination of antibody directed against Lipopolysaccharide (LPS) prepared from compost	X	X	X	X	X
Skin testing with A. fumigatus antigen preparation**				X	
Pulmonary function testing	X	X	X	X	X
Immunochemical determination of CRP, C3, and immunoglobulins (IgM, IgG, and IgA)	X	X	X	X	X
IgE determination					X
Environmental monitoring		X	X	X	X
Renal and liver profiles	X				X

*Aspergillin Meridian Diagnostics, Cincinnati, Ohio.

**A. fumigatus allergic extract for scratch test (1:10 w:v) Hollisten-Stein Lab., Spokane, Washington.

County Municipal Utilities Authority and by the District of Columbia Department of Environmental Services, respectively. In Camden the compost unit is located at the main wastewater treatment plant and in Washington, D.C. at the Blue Plains Wastewater Treatment Plant. Permission was not received from the U.S. Department of Agriculture to recruit their compost-exposed employees at the Beltsville Agricultural Research Center but it was received from Maryland Environmental Services which employs many of the workers involved with the Beltsville compost-related activities. Non-exposed workers in the Camden area were recruited from a small secondary wastewater treatment plant in Camden, the Baldwin Run Plant. In the Washington, D.C.-Beltsville, MD, area, permission was received from the Washington Suburban Sanitary Commission to recruit control workers from their secondary wastewater treatment plant at Piscataway, MD located about 10 miles down the Potomac River from the Blue Plains Plant. The Beltsville compost facility treats a small portion of the sludge produced at the Blue Plains Plant. Another portion of the Blue Plains Plant Sludge is trenched elsewhere in Maryland. The Camden compost facility treats sludge from the treatment plant at which it is located as well as that from the smaller Baldwin Run Plant. The size and type of treatment facilities from which these study participants were recruited are given in Table II. A Total of 170 workers were recruited from these plants for the study that is now being reported. About 100 additional low exposure workers have been recruited from the Blue Plains Plant.

Table II. Size and Type of Treatment Facilities from Which Participants Were Recruited

Location	Wastewater Size (MGD)	Type	Compost Size (dry T/day)	Date Started
Camden, NJ				
Main Plant	30	Primary	15	5/78
Baldwin Run	4	Secondary	-------None--------	
Beltsville, MD	-----None-------		15	3/73
Washington, D.C. (Blue Plains)	300	Secondary	75	2/79
Piscataway, MD	30	Secondary	-------None--------	

Compost workers and control subjects in Philadelphia, PA (30) and Bangor and Portland, ME (20), have also been recruited. Results from these workers will be included in a future report.

METHODS

Clinical and Serological

The study was designed to evaluate the participants for objective and subjective symptoms and signs possibly related to exposure to dust associated with the composting operation. Participants received physical examinations once during the first year of the study. In addition, a detailed medical and occupational history, which included reference to contact with sewage sludge composting operations, or any other occupational dust and/or waste exposure was obtained from each participant. This data base allowed ranking of the participants in this investigation in terms of length and severity of exposure to the environmental dust and other factors associated with composting. A health questionnaire, specifically designed to determine the occurrence of hypersensitivity, respiratory, gastrointestinal and mucocutaneous disorders associated with dust exposure, was obtained from the participants during the first, fifth, seventh, ninth and twelfth months of 1979. LPS (endotoxin) was extracted from bulk compost samples from each composting site by the method of Westphal [23]. The serum from the study population was tested for the presence of antibodies directed against each of the LPS preparations by the enzyme-linked immunosorbent assay (ELISA). Because of the difficulty in demonstrating infection with members of Aspergillus species, four parameters were employed to detect infection with these fungi: 1) determination of precipitating antibody directed against A. fumigatus, A. carneus, A. niger and A. flavus in the serum of the study population, 2) determination of skin test reactivity to an extract of A. fumigatus, 3) semi-quantitative sputum cultures for A. fumigatus and 4) evidence of transient or migratory infiltrates on chest X-ray.

Precipitating antibody to A. fumigatus, A. flavus, A. carneus and A. niger (Aspergillin, Meridian Diagnostics, Cincinnati, Ohio) in the serum of the study population was measured semi-quantitatively by titration of antibody using the counter immunoelectrophoretic technique [24]. Antigens of A. fumigatus used in the determination of specific IgG antibody by the ELISA were prepared by the method of Coleman and Kaufman [25] and further purified by gel filtration on Sephadex G-200. The first peak obtained by gel filtration of the antigen preparations were used for determination of precipitins in the serum samples tested. Sera positive for A. fumigatus were kindly provided by Dr. J. W. Rippon, University of Chicago Hospital, Chicago, IL, and were used as positive controls for precipitin determinations. Serum samples for fungal precipitin determination were obtained from the study population during the first, fifth, seventh, ninth and twelfth months of the first year of the proposed investigation.

Skin reactivity to Aspergillus antigen was determined by scratch test using Aspergillus antigen (Hollisten-Stein Corp., Spokane, WA) in 1:10 weight/volume concentration. The skin test was performed during the twelfth month of the study. Determination of total IgE titers were performed by the ELISA method using sera collected during December, 1979.

Anterior nares and oropharyngeal swabs were obtained from the study population during the fifth, seventh, ninth, and twelfth months of the first year of the investigation and cultured for A. fumigatus. Sputum specimens were obtained during the first month but were discontinued because of the difficulty in obtaining adequate samples. Due to the widespread distribution of A. fumigatus, semi-quantitative determination of A. fumigatus in the samples obtained from the study and control groups was performed. Quantification of the number of A. fumigatus in the samples may provide an index of the degree of colonization between the groups participating in the study.

Transient or migratory pulmonary infiltrates are frequently observed in individuals with hypersensitive lung diseases. A chest X-ray was obtained on participants once during the first twelve-month period and any infiltrative process or other lung pathology was recorded. In the future, prior chest X-rays of the study participants will be reviewed, if possible, and any positive findings will be correlated with the clinical diagnosis made at the time the X-ray was taken and with periods of exposure to the composting operation.

Chronic exposure to microbial antigens present in the dust arising from the composting of sewage sludge may induce a local and/or systemic inflammatory response as well as an elevation of one or more classes of circulating immunoglobulins. C-reactive protein and C3, serum proteins which are often elevated during inflammatory processes, were quantitated by radial immunodiffusion [27]. Hemolytic complement was also assayed [24] since it may be reduced in immunological phenomena involving antigen-antibody complexes. The concentration of total IgG, IgM, and IgA was quantitatively determined in the sera of the study population by radial immunodiffusion [27].

The study population was also evaluated for evidence of hypersensitive pulmonary disease or other lung diseases which may be associated with prolonged exposure to the dust arising from the composting operation. Each participant in the proposed investigation underwent pulmonary function testing. When possible, pre- and post-shift testing was performed. It was felt that pulmonary function testing in this manner would detect bronchial sensitivity associated with exposure to dusts in the work environment.

Environmental Monitoring

Viable particle collection from air samples was performed using the six-stage Andersen cascade impactor designed for that purpose. Air was drawn through the samplers by 12-volt D.C. pumps, which were powered by 12 volt motorcycle batteries. The pumps, attached to an assembled sampler loaded with petri dishes containing agar, were calibrated against a dry gas meter to pull one cubic foot of air per minute (CFM), as recommended by the manufacturer. The samplers were mounted on tripods such that samples were collected at a height of five feet, which approximates the breathing zone of the average worker. Sampling times were adjusted in an attempt to obtain colony numbers which were optimal for counting. The volume of air sampled was calculated from the sampling time and the flow rate of the pumps. Total colony counts for bacterial and fungal plates were obtained by direct colony count, unless excessive colony numbers precluded this procedure. If colony numbers precluded counting all colonies, the colonies on 1/4 or 1/2 the plate were counted and the count was multiplied by the appropriate number to obtain a total plate count.

Bacterial samples were collected on Trypticase soy agar (TSA) manufactured by Baltimore Biological Laboratories (BBL), Cockeysville, MD. The medium was prepared according to the manufacturers instructions, autoclaved at 121°C for 15 minutes, cooled to 45-50°C and dispensed into petri dishes. Prior to use, TSA plates were incubated at 36°C for 24-48 hours to detect contamination and minimize surface moisture due to condensation.

Fungal samples were collected on modified Czapek-Dox medium. The medium was prepared by adding 35 grams of Czapek-Dox broth (Difco Laboratories, Detroit, MI), 20 grams agar (Difco), and 15 grams Oxgall (Matheson, Coleman and Bell, Cincinnati, OH) to 1000 milliliters distilled water, adjusting the pH to 7.3 and autoclaving at 121°C for 15 minutes. Prior to dispensing, the medium was cooled to 45-50°C and 50 μg/ml Streptomycin, 50 μg/ml Chloramphenicol and 20 units/ml Penicillin were added to retard bacterial growth. These plates were incubated at room temperature or occasionally at 36°C to minimize moisture.

Aerosol samples obtained on TSA plates were shipped directly (within 24 hours) to the Research Surgical Bacteriology Laboratory at the University of Cincinnati, where they were incubated at 36°C for 24 hours. At the end of the incubation period, total bacterial counts were made from the TSA plates which were then replicate plated onto three different types of media: KF streptococcal agar (BBL), M - FC agar (BBL), and M-Endo agar (LES) (BBL).

KF streptococcal agar is a selective medium which supports the growth of fecal streptococci (Lancefield's serological groups D and Q) while inhibiting the growth of most gram-negative and other gram-positive bacteria [5]. The presence of pink/red colonies after 48 hours at 36°C is indicative of fecal streptococci. All growth on KF streptococcal agar was re-replicated to Bile-Esculin agar (Difco) and tested for catalase activity. Gram-positive cocci which grew as brown/black colonies on Bile-Esculin agar, and which failed to decompose a three percent solution of hydrogen peroxide, were considered to be confirmed fecal streptococci.

The purpose of replication onto M-FC agar was to determine the presence or absence of fecal coliforms. These plates were incubated at 44.5°C for 24 hours and the blue to blue-green colonies were transferred to a Lauryl sulfate broth (BBL) which was incubated 24-48 hours at 37°C and to EC broth (BBL) which was incubated for 24 hours at 44.5°C. Growth with gas in these media confirmed the presence of fecal coliforms.

In order to determine total coliform counts, the TSA plates were replicated onto M-Endo agar (BBL) and incubated at 37°C for 24 hours. The colonies with a green sheen were picked to Lauryl sulfate broth and Brillant green bile broth (BBL). These media were incubated at 37°C for 48 hours, $\pm$ 3 hours. Growth with gas confirmed the presence of coliforms.

Fungal plates, upon reaching the laboratory, were incubated at 42.5 to 44.5°C for 48 hours and total plate counts were performed. The microorganisms were separated into A. fumigatus and "other thermophilic microorganisms" on the basis of microscopic examination, colony color and morphology.

RESULTS

At this time only a preliminary analysis of a portion of the data from participants at Camden, Beltsville, Washington, D.C. (Blue Plains) and Piscataway is available. Data from workers in Philadelphia, PA, Bangor and Portland, ME and from the additional lower exposed Washington, D.C. volunteers will be presented at a future time.

Populations Recruited

A summary of the 173 workers recruited from Camden, N.J., Beltsville, MD, Washington, D.C. (Blue Plains) and Piscataway, MD is presented in Tables III, IV, V and VI, respectively. Initial recruitment began in late January, 1979 in Camden, N.J. and in March for Beltsville, MD, Washington, D.C. and Piscataway, MD. The participants from each of these areas have been assigned compost exposure categories: high, intermediate or control. Assignment to a particular category was based on interviews and direct observations. Workers who were directly involved in compost operations as their major activity were assigned to the high exposure category. The intermediate category applied to workers either routinely working within about 100 meters of a compost operation or who are sometimes involved with composting operations but not as the major part of their work. These categories are relative ones and are area specific; that is an "intermediate exposure" worker at the Camden compost plant would not necessarily have the same exposure as an "intermediate exposure" worker at Blue Plains.

In Camden, the intermediate exposure category was used for workers at the Main Plant who were not involved in the compost operation. Control workers in Camden were from the Baldwin Run Plant. There were 31 Camden participants in the high exposure category, 22 in the intermediate and 14 in the low exposure group. At Beltsville, MD there were 10 people in the high exposure group working at the compost site and there were eight in the intermediate group who primarily worked with compost samples at a research laboratory some distance from the compost. On occasions they visited the compost site for sample collection and other purposes. Included in the high exposure group was one person whose job was based in a trailer located at the compost site. In Washington, D.C. 14 compost workers have been recruited in the high

Table III. Camden Workers: (Age, Race, Sex and Exposure)

Exp.	Whites No.	Med.	Range	Non-Whites No.	Med.	Range	Total No.	Med.	Range
High	19(1)	28	19-61	2	55	48-61	21(1)	29	19-61
Int.	19	26	19-70	9	48	25-68	28	28	19-70
Cont.	7	27	23-61	7	43	22-64	14	37	22-64

Number of females appears in parentheses.

Table IV. Beltsville Workers: (Age, Race, Sex and Exposure)

Exp.	Whites No.	Med.	Range	Non-Whites No.	Med.	Range	Total No.	Med.	Range
High	9(1)	50	29-59	1	43		10(1)	47	29-59
Int.	8(2)	27	25-65	0			8(2)	27	25-65
Cont.	0			0			0		

Number of females appears in parentheses.

Table V. Blue Plains Workers: (Age, Race, Sex and Exposure)

Exp.	Whites No.	Med.	Range	Non-Whites No.	Med.	Range	Total No.	Med.	Range
High	1	32		13(1)	37	20-55	14(1)	37	20-55
Int.	5	35	26-45	22(6)	31	23-49	27(6)	31	23-49
Cont.	2	36	29-42	9	42	23-51	11	42	23-51

Number of females appears in parentheses.

Table VI. Piscataway Workers: (Age, Race, Sex and Exposure)

Exp.	Whites			Non-Whites			Total		
	No.	Med.	Range	No.	Med.	Range	No.	Med.	Range
High	0								
Int.	0								
Cont.	18(1)	28	22-63	22(6)	31	25-47	40(7)	30	22-63

Number of females appears in parentheses.

exposure group. An intermediate exposure group consisted of 27 workers whose job occasionally required them to repair equipment in the Blue Plains compost pit and on occasion at the Beltsville compost site. A control group of 11 workers was recruited who report to a pumping station several miles from the Blue Plains Plant but spend a portion of their time at the Blue Plains Plant maintaining the pumps in the raw sewage inlet station and in the air blower building, both of which were several hundred meters from the compost pit. The 40 participants recruited at the Piscataway, MD plant were engaged in various operational and laboratory functions. All have been assigned to the control category since no composting operations are involved at their plant. The median age for the Camden groups ranged from 28 to 37 years, at Beltsville, from 27 to 47, at Blue Plains from 31 to 42 and at Piscataway the median age of the one group was 30 years. The gradations of exposure at Camden, Beltsville and Blue Plains allowed for comparison of results between workers at these locations. In addition, Blue Plains workers were compared to the control group at nearby Piscataway, MD.

Air Sampling

During the period May-December, 1979 viable particle air sampling was conducted at several locations at each of the treatment facilities involved in the study. Results are summarized for A. fumigatus and "Other Thermophilic Microorganisms" in Table VII, for Fecal Streptococci and Fecal Coliforms in Table VIII, and for Total Coliform and Total Bacteria in Table IX. At the Blue Plains and Main Camden plants results were summarized for samples collected in the vicinity of the compost area and at locations more distant from the compost areas. Because of the much larger size of the Blue Plains Plant, "Other Areas" for it were at least 150 meters from the compost area while they were only at least 50 meters away at the Camden plant. (At the Camden plant, all areas of the plant were within 150 meters from the compost

Table VII. Summary of 1979 Air Monitoring for Respirable Concentrations of A. fumigatus and Other Thermophilic Microorganisms (colony forming units per cubic meter, CFU/M³)

Area	No. Samples	A. fumigatus Range	A. fumigatus Average	Other Thermophilic Microorganisms Range	Other Thermophilic Microorganisms Average
Metro. Washington, D.C.					
Beltsville, Compost	4	48-131	81	0-34	17
Blue Plains, Compost	25	0-475	52	0-12,000	1060
Blue Plains, Other Areas*	14	0-21	3	0-246	73
Piscataway Plant	15	0-18	3	0-5	<1
Camden, N.J.					
Main Plant, Compost	13	0-2940	918	0-580	146
Main Plant, Other Areas**	6	0-126	24	0-7	2
Baldwin Run Plant	11	0-52	14	0-7	1

*At least 150 meters from compost area.
**At least 50 meters from compost area.

Table VIII. Summary of 1979 Air Monitoring for Respirable Concentrations of Fecal Streptococci and Fecal Coliforms (colony forming units per cubic meter, CFU/M^3)

Area	No. Samples	Fecal Streptococci Range	Fecal Streptococci Average	Fecal Coliforms Range	Fecal Coliforms Average
Metro. Washington, D.C.					
Beltsville, Compost	4	0-23	6	0	0
Blue Plains, Compost	4	11-79	48	0	0
Blue Plains, Other Areas*	0				
Piscataway	6	0-6	1	0-3	1
Camden, N.J.					
Main Plant, Compost	8	0-482	85	0	0
Main Plant, Other Areas**	2	0-5	3	0-5	3
Baldwin Run Plant	6	0-25	8	0-21	5

*At least 150 meters from compost area.
**At least 50 meters from compost area.

Table IX. Summary of 1979 Air Monitoring for Respirable Concentrations of Coliforms and Total Bacteria (colony forming units per cubic meter, CFU/M^3)

		Concentrations			
		Total Coliforms		Total Bacteria	
Area	No. Samples	Range	Average	Range	Average

Metro. Washington, D.C.

Beltsville, Compost	4	0	0	5,500-16,800	10,700
Blue Plains, Compost	4	0	0	8,070-18,100	14,600
Blue Plains, Other Areas*	0				
Piscataway, Plant	6	0-27	12	564-1330	815

Camden, N.J.

Main Plant, Compost	8	0-56	8	590-5170	2720
Main Plant, Other Areas**	2	0-10	5	443-524	484
Baldwin Run Plant	6	0-35	10	196-3380	1460

*At least 150 meters from compost area.
**At least 50 meters from compost area.

area.) Samples were collected at each plant on from three to six days, except for Beltsville where samples were collected on only one day. Only colonies from plates 3-6 (respirable size range) from the Andersen samplers were included in Tables VII- IX. Concentrations of A. fumigatus varied considerably and on average were higher in compost areas than in non-compost areas. Concentrations at areas at least 150 meters from the compost pit at the Blue Plains Plant were about the same as at the Piscataway Plant where no composting occurs. At the Main Camden Plant, A. fumigatus levels in areas at least 50 meters from the compost pit were on average much lower than concentrations in the compost area, but were on occasion higher than those at the Baldwin Run Plant. Concentrations of A. fumigatus at Camden were generally higher than those at Beltsville and Blue Plains. These differences may be due in part to differences in compost activities underway at the time of the sampling or to differences in precipitation levels prior to sampling. Two days prior to one of the Blue Plains sampling days, about four inches of rainfall occurred in the vicinity of the plant as tropical storm David passed through the D.C. area. About six percent of all colonies of A. fumigatus were found on plate 6 and were thus thought to be small enough to reach the alveoli. This compares to less than one percent found to be in that size range by Millner et al. [18]. About 80 percent were found to be in the respirable size range (plates 3-6) which is practically the same as that reported by Millner et al. [18], 79 percent.

In Camden the two treatment plants are clearly distinguishable from each other with respect to A. fumigatus levels. At the main Camden plant A. fumigatus levels were considerably higher at compost areas than elsewhere in the plant where samples were collected. Similarly, A. fumigatus concentrations at the Piscataway Plant and at areas of the Blue Plains Plant remote from the compost pit were clearly lower than levels reached at the Beltsville and Blue Plains composting sites.

Concentrations of "Other Thermophilic Microorganisms" varied widely but were generally higher at the compost sites.

Fecal Streptococci were detected at each of the areas where they were measured and were generally higher at the compost areas (Table VIII). Fecal Coliforms were infrequently detected (Table VIII). Coliforms (Table IX) were not detected in samples collected at Beltsville and Blue Plains. Concentrations were about the same at Piscataway as they were at the Camden sampling locations (averages ranged from 5 to 12 CFU/M^3). Concentrations of Total Bacteria ranged somewhat higher at compost sites than at other sites but were above 10^2 CFU/M^3 in all samples.

A. fumigatus in Study Participants

Culturing for A. fumigatus colonies was performed on anterior nares, and oropharyngeal swabs and sputum specimens obtained from study participants. Between one and five determinations were therefore made for each individual. A summary of the highest value for each individual, according to location and exposure category, is presented in Table X. The number of colonies observed are expressed as either zero or greater than or equal to one. Statistical analyses by the chi-square test were performed between results for the various exposure groups and combinations of exposure groups in each city and between groups for all areas combined. The null hypothesis tested was that the probability of a positive count is independent of exposure category. Statistically significant differences were found between many of the groups with those in exposure category 1 having more positive determinations than either those in category 2 or category 3. Workers in category 2 had more positive values than those in group 3 but the results were not statistically different except when values for all sites were combined.

The differences among the groups in A. fumigatus colonies verified the overall validity of the exposure categories. Even though viable A. fumigatus spores are ubiquitous in the environment, the increase, above ambient levels in their concentration at or near compost sites was apparently large enough to result in a higher degree of colonization among workers with compost exposure than in those without. Future analyses of the data will include the numerical value for each determination performed. Combining all sites, the percentage of study participants exhibiting at least one positive value for AF colonies, distributed by exposure category, was category 1 (high exposure) 70%; category 2 (intermediate exposure) 20%; category 3 (control), 5%.

Liver Function and Total Immunoglobulin Determination

Preliminary analyses of the highest results for each participant on tests for the enzymes serum glutamic oxaloacetic transaminase (SGOT), serum glutamic pyruvic transaminase (SGPT), measures of liver function, and levels of total IgG, IgE and IgM, did not reveal any differences among exposure groups at any of the sites. Combining data by exposure group for all sites again did not reveal any significant

Table X. Distributions and Analyses of Study Participants According to the Highest Value of Aspergillus fumigatus Colonies (AF) Observed on Cultures Taken During 1979

	Exposure Categories			Exposures Compared	Testing of Hypothesis of Independence	
	1	2	3		X2 Test	Fisher's Exact Test
Camden						
AF > 1	16	5	1	1:2	.0005	
AF = 0	5	20	12	1:3	.0005	
				1:2,3	.0005	
				2:3		.314
Blue Plains						
AF > 1	8	7	0	1:2	.05	
AF = 0	5	20	11	1:3		.002
				1:2,3	.005*	
				2:3		.070
Blue Plains & Piscataway						
AF > 1	8	7	2	1:2	.05	
AF = 0	5	20	48	1:3		.0001
				1:2,3	.005*	
				2:3		.007
Beltsville						
AF > 1	7	0		1:2		.004
AF = 0	3	8				
All Sites Combined						
AF > 1	31	12	3	1:2	.0005	
AF = 0	13	48	60	1:3	.0005	
				2:3	.01	

*Small cell expected value renders the validity of the X2 test "borderline."

difference among exposure groups. A more detailed examination of the data may reveal differences in mean values for the various exposure groups, however.

Delayed Hypersensitivity to Aspergillus Antigens

Skin reactivity to Aspergillus antigen was determined on 109 participants in December, 1979. Only five persons tested positive, three compost workers, one intermediate-exposed worker and one in a control group. These numbers were too small for statistical analyses.

C-Reactive Protein (CRP)

Preliminary analysis of the highest CRP value for each individual was performed on two groupings of the data: (1) <1 mg% and >1 mg% and (2) positive and non-reactive. For the first grouping there were no differences among exposure groups at the various sites or for all sites combined. Comparisons of the positive and non-reactive results revealed only one borderline statistically significant result (α = 0.053) by the Fishers exact test: four of nine compost workers at Beltsville had positive values for CRP compared to none of eight intermediate-exposed workers at Beltsville.

Eosinophil Count

Preliminary analyses of the highest absolute eosinophil count for each of the participants did not reveal any differences among the exposure groups at the various sites, separately or combined.

Chest X-Ray Findings

X-rays were obtained from 74 participants and were reviewed by a radiologist at the University of Cincinnati. Although a total of 17 abnormal findings were observed, only two were suggestive of occupational exposure or smoking, one a compost worker and one a control.

Antibody to A. fumigatus

Specific IgG directed towards A. fumigatus, as determined by the ELISA, was compared by exposure group for each site separately. Statistical analyses were performed by the Mann-Whitney one-tailed ranking test. At no site did the compost workers exhibit statistically higher values.

Antibody to Lipopolysaccharide (LPS)

LPS was prepared from compost samples obtained from

the Camden site, and from the Beltsville-Blue Plains sites combined, by the Westphal method and tested against workers sera to determine possible responses to LPS present in the compost material. Specific IgG directed against the LPS was determined by the ELISA. Statistical analysis was by the Mann- Whitney one-tailed ranking test. Antibody titers in the Camden compost workers (group one) directed against LPS prepared from compost obtained from Camden were not higher than in the control workers (group 3). However in Beltsville, the compost workers had higher antibody titers against LPS prepared from compost obtained from the Blue Plains-Beltsville compost sites, than the intermediate group ($\alpha = 0.04$). In Blue Plains, antibody titers in the compost workers were higher than the intermediate group ($\alpha = 0.025$); when group 3 (control) workers at Piscataway and Blue Plains were included, the significance was still $\alpha = 0.02$.

Health Effects Associated with Composting

A review of the medical histories, physical examinations and chest X-rays of compost workers participating in the study has thus far revealed one worker with several health problems which may be associated with exposure to microbial pathogens and dust present at the composting site. Approximately one year after starting employment at a compost site, the worker developed chronic otitis media of the right ear, with sclerosis of the right mastoid and erosion of the posterior wall of the right external canal demonstrated on X-rays of the mastoid. Over a period of several months the infection gradually progressed, with erosion of 70% of the right tympanic membrane, despite topical antibiotics, local debridement, and a right tympanomastoidectomy. Aspergillus niger, a fungus which has been reported present in compost, was the only pathogen isolated from the purulent discharge present in the right ear. This worker also has small irregular shadows in both lungs suggestive of parenchymal disease. Since this individual states that he has never smoked, the pulmonary abnormalities observed on the chest X-ray may be job related. Except for evidence of the right chronic otitis, the physical examination of this worker was normal, as were his pulmonary function studies on three different occasions. The worker also complained of symptoms suggestive of chronic sinusitis, however, the relation of the onset of these symptoms and his exposure to composting could not be definitely determined.

DISCUSSION AND CONCLUSION

Workers at sludge compost sites are exposed to a variety of fungal and bacterial pathogens, as well as toxins of microbial origin, in their work environment. No information is currently available regarding the potential long term effects of this exposure on the health of these workers. This report has reviewed current literature related to host responses to microorganisms and microbial toxins known to be present at compost sites and has presented preliminary results of a study designed to evaluate potential health effects of exposure to microbial factors present at compost sites.

Results of the environmental monitoring demonstrated that workers at each of the compost sites studied were exposed to markedly elevated numbers of A. fumigatus spores. This observation was supported by the high frequency of isolation of A. fumigatus from oropharyngeal and anterior nares cultures obtained from the compost workers when compared to the other two exposure groups. However, the presence of A. fumigatus in the anterior nares and oropharyngeal cultures did not correlate with antibody response directed against A. fumigatus in those workers with positive cultures or among the three exposure groups. One worker in the study, employed at a compost site, was found to have a local infection involving his right ear caused by A. niger and chest X-ray changes compatible with an occupationally-related disorder. It has not been possible to document whether the ear infection or the chest X-ray findings were directly related to exposure to microbial agents present at the compost site.

Due to the large numbers of gram-negative bacteria in the sewage sludge which is composted it was assumed that workers at compost sites would be exposed to relatively large amounts of LPS (endotoxin) in the dust at the compost site. The only means by which this exposure could be documented was by the viable counts of gram-negative microorganisms (Fecal coliforms) collected during environmental monitoring. These levels may be artificially low since a major component of the LPS in dust may be comprised of non-viable bacteria. The LPS (endotoxin) used for determination of antibody titers in the study participants was prepared from compost obtained from the Camden and Beltsville-Blue Plains compost sites. Levels of specific IgG antibody directed against LPS prepared from Beltsville-Blue Plains compost were significantly higher in the Beltsville and Blue Plains compost workers than in the non-exposed controls in the Washington, D.C. area. No difference in specific IgG antibody levels directed against LPS prepared from Camden compost was observed in exposed and unexposed workers in the Camden, NJ area. The immune response observed in the Blue Plains-Beltsville compost workers to challenge with LPS by the respiratory route was similar to that reported by Rylander et al. for workers in cotton mills exposed to gram-negative bacteria in cotton dust and rats challenged with purified E. coli O25:B6 LPS [12,13].

Preliminary analysis of other parameters studied, such as C-reactive protein, total immunoglobulin (IgG, IgM, IgA, IgE) levels, liver function tests, skin test reactivity to A. fumigatus, chest X-rays and absolute eosinophil counts, did not reveal significant differences among the three exposure groups. When the available data is subjected to a more complete analysis, incorporating such factors as age, race, precise length of exposure, the preliminary findings may require revision. In addition, an extended period of observation may reveal abnormalities which are related to the length of exposure. The preliminary data on antibody response of the compost workers at the Beltsville-Blue Plains sites to LPS suggest that it may be advisable to take reasonable precautions to reduce exposure such as the use of respirators by compost workers and periodic water spraying of the compost sites to reduce dust.

ACKNOWLEDGMENTS

Support for the research came in part from Research Grant N. R805445 from the Health Effects Research Laboratory (Cincinnati) of the U.S. Environmental Protection Agency. Appreciation is expressed to the many people who were involved in the collection and analysis of specimens and in other aspects of the study, and in particular, Darryl Alexander, Mel Barber, Jane Onslow, Deanne Hanes, Sue Lewis, Eddie Oliver, Geraldine Perkins, Michael Head, Vickie Gillespie and Sharon Humiston. The cooperation of the participants and their organizations is greatly appreciated.

REFERENCES

1. "Symposium on Wastewater Aerosols and Disease," Health Effects Research Laboratory (Cincinnati), U.S. Environmental Protection Agency, Office of Research and Development, Cincinnati, Ohio, September 19-21, 1979.

2. Kominsky, J. and M. Singal. "Non-Viable Contaminants from Wastewater Aerosols and Disease, Cincinnati, Ohio, September 19-21, 1979.

3. Elia, V. J. et al. "Worker Exposure to Organic Chemicals at an Activated Sludge Plant," Paper presented at the Symposium on Wastewater Aerosols and Disease, Cincinnati, Ohio, September 19-21, 1979.

4. Sekla, L. H. et al. "Sewage Treatment Plant Workers and Their Environment: A Health Study," Paper presented at the Symposium on Wastewater Aerosols and Disease, Cincinnati, Ohio, September 19-21, 1979.

5. Clark, C. S. et al. "Health Effects of Occupational Exposure to Wastewater," Paper presented at the Symposium on Wastewater Aerosols and Disease, Cincinnati, Ohio, September 19-21, 1979.

6. "Process Design Manual Sludge Treatment and Disposal," Municipal Environmental Research Laboratory, USEPA Report - 625/1-79-001 (1979).

7. Environmental Resources Company. "Workshop on the Health and Legal Implications of Sewage Sludge Composting" (Cambridge, Massachusetts, December 18-20, 1978).

8. Detroy, R. W. et al. "Aflatoxin and Related Compounds" in *Microbial Toxins: A Comprehensive Treatise*, Vol. VI, *Fungal Toxins*, A. Ciegler, S. Kadis and S. J. Aji Eds. (New York: Academic Press, Inc., 1971).

9. Rylander, R., K. Andersson, L. Belin, G. Berglund, R. Bergstrom, L. Hanson, M. Lundholm and I. Mattsby. "Studies on Humans Exposed to Airborne Sewage Sludge," *Schweir. Med. Wschr.* 107:182-184 (1977).

10. Mattsby, I. and R. Rylander. "Clinical and Immunological Findings in Workers Exposed to Sewage Dust," *Jour. of Occupt. Med.* 20(10):690-692 (1978).

11. Rylander, R., K. Andersson, C. Belin, G. Berglund, R. Bergstrom, C. Hanson, M. Lundholm and I. Mattsby. "Sewage Worker's Syndrome," Lancet 28:478-479 (1976).

12. Rylander, R. et al. "Local and Systemic Immune Response to Cotton Dust Bacteria," Paper presented at the 3rd Special Session on Cotton Dust, The Beltwide Cotton Conference, Phoenix, Arizona, 1979.

13. Rylander, R. "Exposure to Gram-Negative Bacteria and the Development of Byssinosis," Paper presented at the XIX International Congress on Occupational Health, Dubrounik, Yugoslavia, 1978.

14. McGuire, W. et al. "Activation of Plasma Contact System in Inflammatory Lung Disease." Federation Proceedings, 39, 906 (1980).

15. Dutkiewiez, J. "Exposure to Dust-Borne Bacteria in Agriculture. I. Environmental Studies," Arch. Environ. Health 33(5):250-259 (1978).

16. Dutkiewiez, J. "Exposure to Dust-Borne Bacteria in Agriculture. II. Immunological Survey," Arch. Env. Health 33(5):260-270 (1978).

17. Lundholm, B. and R. Rylander. "Occupational Symptoms Among Compost Workers," J. Occup. Med. 22(4):256-257 (1980).

18. Millner, P. D., D. A. Bassett and P. B. Marsh. "Dispersal of Aspergillus fumigatus from Sewage Sludge Compost Piles Subjected to Mechanical Agitation in Open Air," Appl. Env. Microbiol. 39(5):1000-1009 (1980).

19. May, K. R. "Calibration of a Modified Andersen Bacterial Aerosol Sampler," Appl. Microbiology 12(1):37-43, (1964).

20. Passman, F. J. Energy Resources Co., Personal Communication, November 12 (1979).

21. Millner, P. D., P. B. Marsh, R. B. Snowden and J. F. Parr. "Occurrence of Aspergillus fumigatus During Composting of Sewage Sludge," Appl. Env. Microbiol. 34(6): 765-772, (1977).

22. Rippon, J. W. *Medical Mycology: The Pathogenic Fungi and the Pathogenic Actinomycetes* (W. B. Saunders Co., Philadelphia, Pennsylvania, 1974).

23. Westphal, O. and K. Jann. "Bacterial Lipopolysaccharides. Extraction with Phenol-Water and Further Applications of the Procedures," in *Methods in Carbohydrage Chemistry, Vol. 5*, R. L. Whistler and M. L. Wolfrom, Eds. (New York: Academic Press, Inc., 1965) pp. 83-91.

24. Gewurz, H. and L. A. Sujehira. "Complement," in *Manual of Clinical Immunology*, N. R. Rose and H. Friedman, Eds. (American Society for Microbiology, Washington, D.C., 1976) p. 36.

25. Coleman, R. M. and L. Kaufman. "Use of the Immunodiffusion Test in the Serodiagnosis of Aspergillus," *Appl. Microbiol.* 23(2):301-308 (1972).

26. Engrall, E. and P. Perlmann. "Enzyme-Linked Immunosorbent Assay. III. Quantitation of Specific Antibodies by Enzyme-Labled Anti-Immunoglobulin in Antigen Coated Tubes," *J. Immunol.* 109:129-134 (1972).

27. Mancini, G., A. D. Carbonera and J. F. Heremans. "Immunochemical Quantitation of Antigens by Single Radial Immunodiffusion," *Immunochemistry* 2:235-239 (1965).

CHAPTER 11

HEALTH ASPECTS OF COMPOSTING: Primary and Secondary Pathogens

W. D. Burge., P. D. Millner. Biological Waste Management and Organic Resources Laboratory, Science and Education Administration, U. S. Department of Agriculture, Beltsville, Maryland.

INTRODUCTION

High temperature composting, which is a result of the activity of oxidative thermophilic microorganisms, is increasingly becoming an accepted method of processing sewage sludges. If properly managed, it can convert sewage sludge to an esthetically acceptable humus-like product free of enteric pathogens. As the composting process proceeds, a succession of microorganisms develops, adapting to the increasingly elevated temperatures resulting from the heat released by their activities.

Several methods of high temperature composting have been developed since the pioneering efforts of Howard [1] and Archarya [2] in India, van Vuren in South Africa [3] and Scott [4] in China. These methods include enclosed mechanical systems that are more popular in Europe than in the United States. In this country, the Beltsville method of forced aeration composting [5] is gaining acceptance.

In considering disease hazards connected with composting, it becomes necessary to take into account the pathogens present in sewage sludges and those that might grow during the composting process. For convenience, pathogenic organisms that might be encountered in sewage-sludge composting operations can be somewhat arbitrarily divided into two groups. Organisms that can initiate an infection in apparently healthy individuals will be called primary pathogens, whereas those that usually attack only hosts whose defense systems have been weakened by certain prior diseases or therapies will be called secondary pathogens. Both primary and secondary pathogens can be found in sewage sludges before composting. Secondary pathogens that concern us most, however, are those that actually may grow as a result of the composting process. In addition, certain organisms that grow in compost may incite allergic responses.

Primary Pathogens

   Four groups of primary pathogens occur in sewage wastes: viruses, bacteria, cysts of protozoans, and ova of helminths. The major pathogenic organisms in these four groups and the diseases they cause have been discussed by B.P. Sagik in a paper presented at this symposium. Most of these organisms are incorporated into the sludge during its formation in wastewater processing [6]. Their survival depends upon the treatment the sludge receives before it reaches the compost site. Sludge digestion can inactivate most of the protozoan cysts and greatly lower the numbers of the other pathogens [7]. However, digested sludge composts more slowly than raw sludge because some of the microbial substrate present in raw sludge is utilized during digestion. Liming sludge to a high pH will inactivate most organisms, except ascaris [8]. However, in the future, for economy, mixtures of primary, secondary and chemically precipitated tertiary sludges that contain relatively high levels of pathogens will probably be composted. Since composting itself can eliminate the primary pathogens, the presence of these organisms in sludge does not justify using lime or other methods when composting is to be done.

Secondary Pathogens and Allergens

   Many microorganisms can function as secondary pathogens, but composting conditions favor the growth of some more than others. The fungus Aspergillus fumigatus Fres. (Af) has been isolated at relatively high concentrations from finished compost and from compost-pile zones at less than 60°C [9]. Other secondary fungal pathogens isolated occasionally from compost are Mucor pusillus and M. miehei. Serratia marcescens and Pseudomonas spp. are secondary bacterial pathogens isolated from compost, but their heat lability and inability to form spores seem to limit their occurrence. Also some of the thermophilous actinomycetes that proliferate during composting incite respiratory allergic responses in certain sensitized people.
   Aspergillus fumigatus growing in the compost produces spores that may become airborne and then be inhaled. Upon inhalation, these spores, 2-3 µm in diameter, may penetrate to the secondary bronchi and alveoli and germinate [10]. After germination, growth, with the formation of fungus balls may be confined to the lung airways. Coughing up of mucus and blood may become chronic and debilitating. Growth may, however, penetrate lung tissue; from there the fungus may disseminate via the circulatory system to other critical body organs. Successful treatments of such extreme conditions are unusual.

The factors leading to susceptibility to infection by Af were discussed in a recent review [11]. Previous or existing lung disorders, such as tuberculosis, sarcoidosis, histoplasmosis, chronic bronchitis, asthma, and emphysema, and defective host defenses associated with several factors, including aging, can predispose hosts to infection [12]. In addition, therapies, including the use of cortisone and corticosteroids, for various diseases may predispose individuals to infection [13].

In addition to infectious types of aspergillosis, there is also a respiratory disorder known as allergic bronchopulmonary aspergillosis (ABPA), a response of the immune system to allergenic components of _Aspergillus_ species, especially Af. Like other allergic, hypersensitivity responses, a period of sensitization, or induction of hypersensitivity, usually requiring one or more uneventful exposures to the agent followed by a latent period, precedes an extreme reaction. Allergenic agents, of which Af is one, apparently have little if any sensitizing effect on most individuals. Of those who do become sensitized, some have family histories indicating hypersensitivity to a variety of agents [14].

Pulmonary extrinsic allergic alveolitis (PEAA) is another respiratory immunologic response to microbial or organic materials. Some examples of organic materials and the associated respiratory disorders are listed in Table I. The disease usually develops acutely, but also may develop insidiously [31]. The most commonly known form of PEAA is farmer's lung, reportedly more prevalent in England than in the USA. This disorder has been designated as an industrial disease in Great Britain since 1965 for purposes of workman's unemployment compensation [31].

Although the host condition is an important factor in the development of the several forms of aspergillosis and of PEAA, exposure dose must be a significant factor also. Unfortunately, dose-response data are not available for either healthy or predisposed hosts. Without such data, the use of the concept of "exposure relative to natural background" [32] is of some use as a guideline until dose-response data become available. Application of this natural background exposure concept requires that aerial concentrations of the critical microbes present at and around a composting site be compared with those naturally present in the environmental background. Arbitrary tolerable levels could be established based on those background levels.

ON-SITE HAZARDS

The hazard to on-site personnel relative to infection by primary or secondary pathogens can be evaluated definitively

TABLE I. Hypersensitivity Pneumonitis Associated with Exposure to Various Organic Materials and Microbial Agents.

Hypersensitivity	Organic Material	Microbial
Bagassosis [15, 16]	Sugar cane bagasse	Thermoactinomyces sacchari
Farmer's lung [18,19,20]	Moldy hay	Micropolyspora faeni, T. vulgaris, A. fumigatis
Fermentor's lung [21]	Contaminated citric acid fermentation	A. fumigatus Penicillium spp.
Humidifier lung [22-24]	Dust in air handling systems	Neisseria, Acanthamoeba spp. T. vulgaris, M. faeni, A. fumigatus
Maltworker's lung [25]	Malted barley	A. clavatus, A. fumigatus
Maple bark splitter's lung [26]	Maple bark	Cryptostroma corticale
Mushroom worker's lung [27,28]	Compost	M. faeni, T. vulgaris and A. fumigatus
Sewage worker's lung [29]	Sewage dust	Possibly microbial endotoxin
Suberosis [30]	Moldy cork	Penicillium frequentans

only from actual experience reinforced by epidemiological studies involving serology. Reinforcement is necessary because sick-leave records, even if kept carefully and made available for comparison with those of workers suitable for controls, will not reveal subclinical infection or the risks that this kind of infection may constitute for members of the worker's family and others contacted outside the work area. We know of only one such epidemiological study, and since it is in an early stage no data have been made available. Judgments about the hazards to compost site personnel can, therefore, be made only from an analysis of what must be considered hypothetical threats and what data are available from studies of working groups undergoing similar or more intense hazards.

Primary Pathogens

The degree of hazard for workers at compost sites will be influenced presumably by how much contact the workers have with sewage sludges and pathogen-carrying aerosols generated during the process, and the effectiveness of the process in eliminating pathogens. Precautions taken to protect workers from these hazards will, of course, reduce the extent of these hazards. The most labor-intensive processes, that is, the open-air pile or windrow, may be expected to bring the workers into the greatest direct bodily contact with sludge, mainly through possible hand contact and eventual transfer of infectious materials to the mouth or through splatter contact to the face causing risk of infection through the eyes or the mouth. Production of liquid aerosols seems a relatively negligible hazard at compost sites as compared with that at wastewater treatment plants. Hazards from particulate aerosols may be greater at compost sites, especially around screening operations and when the compost pad is dry and dust is consequently generated.

The available epidemiological data concerning health hazards for wastewater treatment plant, sewer maintenance, and sewage-farm workers are insufficient for definite conclusions, but they do not seem to indicate any acute hazards [33]. Clark et al. [34] cited evidence from an extensive study indicating that sewer workers in West Berlin did not experience any unusual health difficulties. Other less extensive studies they cited, however, showed a connection between work with sewage and disease incidence, but these data were inconclusive. The lack of hazard indicated in the West Berlin study was also shown in a recent sero-epidemiologic study carried out on more than 100 workers in a Cincinnati activated-sludge plant [35]. This study failed to demonstrate any increased risk of infection for wastewater workers or elevation of antibody titers that had been expected in view of the presumed exposure of the workers to pathogens in the work environment. Unless particulate aerosols are found to be more hazardous than liquid aerosols, the above experiences would lead us to believe that compost-site workers would not be subjected to any appreciable risk from infection by primary pathogens.

Secondary Pathogens And Allergens

The nature of secondary pathogens and the diseases they cause offer compost-site operators some reasonable guide concerning worker exposure. The overwhelming association between host predisposition and increased susceptibility to infectious forms of aspergillosis suggests that individuals whose health predisposes them to infection would do well to avoid exposures to composts or other materials known to contain high concentrations of opportunistic microorganisms. In a practical sense, several predisposing factors would seem to be associated with disorders in which the individuals would be so debilitated that they could not perform the rigorous physical activities required in compost work. Thus, compost-site operators may find that they are not confronted with obviously predisposed employees. On the other hand, some predisposing conditions, such as a prior or existing case of tuberculosis or sarcoidosis, in which scar tissue developed in the lung, would not necessarily render individuals physically incapable of performing the tasks of a compost-site job. However, such predisposing conditions could be identified by using preemployment health histories for compost facility workers. Periodic medical checkups should be required for all employees.

Exposure to microbial dusts at a compost site will depend largely on the proximity of a worker to particular parts of the operation. A worker standing next to an operating open-air screener or a front-end loader dumping compost or woodchips will be exposed to higher concentrations of thermophilous microbes than a worker located some distance away from the central operational area or upwind of dust-generating operations. Exposures may also be high for equipment operators who leave doors to their cabs open while dumping woodchips or compost.

Some examples of viable particle concentrations of thermophilous microbes that have been measured in air at the Beltsville composting site [36] are shown in Table II. Downwind concentrations of Af are generally higher than upwind (9 to 155 Af/m^3) and noncompost (0 to 24 Af/m^3) site concentrations.

Woodchips stored outdoors at the compost site, when agitated, release into the air high concentrations of viable Af and very few thermophilous actinomycetes. In contrast,

Table II. Aerial Concentrations of Viable A. fumigatus and Thermophilous Actinomycete Particles at the Beltsville Sewage-sludge Compost Site[1/]

Activity	Distance Downwind (m)	A. fumigatus no./m^3	Actinomycetes no./m^3
Dumping Compost	3	3,860	9,510
	60	3,590	4,150
Screening	3	155	1,690
	76	407	77
	152	24	9
Dumping Woodchips	60	37,000	28
	122	19,000	14
	213	3,220	9

[1/] Based on collections using an Andersen 6-stage viable sampler [36].

the composted sludge/woodchip mixture when agitated releases fewer Af than do woodchips but many more thermophilous actinomycetes. Preliminary identifications of these actinomycetes indicated that the following are present in compost dust: Actinobifida dichotomica, Micropolyspora faeni, Thermoactinomyces vulgaris, Saccharomonospora viridis, and Streptomyces spp. Further work on the concentrations of these microbes in compost dust is underway in this laboratory.

In view of the demonstrated role that the above-named microbes have in sensitizing individuals and in eliciting hypersensitivity responses that affect respiratory function (Table I), it seems prudent to avoid compost facility designs that enclose a worker in a dust-generating part of the operation. Such enclosures would create exposure conditions that could lead to development of allergic alveolitis syndromes, such as those experienced by mushroom-house and sewage-drying workers.

In some mushroom houses, workers exhibited symptoms of PEAA [37]. In one such mushroom house (unpublished data), 88,500 Af/m^3 and 661 thermophilous actinomycetes/m^3 were measured during the dusty clean-out operation that involves the manual removal of spent mushroom compost from traybenches. In comparison, Af concentrations in open air at the Beltsville compost site have reached 55,500/m^3 and thermophilous actinomycetes, 15,300/m^3 [36].

Thus, it seems reasonable to conclude that enclosing compost working areas would increase the respirable dose of Af and thermophilous actinomycetes to a level that could elicit a pulmonary hypersensitivity response in sensitized individuals.

Noninfectious responses to microbial aerosols generated during the composting of garbage sewage-sludge mixtures [38] in Sweden included nausea, headache, diarrhea, fever, and eye irritation. Because the air in both work places contained a predominance of gram-negative bacteria, and because of reports on the effects of gram-negative bacterial endotoxin on man, the investigators suggested that the work-related symptoms may be due to endotoxin. At present, no studies of this phenomenon have been reported for compost workers in the USA.

The above-described hypersensitivity responses associated with contact of microbial agents known to be in sewage-sludge compost suggest that pre-employment medical histories should include information about prior or existing asthmatic or allergic problems and that workers should be provided with protection devices, such as dust masks or respirators.

Because there are no definitive tests that identify individuals who can become sensitized to these microbial agents, site operators should inform compost workers about the possibility for development of pulmonary hypersensitivity. Workers should be encouraged to report unusual health problems to supervisors and medical officers.

OFF-SITE HAZARDS

Hazards off-site may occur from particulate or liquid aerosols carrying pathogenic organisms or from contact of consumers with the compost as a product. The quantity of liquid aerosols generated at a compost site seems extremely small as compared with those generated at wastewater treatment plants. Although filter-cake sludges may be 80 percent or more water, the water is held in the sludge matrix by capillary forces and adsorption. It seems likely that the force required to dislodge water as liquid aerosol droplets from sludge exceeds that applied in handling. The clouds of moisture droplets frequently seen emerging from compost piles result from water leaving the sludge matrix as gas and condensing into visible droplets that should essentially be free of microorganisms as well as salts.

Microorganisms may be transported aerially to off-site locations either as microbial propagules, such as vegetative cells or spores, or in dust that carries microbial propagules.

Primary Pathogens

    Essentially no data are available on the dispersion of primary pathogens as particulate aerosols from compost sites or any other kinds of facilities. Possible hazards can only be evaluated by consideration of hazards associated with liquid aerosols. According to Hickey and Reist [39], it has been thoroughly documented that liquid aerosols from wastewater treatment plants can carry primary pathogens. They stated that environmental monitoring has shown that people residing near treatment plants are exposed to pathogen-containing aerosols, but health effects are undetermined to date. In an attempt to determine health effects, epidemiological studies were made recently of urban populations surrounding wastewater treatment plants near Chicago, Illinois [40], in Tecumseh, Michigan [41], and in Skokie, Illinois [42]. Also, attendance at an elementary school was monitored for periods before and after construction of a wastewater treatment plant adjacent to the school grounds [43]. Evidence that the students were exposed to pathogen-containing aerosols was obtained by monitoring air concentrations close to aerosol sources for short periods using high volume aerosol samplers and extrapolating these data to the air concentrations students would encounter in classrooms and on the school playground. Calculations indicated that students likely would respire pathogens. In the Chicago and Skokie, Illinois studies, serological techniques and health questionnaires were used to determine exposure. For the Tecumseh, Michigan study, only a health questionnaire was used. In all of these studies, there was no evidence of serious adverse effects on the health of the various populations; however, some relatively minor adverse effects were noted. For the Chicago, Illinois study, there was evidence of some increased skin disease, nausea, weakness, diarrhea, and breathing pain as compared with the control populations. Alpha- and gamma-hemolytic streptococcal throat isolations were higher for the aerosol exposed populations, but this was regarded as having no health significance. For the Tecumseh, Michigan study, disease incidence was increased for people living within 600m of the sewage plant, but this was thought to be related to low income and housing density, rather than to nearness to the plant. In the Skokie, Illinois study, the authors cautioned that, although there were no obvious adverse health effects, the population receiving the highest dosage levels was very small, making these conclusions somewhat tenuous.

    Although these studies cannot be taken as conclusive evidence that populations living adjacent to compost sites are not subject to infection from primary pathogens, it does appear that there is no reason to conclude that the danger is extreme. Of course, if pathogens are found to be more infectious when associated with particulate aerosols than when associated with liquid aerosols, the data for communities around wastewater treatment plants will have little relevance for those around compost sites.

Secondary Pathogens and Allergens

Concerns about the location of a compost site focus on the potential for increasing aerial Af-spore concentrations substantially above the communities' precompost-site levels. In an attempt to describe the aerial concentrations that can be expected around a compost site, Millner et al. [36] studied the rate of spore emission during the turbulent movement of compost by front-end loaders. Using Pasquill's atmospheric dispersion model, they found that emission rates from six trials ranged from 1.1×10^5 to 4.6×10^6 Af/sec. Using the latter rate, they estimated aerial Af concentrations at unobstructed distances downwind from an agitated compost pile for different atmospheric conditions. These estimations (Table III) showed that 0.6 km downwind from the source,

Table III. Estimated Centerline Concentrations of A. fumigatus at Unobstructed Distances Downwind from Agitated Compost[a]

Downwind Distance (km)	Atmospheric Condition	
	Neutral[b] (Af/m^3)	Unstable[c] (Af/m^3)
0.1	6.1×10^3	1.1×10^3
0.2	2.6×10^3	3.3×10^2
0.3	1.3×10^3	1.4×10^2
0.4	8.8×10^2	7.2×10^1
0.5	6.1×10^2	4.1×10^1
0.6	4.2×10^2	2.4×10^1
0.7	3.3×10^2	1.5×10^1
0.8	2.6×10^2	1.0×10^1
0.9	2.2×10^2	7.1×10^0
1.0	1.9×10^2	5.2×10^0

[a] Calculated using an atmospheric dispersion model with emission rate 4.6×10^6 Af/sec, source height 5.0 m, and receptor height 2.1 m [36].

[b] Moderately turbulent with windspeed 3.5 m/sec.

[c] Very turbulent with windspeed 3.1 m/sec.

under very turbulent atmospheric mixing conditions, Af concentrations along the centerline of the dispersing spore cloud would not be higher than background levels. Under somewhat less turbulent conditions, the Af concentrations at 1 km downwind would be 190 Af/m^3, higher than usual background levels.

The median diameter of Af particles collected in the dispersing aerosols was 3.8 μm. The cumulative distribution plot of the collected viable Af particles indicated that 87 percent were respirable (<7 μm); about 70 percent could pass the trachea and primary bronchi, 5 percent could reach the terminal bronchi and less than 1 percent could reach the alveoli [10]. Thus, Af particles in the aerosols could, when inhaled, reach points in the respiratory system critical for an infectious process.

From these studies, they concluded that background airspora concentrations around proposed compost sites should be evaluated on a case-by-case basis, because the airspora concentrations can be substantially influenced by local conditions. Furthermore, it is important to determine the frequency of occurrence of meterological conditions that favor low spore dispersal during transport. The latter should be considered because of the impact on sensitive sites, such as hospitals, in surrounding communities. Obviously, airspora concentrations can increase very substantially 0.1 to 0.3 km downwind from the compost-unloading activity. Thus, the location of large-scale composting facilities should be prudently remote from health-care facilities where predisposed individuals are housed.

PRODUCT USERS

Primary Pathogens

Hazards to consumers using the compost will depend upon whether the organisms have been destroyed by the composting. Several studies have shown that composting can destroy pathogens. Wiley and Westerberg [44] composted sewage sludge by mixing it with augers and force-aerating it in a bin. Poliovirus type 1, the bacterium Salmonella newport, and ascaris ova mixed into the sludge were destroyed. They concluded that temperatures ranging from 60-70°C maintained for 3 days of composting would destroy all pathogens. Gaby [45] followed the destruction of inserted and naturally present pathogens during the windrow composting of mixed refuse and sewage sludge at Johnson City, Tennessee. The organisms were inserted in test tubes or in small packets of material to be composted. Salmonellae and shigellae, present or inserted, disappeared within 7 to 21 days. Inserted poliovirus type 2 was inactivated within 3 to 7 days. The spirochaete <u>Leptospira philadelphia</u>, artificially inserted, was destroyed within 2 days. Human parasitic cysts and ova disintegrated after 7 days.

Morgan and Macdonald [46] concluded on the basis of insertion experiments at Johnson City that the bacterium causing tuberculosis would be destroyed in 14 days if the temperature averaged at least 65°C. The organism, however, survived indefinitely when composting was inhibited by low ambient temperatures of about -2°C.

Although the work cited has shown that pathogenic organisms are destroyed by properly conducted composting systems, no set of criteria has been advanced so that personnel operating compost sites can determine under varying conditions whether pathogen destruction has indeed been achieved or so that health authorities or others obligated to protect public health can set standards for quality control. A temperature-by-time procedure for determining criteria will be discussed in a later section.

Secondary Pathogens and Allergens

Any hazard associated with the use of compost or similar products would relate to the manner in which the product is handled. In a survey of 21 commercially available potting soils, manures, and mulches, Millner et al. [9] found that most contained more than 100 Af/g (dry wt.) and five products contained more than 1.3×10^4 Af/g, the amount in Beltsville compost cured for 4 months. Dumping bags of such organic materials or agitation during potting of plants might expose the user to high spore concentrations, presenting a special problem for sensitized or predisposed individuals.

CONTROL OF PATHOGENS DURING COMPOSTING

Primary Pathogens

In seeking a method of quality control, researchers have often suggested using fecal coliforms as indicator organisms to determine whether pathogens have been eliminated from the final product. Our experience has been that they can grow in finished compost and that it is impossible to prevent reinoculation of the product at outdoor compost sites, as they are presently operated. Therefore, the presence of fecal coliforms is no indication of the history of the compost relative to inactivation of pathogens. Because direct determination of pathogens is not feasible, Burge et al. [47] proposed using temperature-by-time criteria to determine the level of inactivation achieved. To develop these criteria, coliphage f2 was used as the model organism because a search of the literature for inactivation rate constants, as influenced by temperature, showed a relatively high degree of thermal stability for this organism. The rates of inactivation for pathogenic helminth ova, protozoan cysts, enteric viruses and enteric bacteria were higher than for f2 in the temperature range of 50 to 65°C. Thus, inactivation of any concentration of f2 should mean inactivation of any enteric primary pathogen at that concentration or below. Using the first-order-rate equation and the Arrhenius equation, they derived the following expression relating number of logs (L) of inactivation to the period of time (t) in days for which any particular temperature (T) in °K was achieved during composting:

$$L = 2.589 \times 10^{38} \, t \, \exp(-28,349 \, 1/T).$$

Using this expression or a graph produced from it, one can determine the time needed at a particular temperature for any specified number of logs of inactivation. Whether the temperature-by-time regimes have been achieved can be determined by monitoring the compost pile temperature.

Once a temperature-by-time regime has been adequate for a desired level of pathogen inactivation, then pathogen control can be considered to have been achieved, and other goals, such as stabilization or achievement of a desirable carbon:nitrogen ratio, can become the controlling criteria. Burge et al. [47] suggested 15 logs for the pathogen inactivation criterion. This can be achieved in 2 days at 55°C.

A weakness of this proposal is that the inactivation rates obtained from the literature came from studies in which media other than sewage sludge were used. To put the proposal on a sounder basis, studies should be made to determine actual pathogen inactivation rates during composting. Because it is difficult to find enteric pathogens at concentrations in sewage sludge that would permit the measurement of their inactivation rates during composting, we suggest that the studies be made utilizing small batches of pathogen-seeded sludge. The composting could most conveniently be carried out in self-heating laboratory composters.

To utilize the temperature-by-time system, one must establish the correct temperature-monitoring procedure for any composting operation. Experience at Beltsville with single compost piles has indicated that the outside lower corners of a compost pile are the lowest temperature zones and, therefore, the ones that should be monitored. Once monitoring has shown that the criterion has been met in these zones, then the rest of the pile should be reasonably free of pathogens.

Secondary Pathogens and Allergens

Both Af and thermophilous actinomycetes typically appear as part of the microbial succession occurring in self-heating piles of organic matter. The growth and sporulation of Af in the pile and blanket are influenced by physical and nutritional factors. Significant among these is the availability of suitable carbon substrate; Af readily decomposes cellulosic materials [48]. Our preliminary (unpublished) studies indicated that one practical way to substantially reduce Af growth in open-air composting is to use noncellulosic bulking agents, rather than woodchips or other currently used cellulosic bulking agents.

The influence of physical and nutritional factors in sewage-sludge composting on thermophilous actinomycete growth has not been studied. These should be examined because of high potential for exposure of compost workers to potentially allergenic compost dust.

SUMMARY

Primary and secondary pathogens, as well as allergenic microorganisms and substrates, are encountered during the composting of sewage sludge. Primary pathogens may infect apparently healthy people and include bacteria viruses, cysts of protozoans, and ova of intestinal helminths. These organisms become trapped in the sludge during its formation in the wastewater-treatment process. Secondary pathogens infect only people whose disease defense systems have been compromised. Currently, it appears that the most abundant secondary pathogen encountered during composting is Af that grows in compost-pile zones whose temperatures do not exceed $60^{\circ}C$. It can cause noninvasive and invasive pulmonary, and sometimes, disseminated infections. Thermophilous actinomycetes that grow profusely during composting can produce allergic responses in sensitized people. Other, but not as yet positively identified, allergenic substances are the thermo-stable endotoxins present in the cell walls of enterobacteriacae, organisms present in high concentrations in sewage sludges.

Hazards from infection by primary pathogens to compost-site workers, communities surrounding the compost site, and people utilizing the compost seem to be low. Compost-site workers could possibly become infected through hand-to-mouth transfer of the sludge or through splatter to the workers' faces, but methods of prevention of infection by these routes seem obvious and simple. Generation of liquid aerosols during composting is extremely small, essentially eliminating liquid aerosol transport as a source for infection of workers or surrounding communities. Transport of pathogens via particulate aerosols and their infectivity in this form have not been evaluated. Proper composting procedures are capable of destroying primary pathogens present in the sludge, thus, eliminating them as a possible source of infection for users.

Thermophilous actinomycetes and Af spores are frequent components of respirable airborne particulates at a compost site. They may be transported off the site, producing, in some circumstances, higher than normal background concentrations. Because inhalation of these microbes can adversely affect sensitized or predisposed individuals or even those not predisposed if concentrations are high enough, compost-site planners and operators should consider taking the following precautions:

1. Locate a compost site at a prudent distance from hospitals or other health-care facilities.

2. Use preemployment health histories to screen out employees who would be at some risk from Af inhalation.

3. Distribute preemployment health-safety information relevant to the compost-site environment.

4. Provide workers in enclosed dusty areas or other high-spore concentration areas with respirators and air-handling equipment.

For users of composted sewage-sludge, Af concentrations are comparable to those found in several commercially available organic soil amendments. All of these and sewage-sludge compost contain substantial amounts of cellulosic materials. The latter favorably support Af growth. Using noncellulosic bulking materials should reduce the proliferation of Af during the composting process and greatly lower if not eliminate its concentration in the final product.

REFERENCES

1. Howard, A. 1935. The manufacture of humus by the indore process, J. Roy. Soc. Arts, 84:25.

2. Archarya, C. N. 1949. Preparation of compost manure from town wastes, Miscellaneous Bulletin No. 60, The Indian Council of Agricultural Research, Simla.

3. van Vuren, J.P.J. 1949. Soil Fertility and Sewage, Faber and Faber Limited, London.

4. Scott, J.C. 1952. Health and Agriculture in China, Faber and Faber Limited, London.

5. Epstein, E., G. B. Willson, W. D. Burge, D. C. Mullen, and N. K. Enkiri. 1976. A forced aeration system for composting wastewater sludge. J. Water Pollut. Control Fed. 48, 688.

6. Kabler, P. 1959. Removal of pathogenic microorganisms by sewage treatment processes. Sewage Ind. Wastes. 31, 1373.

7. Cliver, D. O. 1976. Surface application of municipal sludges. pp. 77-83. In Virus aspects of applying municipal wastes to land. L. B. Baldwin, J. M. Davidson, and J. F. Gerber (eds.), Center for Environ. Programs, Institute of Food and Agricultural Sciences, Univ. of Florida, Gainesville.

8. Farrell, J. B., J. E. Smith, Jr., S. W. Hathaway, and R. B. Dean. 1974. Lime stabilization of primary sludges. J. Water Pollut. Control Fed. 46, 113.

9. Millner, P. D., P. B. Marsh, R. B. Snowden, and J. F. Parr. 1977. Occurrence of Aspergillus fumigatus during composting of sewage sludge. Appl. Environ. Microbiol. 34, 765.

10. Hatch, T. F. and P. Gross. 1964. Pulmonary deposition and retention of inhaled aerosols. Academic Press. New York.

11. Marsh, P. B., P. D. Millner, and J. M. Kla. 1979. A guide to the recent literature on aspergillosis as caused by Aspergillus fumigatus, a fungus frequently found in self-heating organic matter. Mycopathologia 69, 67.

12. Allen, J. C. 1976. Infection and the compromised host. The Williams and Wilkins Co., Baltimore.

13. Sidransky, H. 1975. Experimental studies with aspergillosis. In Chick, E.W., A. Balows, and M. L. Furcolow (eds.) Opportunistic fungal infections. Amer. Lecture Ser. Pub. 974, pp. 165-176. Chas. C. Thomas, Springfield, Ill.

14. Coca, A. F. and R. A. Cooke. 1923. On the classification of the phenomena of hypersensitiveness. J. Immunol. 8, 163.

15. Lacey, J. 1971. Thermoactinomyces sacchari sp. nov., a thermophilic actinomycete causing bagassosis. J. Gen. Microbiol. 66:327.

16. Khan, Z.U., R.S. Sandhu, H.S. Randhawa, and D. Parkash. Allergic bronchopulmonary aspergillosis in a cane sugar Scand. J. Resp. Dis. 58, 129.

17. Hargreave, F. E., K. F. Hinson, L. Reid, G. Simon, and D. S. McCartney. 1972. The radiological appearances of allergic alveolitis due to bird sensitivity (bird fancier's lung) Clinical Radiology 23, 1.

18. Pepys, J., P. A. Jenkins, G. N. Festenstein, P. H. Gregory, M. E. Lacey, and F. A. Skinner. 1963. Farmer's Lung. Thermophilic actinomycetes as a source of "farmer's lung hay" antigen. Lancet ii, 607.

19. Lacey, J. 1974. Thermophilic actinomycetes associated with farmer's lung pp. 155-163. In R. de Haller and F. Suter (eds.) Aspergillosis and farmer's lung in man and animal. Hans Huber Publishers, Bern.

20. Katila, M. L., and R. A. Mäntyjärvi. 1978. The diagnostic value of antibodies to the traditional antigens of farmer's lung in Finland. Clinical Allergy 8, 581.

21. Horejsi, M., J. Sach, A. Tomsiková, and A. Mecl. 1960. A syndrome resembling farmer's lung in workers inhaling spores of aspergillus and penicillia moulds. Thorax 15, 212.

22. Banaszak, E. F., W. H. Thiede, and J. N. Fink. 1970. Hypersensitivity pneumonitis due to contamination of an air conditioner. New Engl. J. Med. 283, 271.

23. Seabury, J., J. Salvaggio, J. Domer, J. Fink, and T. Kawai. 1973. Characterization of thermophilic actinomycetes isolated from residential heating and humidification systems. J. Allergy Clin. Immunol. 51, 161.

24. van Assendelft, A., K. O. Forsen, H. Keskinen, and K. Alanko. 1979. Humidifier associated extrinsic allergic alveolitis. Scand. J. Work Environ. Health 5, 35.

25. Riddle, H.F.V., S. Channell, W. Blyth, D. M. Weir, M. Lloyd, W. M. G. Amos, and I.W.B. Grant. 1968. Allergic alveolitis in a maltworker. Thorax 23, 271.

26. Emanuel, D. A., B. R. Lawton and F. J. Wenzel. 1962. Maple-bark disease: Pneumonitis due to _Coniosporium corticale_. New Engl. J. Med. 266, 333.

27. Jackson, E. and K.M.A. Welch. 1970. Mushroom worker's lung. Thorax, 25, 25.

28. Lacey, J. 1974. Allergy in mushroom workers. Lancet i, 366.

29. Mattsby, I. and R. Rylander. 1978. Clinical and immunological findings in workers exposed to sewage dust. J. Occupat. Med. 20, 690.

30. Avila, R. and T. G. Villar. 1968. Suberosis respiratory disease in cork workers. Lancet i:620.

31. Boyd, G. 1978. Clinical and immunological studies in pulmonary extrinsic allergic alveolitis. Scot. Med. J. 23, 267.

32. Lowrance, W. W. 1976. _Of Acceptable Risk_. William Kaufman, Inc. Los Altos, Calif.

33. Dixon, F. R. and L. J. McCabe. 1964. Health aspects of wastewater treatment. J. Water Pollut. Control Fed. 36:984-989.

34. Clark, C. S., E. J. Cleary, G. M. Schiff, C. C. Linnemann, Jr., J. P. Phair, and T. M. Briggs. 1976. Disease risks of occupational exposure to sewage. J. Environ. Eng. Div. 102, 375.

35. Clark, C. S., G. L. VanMeer, C. C. Linnemann, A. B. Bjornson, P. S. Gartside, G. M. Schiff, S. E. Trimble, D. Alexander, E. J. Cleary, and J. P. Phair. 1980 Health effects of occupational exposure to wastewater. U.S.E.P.A. Symp. Wastewater Aerosols and Disease Proc. September 19-21, 1979.

36. Millner, P. D., D. A. Bassett, P. B. Marsh. 1980. Dispersal of *Aspergillus fumigatus* from sewage sludge compost piles subjected to mechanical agitation in open air. Appl. Environmental Microbiol. 39:1000.

37. Bringhurst, L. S., R. N. Byrne, and J. Gershon-Cohen. 1959. Respiratory disease of mushroom workers (Farmer's lung). J. Amer. Med. Assoc. 171, 15.

38. Lundholm, M. and R. Rylander, 1980. Occupational symptoms among compost workers. J. Occupat. Med. 22, 256.

39. Hickey, J.L.S. and P. C. Reist. 1975. Health significance of airborne microorganisms from wastewater treatment processes. II. Health significance and alternatives for action. J. Water Pollut. Control Fed. 47, 2758.

40. Johnson, D. E., D. E. Camann, J. W. Register, R. J. Prevost, J. B. Tillery, R. E. Thomas, J. M. Taylor, and J. M. Rosenfeld. 1978. Health Implications of Sewage Facilities. Health Effects Research Laboratory. Office of Research and Development, U.S.E.P.A. Report 600/1-78-032.

41. Fannin, K. F., K. W.Cochran, H. Ross and A. S. Monto. 1978. Health effects of a wastewater treatment system. Health Effects Research Laboratory. Office of Research and Development, U.S.E.P.A. Report 600/1-78-062.

42. Carnow, B., R. Northrop, R. Wadden, S. Rosenberg, J. Holden, A. Neal, L. Sheaff, and S. Meyer. 1979. Health effects of aerosols emitted from an activated sludge plant. Health Effects Research Laboratory, Office of Research and Development, U.S.E.P.A. Report 600/1-79-019.

43. Johnson, D. E., D. E. Camann, H. J. Harding and C. A. Sorber. 1979. Environmental monitoring of a wastewater treatment plant. Health Effects Research Laboratory, Office of Research and Development, U.S.E.P.A. Report - 600/1-79-027.

44. Wiley, J. S. and S. C. Westerberg. 1969. Survival of human pathogens in composted sewage sludges. Appl. Environ. Microbiol. 18, 994.

45. Gaby, L.W. 1975. Evaluation of health hazards associated with solid waste/sewage sludge mixtures. U.S.E.P.A. Report 670/2-75-023.

46. Morgan, M. T. and F. W. Macdonald. 1969. Tests show MB tuberculosis doesn't survive composting. J. Environ. Health. 32, 101. 47.

47. Burge, W. D., D. Colacicco and W. N. Cramer. 1980. Control of pathogens during sewage-sludge composting. Proc. Natl. Conf. Munic. Indus. Sludge Composting, Nov. 14-16, 1979. Inform. Transfer, Inc.

48. White, W. L., R. T. Darby, G. M. Stechert and K. Anderson. 1948. Assay of cellulolytic activity of molds isolated from fabrics and related items exposed in the tropics. Mycologia 40, 34.

SECTION IV

MAJOR FINDINGS FROM THREE SLUDGE PROJECTS

CHAPTER 12

OBSERVATIONS ON THE HEALTH OF SOME ANIMALS EXPOSED TO ANAEROBICALLY DIGESTED SLUDGE ORIGINATING IN THE METROPOLITAN SANITARY DISTRICT OF GREATER CHICAGO SYSTEM

Paul R. Fitzgerald. College of Veterinary Medicine, University of Illinois, Urbana, Illinois 61801

INTRODUCTION

The use of liquid sludge as an irrigant-fertilizer-soil conditioner for forage crops is one of importance. Anaerobically digested sludge containing 4 to 5 percent solids is beneficial and useful in the reclamation of strip-mined and other soils. For example, rough strip-mined land is rough graded and anaerobically digested sludge is applied to the soil by overhead sprinklers, injection, or by flood irrigation. Incorporation of the organic material in the sludge improves the tilth and increases the organic components of the soil. One way of utilizing the disturbed land, in the early stages of reconstruction, is to plant forage and allow animals to utilize the forage. Corn or other cereals are often grown in the sludge treated soils and the grains and the residual stubble may be used for forage by ruminant animals. The grains may be harvested by conventional methods or livestock, such as cattle or hogs, may be allowed to forage, feeding upon the grass, legumes, stalks, ear corn, or kernels which may have been left after the harvest.

Origin of Sludge and Potential Animal Exposure

Typically, anaerobically or aerobically digested sludges, originating from different communities, vary in their contents of nutrients, biological organisms, heavy metals, organic insecticides, other organic materials, chemicals, etc. The value of aerobically or anaerobically digested sludges is dependent upon the individual quantities of each component in the sludge that may be dispensed to the soil. For example, anaerobically digested sludge originating from the Metropolitan Sanitary District of Greater Chicago contains sedimented and digested materials from 5 million people and industrial wastes from the equivalent of an additional 5 million people. For this reason, sludge from this system may be entirely different from sludge which might originate from a small rural system limited to 100,000 people producing only waste from a residential community. Table 1 shows an analysis of anaerobically digested sludge from the MSDGC system listing metals, nutrients, etc. Sludges examined from other municipalities indicate, for example, a range of 6-444 ppm Cd, 100-7600 ppm Pb, 600-7000 ppm Zn and 0.01-23 ppm PCB (1,12).

Soils and Plants

With each application of sludge, organic and inorganic components (including chemicals and heavy metals) in the soil change as has been previously indicated [2,3]. Table 2 shows the changes in heavy metals content of soils repeatedly treated with anaerobically digested sludge originating from the MSDGC system. Depending upon the nature of the soil, such additions may be beneficial or harmful. Some organic and most inorganic additions remain unchanged in the soil for long periods.

Plants grown in the treated soils, therefore, are sometimes exposed to unusual quantities of materials like heavy metals. Plants take up the nutrients, as well as other materials, such as heavy metals, and incorporate them in the plant tissues. Forage plants and cereals of various kinds may incorporate some of the materials into the plant tissues or the grains. These in turn may be ingested by animals and a portion of whatever is present may be incorporated in the tissues of the animals. Table 2 also shows the levels of some heavy metals in forage grown in soils treated with anaerobically digested sludge. The figures shown are averages calculated from monthly samplings of soil and plants taken from the same plots. Control plants were grown in soils not treated with sludge but in the same general area as the experimental soils.

Aerosols and Water

A minor exposure of animals to sludge could come as a result of breathing small quantities of aerosol mist during overhead sprinkler application of sludge. This would appear to be minor because of the mechanics of the application system. Another possible source of contamination to animals could be the water supply. Often water for watering livestock accumulates in small farm ponds or lakes situated at the lowest site on the farm. As a

TABLE I.   Analysis of Liquid Fertilizer (Anaerobically Digested Sludge) Applied to Cropland in Fulton County During July 1978. Originated from Lagoon Storage.

		Application Dates		
		7/10-14/78	7/16-20/78	7/23-29/78
pH		7.6	7.5	7.6
Total Solids		5.34	5.56	5.55
Total Volume Solids		47.00	49.20	48.90
Total P	mg/L	1470.00	1890.00	1220.00
N-Kjeldahl	mg/L	2556.00	2905.00	2545.00
N-NH$_3$	mg/L	1092.00	1072.00	1053.00
Alk. as CaCO$_3$	mg/L	3100.00	3900.00	3220.00
Cl	mg/L	256.00	584.00	360.00
E.C.	µmmhos/cm	3400.00	3000.00	2500.00
Fe	mg/L	2300.00	2290.00	2300.00
Zn	mg/L	208.00	192.00	165.00
Cu	mg/L	93.50	87.60	88.40
Ni	mg/L	21.00	20.00	20.00
Mn	mg/L	22.60	21.90	24.90
K	mg/L	170.00	160.00	180.00
Na	mg/L	90.00	80.00	80.00
Mg	mg/L	460.00	440.00	400.00
Ca	mg/L	1300.00	1340.00	1240.00
Pb	mg/L	40.80	42.90	42.90
Cr	mg/L	513.00	154.00	181.00
Cd	mg/L	14.50	13.30	13.30
Al	mg/L	800.00	620.00	780.00
Hg	µg/L	146.00	825.00	572.00

TABLE II. Summary of Heavy Metals in Soils and Plants Irrigated or nonirrigated with Anaerobically Digested Sludge. Mean Values in µg/g (ppm) in Experimental and Control Groups. Values Shown were Calculated From Monthly Samplings.

	Soils				Plants			
	1975	1976	1977	1978	1975	1976	1977	1978
Cd								
Exp $\bar{x}$	22	36	35	27	12	15	9	16
Cont $\bar{x}$	3	2	2	2	6	3	2	3
Cr								
Exp $\bar{x}$	268	382	391	340	47	136	24	23
Cont $\bar{x}$	16	19	18	7	4	8	2	1
Cu								
Exp $\bar{x}$	155	224	192	175	30	75	18	21
Cont $\bar{x}$	17	19	31	22	8	9	7	8
Ni								
Exp $\bar{x}$	47	73	59	21	8	19	8	10
Cont $\bar{x}$	24	19	22	46	3	8	2	3
Pb								
Exp $\bar{x}$	79	126	141	135	13	45	13	< 8
Cont $\bar{x}$	16	18	20	23	5	8	4	< 8
Zn								
Exp $\bar{x}$	368	479	485	469	114	198	52	113
Cont $\bar{x}$	46	53	55	50	28	32	27	25
Hg(NG/G)(PPB)								
Exp $\bar{x}$	660	590	361	-----	19	115	68	23
Cont $\bar{x}$	405	92	27	-----	28	49	24	30

result drainage water and "runoff" tends to flow into the pond. If cropland or pastures are treated with sludge, some drainage or "runoff" could include some of the materials present in the sludge. Table 3 shows the results of chemical analyses for heavy metals of some lake waters used as sources of water for livestock in our study conducted in Fulton County, Illinois. These lakes served as "catch" basins for "runoff" and drainage from fields treated with anaerobically digested sludge.

TABLE III. Selected Metals Analyses of Fulton County Pond/Lake Water Samples and One Sludge Sample. The Water Samples were Collected November 30, 1978, and the Sludge Sample was Collected July 13, 1978.

Source	Cd	Cr	Cu	Fe	Hg	Ni	Pb	Zn
				µg/g				
Experimental lakes								
Field 4 at road	.004	.005	.000	.050	.001	.002	.000	.000
Field 4 —East	.035	.001	.000	.120	.000	.003	.000	.000
S.E. Field 17	.005	.003	.000	.060	.001	.008	.000	.000
S.W. Field 17	.003	.004	.010	.540	.001	.001	.000	.000
$\bar{x}$	.012	.003	.003	.190	.001	.004	.000	.000
Control lakes								
Livingston—North	.010	.001	.010	.160	.001	.050	.000	.000
Livingston—East	.007	.001	.010	.130	.001	.009	.000	.000
Gale	.016	.001	.00	.130	.001	.002	.000	.000
$\bar{x}$	.011	.001	.010	.140	.001	.030	.000	.000
Sludge	19	212	108	2800	.100	25	50	218

Dust From Dried Sludge and Ingestion of Soil

Some sludge, applied to the growing forages, may adhere to the plants and result in direct ingestion by animals consuming the forage. Ingestion of sludge-amended soil is a source of exposure for grazing animals [2,4,5] as well as rooting animals [6].

Direct Grazing by Ruminants on Forage Grown in Sludge Amended Soils

Foraging animals react differently to sludge treated forage. In studies conducted cooperatively by the MSDGC and University of Illinois at Fulton County, Illinois it was found that ruminants (cattle) ingested sludge amended forage readily and there was no exclusion because of palatability.

In a study which has been extant for 6 years, we selected beef-type cows and their calves and randomly placed them on pastures treated with sludge or in similar pastures not exposed to sludge. Insofar as possible, 60 mature to aged experimental cows and 20 mature to aged control cows were allowed to graze on Sudax grass, corn stubble, ear corn, grass-legume pastures, etc., for the same lengths of time although in different pastures. The study was conducted as a "field study" and animals were cared for and handled in a manner similar to that which would occur on a farm or ranch operated by a farmer or rancher. All of the animals in each of the two groups were examined twice yearly and observations were made daily in an attempt to detect any unusual illnesses or health defects. Cows and calves were bled in the spring for serological studies and calves were vaccinated, ear tagged, castrated, etc. Insofar as possible, milk samples were obtained from cows during the spring examinations. During the summer, experimental cows, with their calves, grazed in experimentally treated pastures and control cows, with their calves, grazed in untreated pastures. In the fall, the animals were again examined and groups of cows and calves were selected for necropsy.

Under the conditions of our studies, animals never refused to eat forage which had been exposed to anaerobically digested sludge. Although sometimes vegetation had been heavily coated with sludge by overhead irrigation, most of the sludge dried within 24 hours and fell off the plants to the ground. Some animals acquired sludge by licking the ground, the hair coat, or by breathing dust that was stirred up during feeding activities. In some instances, ear corn or seed "heads" of sudax grass contained dried sludge following application of anaerobically digested sludge by overhead irrigation. Water was probably not an important source of any unusual materials that might have been present in sludge (Table 3).

Thus far we have been unable to associate illness of cattle with quantities of sludge ingested on forage. No gross changes in feeding activities have been observed and animals have all appeared healthy and strong as long as an adequate food supply was available.

Changes in Heavy Metal Composition of Animal Tissues Associated with Sludge Residues.

One of the major problems associated with the utilization of sludges as soil conditioners or fertilizers has been concern for the potential transmission of heavy metals, organic compounds or pathogens into the environment. Our direct grazing study was concerned with the potential transmission of heavy metals into the food chain as well as transmission of pathogens. Each year a number of animals from the experimental and control herds were slaughtered with the intent of examining specific tissues for the presence of heavy metals, organic chemicals, or disease. Slaughtered cattle were grouped into two age groups consisting of calves less than 12 months of age born in the project, and of mature cattle, usually 5-15 years old. Some of the latter having been born in the project. Most of the mature cattle had been in the control and experimental groups since the inception of the study (six years).

Tables 4 and 5 show the occurrences of some heavy metals in vital tissues of cows or calves in the experimental and control groups. The figures shown represent the average PPM of heavy metals for the number of animals shown in parenthesis. These results are based upon freeze-dried tissues, a process which extracts essentially all of the water from the tissues. For example, lyophilization for 48 hours removes water so that kidney tissue is reduced to 21 to 23% of its original wet weight. Bone weight is reduced by 55%. Therefore, the PPM shown in the tables needs to be decreased by nearly 5 times (kidney and other soft tissues) and 2 times (bone) to ascertain the approximate live wet weight values in the tissues.

Table 4 also shows the quantity of heavy metals present in milk and blood from cows near the beginning of the study in 1974 and two years later after they had been on pastures exposed to anaerobically digested sludge.

In a somewhat similar grazing study, Kienholz, et al. [7], in Colorado, reported significant increases in Zn and Cd in kidneys and livers of old range cows grazing forages

TABLE IV. Summary of Heavy Metals in Tissues of Adult Cows Ingesting Forage Irrigated with Sludge (ex) Compared to Tissues from Cows Ingesting Forage not Irrigated with Sludge (con). 1975-1979.

Dry weight basis. Cows 8-15 years old.

	μg/g						
	Zn	Cu	Cd	Cr	Ni	Pb	Fe
Diaphragm							
$\bar{x}$ Ex (39)	134	5	0.076	0.120	0.270	<1	120
$\bar{x}$ Con (14)	151	7	0.045	0.164	0.198	<1	136
Heart							
$\bar{x}$ Ex (39)	68	16	0.193	0.086	0.238	<1	242
$\bar{x}$ Con (14)	64	16	0.068	0.611	0.288	<1	216
Liver							
$\bar{x}$ Ex (39)	192	164	7.061	0.084	0.240	1.109	210
$\bar{x}$ Con (14)	117	60	1.800	0.472	0.415	<1	189
Kidney							
$\bar{x}$ Ex (39)	114	21	40.690	0.150	0.328	<1	306
$\bar{x}$ Con (14)	93	22	14.370	0.522	0.471	<1	315
Brain							
$\bar{x}$ Ex (39)	51	13	0.052	0.116	0.411	<1	93
$\bar{x}$ Con (9)	47	11	0.060	0.165	1.330	<1	137
Bone							
$\bar{x}$ Ex (40)	58	2	0.092	0.400	0.803	.736	2.413
$\bar{x}$ Con (14)	68	8	0.441	0.990	3.300	.940	14
Blood (wet wt.)							
$\bar{x}$ Ex (78)	3.1	.99	0.002	0.025	0.007	.084	NA
$\bar{x}$ Con (18)	3.1	.63	0.007	0.010	0.009	.026	NA
Milk (wet wt.)							
1974							
$\bar{x}$ Ex	2.1	.17	0.003	0.012	0.005	0.001	NA
$\bar{x}$ Con	2.3	0.19	0.003	0.009	0.008	0.001	NA
1976							
$\bar{x}$ Ex	3.7	0.10	0.002	0.483	0.166	0.016	NA
$\bar{x}$ Con	2.8	0.06	0.000	0.311	0.091	0.008	NA

TABLE V. Summary of Heavy Metals in Tissues of Calves Ingesting Forage Irrigated with Sludge (ex) Compared to Tissues from Calves Ingesting Forage not Irrigated with Sludge (con). 1975-1979.

Calves less than 10 months old. Dry weight basis.

	μg/g						
	Zn	Cu	Cd	Cr	Ni	Pb	Fe
Diaphragm							
$\bar{x}$ Ex (43)	131	6	0.094	0.309	0.296	<1	87
$\bar{x}$ Con (20)	141	6	0.067	0.618	0.385	0	NA
Heart							
$\bar{x}$ Ex (43)	71	17	0.111	0.220	0.428	<1	242
$\bar{x}$ Con (20)	65	16	0.119	1.215	0.931	<1	NA
Liver							
$\bar{x}$ Ex (43)	114	117	2.301	0.205	6.052	<1	193
$\bar{x}$ Con (20)	118	113	0.688	0.360	0.606	<1	NA
Kidney							
$\bar{x}$ Ex (43)	88	20	12.302	0.188	1.566	<1	351
$\bar{x}$ Con (20)	81	20	3.875	0.357	2.376	<1	NA
Brain							
$\bar{x}$ Ex (3)	47	10	0.062	0.121	0.290	<1	66
$\bar{x}$ Con (20)	51	11	0.032	0.088	0.418	<1	NA
Bone							
$\bar{x}$ Ex (43)	48	3	0.416	0.574	1.670	1.530	19
$\bar{x}$ Con (20)	49	6	0.731	0.907	2.249	1.286	NA

grown in soils in which Denver Metropolitan Sanitary District sludge had been incorporated. None of the animals showed any gross or histopathological abnormalities at necropsy. The levels of Zn and Cd were lower in both liver and kidney than those shown in Table 4 of the present report. This may be partly due to their use of fewer animals, but it may also reflect differences in sludges, soil conditions, kinds of forage or lack thereof, etc.

Table 6 shows a comparison of heavy metals in bones of calves approximately 3 months old compared with bones taken from the same animals when they were 15 months old (10 experimental, 5 control). Up to 3 months of age the principal diet of the calves was milk from the mothers foraging in control or experimental pastures. The bones from the 15 month old animals show the values after approximately 12 months of foraging in sludge treated or nontreated pastures (same animals). There was a striking difference in the lead content of the bones of the experimental calves when compared to the controls. Six times as much lead was present in the bones of the experimental calves as was present in bones of the control calves. Twelve months later the lead content was reduced to slightly less than twice as much even though the young experimental animals ate forage grown in sludge treated soils. No evidence of disease was detectable at any time in either group.

TABLE VI. Chemical Analysis of Ribs Taken From 3-Month-Old Nursing Calves and From 15-Month-Old Heifers (Same Animals).

Ribs	Cd '76	Cd '77	Cr '76	Cr '77	Cu '76	Cu '77
$\bar{x}$ Exp (10)	1.2	0.03	0.08	0	4	3
$\bar{x}$ Cont (5)	1.3	0.03	0.02	0	3	3

Ribs	Ni '76	Ni '77	Pb '76	Pb '77	Zn '76	Zn '77
$\bar{x}$ Exp (10)	2.2	0.06	14.9	4.0	71	59
$\bar{x}$ Cont (5)	1.6	0.04	2.3	2.5	78	52

Two suggestions for the reduction in the level of lead are apparent. First, a dilution by growth and second, reduction by normal metabolism. The young experimental cattle continued to graze in pastures treated with sludge and therefore could be expected to continue to ingest sludge containing lead which would tend to accumulate in the bones. For the same reason one would not expect the lead to be eliminated from the bone when the diet contained abnormally high levels of lead.

These results suggest that during the nursing period, there may be an unusual deposition of lead intricately complexed with calcium in the milk, which ultimately results in a higher concentration of lead in the bones of the young calves nursing mothers grazing forage grown in sludge treated soils.

Changes Due to Direct Sludge Feeding

The quantity of sludge that is ingested by animals appears to have little, or no, short term effect upon the general health of ruminants. Keinholz, et al., [7] found that cattle fed diets consisting of up to 11.6% sludge, originating from the Denver Metropolitan Sanitary District for up to 94 days, did not show harmful effects. They were unable to detect significant changes in any of the tissues examined even though kidneys of cattle, fed the sludge, showed 2 fold increases in Cd (2.4 ppm) and 18 fold increases in Pb (15.8 ppm). There were lesser increases in the livers of the same animals. These levels are less (Cd) and greater (Pb) than our observations in 9 month old calves and mature cows ingesting sludge naturally when foraging in sludge treated pastures.

Smith, et al. [8] indicated that sludge sterilized by irradiation, was a useful nutritional substitute that could be used, in part, to replace cottonseed meal in the diet of ruminants. They pointed out that rumen fermentation may alter some toxic materials or pathogens so that their effects are weakly expressed. Osuna and Edds [9] reported that MSDGC sludge, containing 83 ppm Cd, fed to weanling pigs as 50% dried activated sludge in a starter ration, depressed growth and feed consumption.

Non-ruminants such as rats and Japanese quail are not adversely affected by including 20% sun-dried and autoclaved activated sewage sludge in their diet [10]; however, the digestability is poor and protein quality is deficient so that it cannot be used as a sole source protein in non-ruminant diets.

Changes in Heavy Metals in Tissues of Swine Due to Exposure to Sludge Amended Soils

Table 7 shows the average levels of some heavy metals from tissues of swine exposed to sludge in confined pens. Pigs (35 lbs) were confined to pens 9 m x 14 m and the soil was treated with sludge equivalent to applications of 12, 26, or 50 dry metric tons per hectare. They were fed standard, commercially prepared rations for approximately 4 months and then were necropsied. Their exposure to sludge was by ingestion of plants growing naturally in the pens and by contact with treated soil in the pens. In general, there were linear increases in some heavy metals in some tissues, depending upon the rates of application of sludge to the soil in the pens. Cadmium, for example, varied from 0.54 ppm in the kidneys of unexposed control pigs to 1.32, 4.38 and 4.49 ppm in the kidneys of experimental pigs exposed to soils treated with 12, 26 or 50 dry metric tons of sludge per hectare. Cadmium appears lower than expected in kidneys of the two pigs exposed to the maximum levels of sludge, but the analysis is based on two animals and probably is not indicative of the actual condition.

Health Effects

In all of our studies, as well as those cited, there has been no indication of disease in experimental animals variously exposed to sludges. The transmission of pathogens has not been a problem. Disease could have occurred as a result of the presence of some organics or heavy metals translocated to vital tissues such as kidneys and livers where some heavy metals accumulate, but these tissues have not shown significant alteration. No pathogenic disease organisms have appeared; therefore, there has been no reason to believe that pathogens affecting cattle were present in sludge. Parasitic organisms such as Ascaris lumbricoides of children have been shown to be transmitted through sludge to swine but no similar infections have occurred in cattle.

Serological examinations of blood twice yearly have failed to detect significant levels of pathogens such as Leptospira sp. and Brucella sp. No attempts have been made to isolate specific viruses because no disease suggesting virus origin has appeared.

Performance

There have been no reports of morbidity or mortality in animals consuming sludge-fertilized (SF) products in a natural environment.

TABLE VII. Mean Levels of Heavy Metals in Selected Tissues Taken From Pigs Exposed to Different Levels of Anaerobically Digested Sludge (Dry Equivalents in Metric Tons/Hectare) in Soil During a 4-Month Feeding Period. Dry Weight Basis.

	Animals	Cd	Cr	Cu	Ni	Pb	Zn
			Diaphragm				
Control	8	0.027	0.217	4	NA	< 1	93
12 MT/H*	4	0.047	0.235	4	NA	< 1	79
26 MT/H	8	0.368	0.425	6	NA	< 1	82
50 MT/H	2	0.026	0.469	4	0.100	< 1	59
			Heart				
Control	8	0.040	0.697	18	NA	< 1	74
12 MT/H	4	0.057	0.472	15	NA	< 1	66
26 MT/H	8	0.032	0.288	15	NA	< 1	74
50 MT/H	2	0.050	1.269	14	0.388	< 1	52
			Liver				
Control	8	0.074	0.111	17	NA	< 1	234
12 MT/H	4	0.369	0.160	38	NA	< 1	367
26 MT/H	8	0.822	0.583	18	NA	< 1	247
50 MT/H	2	0.451	0.463	9	0.087	< 1	108
			Kidney				
Control	7	0.544	0.442	45	NA	< 1	100
12 MT/H	4	1.322	0.194	45	NA	< 1	128
26 MT/H	8	4.375	0.322	32	0.125	< 1	116
50 MT/H	2	4.488	0.088	22	NA	< 1	66
			Bone				
Control	8	0.059	0.442	1	NA	< 1	112
12 MT/H	4	0.035	0.797	1	NA	< 1	121
26 MT/H	8	0.039	1.427	1.6	0.132	1.9	116
50 MT/H	2	0.051	0.594	3	0.188	1.25	95

*H = hectare.

Although performance data for growth of adult cattle discussed earlier are not available, weights of 10 month old calves from control and experimental herds are presented. The average dressed weight of all calves was 155 kg; the live weight average was 282 kg. There were no significant weight differences between control and experimental calves. Projected calculations to a yearling basis indicate that calves foraging on sludge treated pastures would be near 350 kg (770 lbs).

Reproduction

Ingestion of forages and grains treated with anaerobically digested sludges or exposure to soils treated with sludges does not appear to adversely affect reproduction in cattle. The Cd intake did not appear to be enough for significant placental transfer and there was no detectable effect on the developing fetus.

Cows foraging on SF pastures have performed as well as those foraging on control pastures. In beef type cattle, the rates of reproduction in control and experimental herds has fluctuated between 85% and 90% during each of the five years of study. Calves born to cows in the experimental group, grazing sludge treated pastures, have been equal to or superior in quality to control calves. The slight increases in some heavy metals (e.g., Cd in Table 5) had no discernible pathological effect on calves.

Gross and Microscopic Pathology

For the most part, animals fed sludge fertilized products perform well and no gross lesions are demonstrable [2,11-13]; therefore, there has been little incentive for detailed microscopic examination of tissues. Histopathological examinations made on tissues taken from 60 of the 170 cattle necropsied in our studies indicate no specific pattern of tissue abnormalities due to exposure to sludge by ingestion or contact. Since Cd accumulation in the kidney is of most concern, one would expect to find abnormalities in that organ if there was a harmful effect. In no instance have we been able to detect significant renal disease in either young or old animals exposed to sludge. In old cows (12 + yrs.) Cd content of the kidney was as high as 46 PPM; in 10 month old calves 37 PPM. The maximum content in comparable control cows and calves was 19 and 2 PPM. Table 8 summarizes the histopathological observations on 10 experimental and 5 control calves and 10 experimental and 5 control cows and bulls. All animals were killed by stunning with subsequent exsanguination. Tissues were quickly removed and were routinely fixed in

TABLE VIII. PATHOLOGICAL CHANGES OBSERVED IN EXPERIMENTAL AND CONTROL COWS AND CALVES EXPOSED OR NOT EXPOSED TO ANAEROBICALLY DIGESTED SLUDGE IN/ON SOIL IN WHICH FORAGE OR CEREALS FOR THE ANIMALS FEED WAS GROWN.

Tissue examined	Focal fatty repl. fibers	Congestion	Degeneration	Neuronal degeneration	Spongiosis	Cellularity Vir-Rob spaces	Early necrosis	Peri-vascular infilt.	Focal portal infilt.	Arteriolar thickening	Mononuclear inter. inf.	Glomerular chg/dis.	Glomerulitis	Focal eosino. infilt.	Portal cellularity	Collegenation	Intersti. infil. foci	Sarcolemma nuclei	Mixed Interlob. ex.	Swollen hepatocytes	Portal fibroplasia	Infiltrates	Frayed	Sarcocystis
Experimental calves (10)																								
Brain		8	1	4	1	1	2																	
Heart		3	3				6	1									1							1
Muscle							4																	
Liver		10	10				4		6															
Kidney		9								9	3	6												
Control calves (5)																								
Brain		5	2				1																	
Heart		4	5																1					1
Muscle		3	5															1					1	
Liver		5	5				1	4																
Kidney		5	5						4	5				2										
Experimental cows (10)																								
Brain		10																						
Heart	1	8	8											3										3
Muscle	9	3	3			1																		4
Liver		10	10																1					
Kidney		10	9						10	10	1			5										
Control Cows (5)																								
Brain		2					2																	
Heart		1	1																				1	4
Muscle	1	2					1										1							1
Liver		5	4				2	1						2										
Kidney		2							2	2	4			2										

10% buffered formalin. Preparation and staining of tissues was carried out by routine procedures and all tissues were examined by the same pathologist.

The pattern of histopathologic lesions was inconsistent in cattle. No pattern emerges although "early necrosis" of brain, heart, muscle, and liver appeared to be more common in experimental calves than in other groups. Observations on necropsied calves from previous years did not confirm this observation. Most tissues in both control and experimental animals showed some degree of degeneration and congestion. No explanation can be given, but it does not appear to be specifically associated with the ingestion of sludge.

CONCLUSIONS

Accumulation of heavy metals in soil, plants and animal tissues appears to be of most concern to those charged with disposing of sewage wastes. Not all heavy metals are of equal concern. Currently, cadmium has received the greatest amount of attention. Whether it justifiably deserves this attention has not yet been confirmed.

Cadmium is accumulative in body tissues and is said to be a cause of failure of some organs. In warm-blooded animals, it accumulates in greatest concentrations in kidneys and livers. In our studies Cd concentrations were always highest in these organs. Kidney and liver normally could be expected to have higher levels of metals like Cd because it is their function to remove excess or harmful materials that may be excessive to the needs of the body. However, Cd is not known to be a required element and therefore any Cd is excessive to body needs. Other organs and tissues do not accumulate quantities of Cd even remotely approaching the levels found in the kidney and liver tissue. It has been suggested that accumulation of Cd in excess of 200 µg/g in human kidney interferes with normal activity. Such a level is "implied" because it is primarily based upon necropsy findings.

Other tissues such as somatic muscle, heart, brain and blood have been shown to acquire comparatively small quantities of heavy metals. Bone tends to accumulate lead, but as shown in Table 4, it may be metabolized and pass out of the body of some animals rather quickly. With the exceptions of Cd and Pb, none of the other heavy metals were significantly increased in various tissues and organs examined in our studies. Under the conditions of exposure herein described copper, chromium, iron, nickel and zinc did not appear to be a threat to the animals well-being.

REFERENCES

1. Furr, A. K., A. W. Lawrence, S. S. C. Tong, M. C. Grandolfo, R. A. Hofstrader, C. A. Bache, W. H. Gutenmann, and D. J. Lisk. "Multielement and chlorinated Hydrocarbon Analysis of Municipal Sewage Sludges of American Cities." Environ. Sci. Technol. 10:682-687. 1976a.
2. Fitzgerald, P. R. "Toxicology of Heavy Metals in Sludges Applied to the Land." Fifth Nat'l. Conf. on "Acceptable Sludge Disposal Techniques". Orlando, FL.: Pp. 106-116. 1978.
3. _____. "Recovery and Utilization of Strip-mined Land by Application of Anaerobically Digested Sludge and Livestock Grazing." In Proc. of Symp. on "Utilization of Municipal Sewage Effluent and Sludge on Forest and Disturbed Land". Penn. State Univ. Press. University Park, Pa. P. 497-506. 1979.
4. Healy, W. B. "Ingested Soil as a Source of Elements to Grazing Animals. Cited by Hockstra, et al., Trace Element Metabolism in Animals." 2, pp. 448-450. Baltimore, Md.: Univ. Park Press. 1974.
5. Thorton, I. "Biogeochemical and Soil Ingestion Studies in Relation to the trace element Nutrition of Livestock." Cited by Hockstra, et al., Trace Element Metabolism in Animals - 2, pp. 451-454. Baltimore, Md.: University Park Press. 1974.
6. Hansen, L. G. and T. D. Hinesly. "Cadmium from Soil Amended with Sewage Sludge: Effects and Residues in Swine." Environ. Health Perspectives. 28:51-57. 1979.
7. Kienholz, E., G. M. Ward, D. E. Johnson, and J. C. Baxter. "Health Considerations Relating to Ingestion of Sludge by Farm Animals." In Proc. of 3rd Nat'l Conf. on Sludge Mgm't. Disposal. Information Transfer, USEPA. Rockville, Md. pp. 128-134. 1976.
8. Smith, G. S., H. E. Kiesling, and E. E. Ray. "Use of Raw Sludge as a Feed Supplement." Proc. of 8th Nat'l. Conf. on Municipal Sludge Mgm't. Information Transfer Inc., Rockville, MD., p. 11. 1979.
9. Osuna, O. and G. T. Edds. "Feeding Trials of Dried Urban Sludge and the Equivalent Cadmium Level in Swine." Proc. of 8th Nat'l. Conf. on Municipal Sludge Mgm't. Information Transfer Inc. Rockville, MD., p. 12. 1979.
10. Cheeke, P. R. and R. O. Myer. "Evaluation of the Nutritive Value of Activated Sewage Sludge with Rats and Japanese Quail." Nutr. Rpts. Intern. 8:383-392. 1973.

11. Chaney, R. L., G. S. Stoewsand, C. A. Bache, and D. J. Lisk. "Cadmium Deposition and Hepatic Microsomal Induction in Mice Fed Lettuce Grown on Municipal Sludge-amended Soil". J. Agric. Food Chem. 26:4, 992-994. 1978a.
12. _____, G. S. Stoewsand, A. K. Furr, C. A. Bache, and D. J. Lisk. "Elemental Content of Tissues of Guinea Pigs Fed Swiss Chard Grown on Municipal Sewage Sludge-amended Soil." J. Agric. Food Chem. 26:4, 994-997. 1978b.
13. Hinesly, T. D., E. L. Ziegler, and J. J. Tyler. "Contents of Selected Chemical Elements in Tissues of Pheasants Fed Corn from Sewage Sludge Amended Soil." Agro-Ecosystems. 3:11-26. 1976.

CHAPTER 13

UPTAKE OF TRACE METALS AND PERSISTENT ORGANICS INTO
BOVINE TISSUES FROM SEWAGE SLUDGE - DENVER PROJECT*

John C. Baxter. Metropolitan Denver Sewage Disposal
District No. 1, Denver, Colorado

Donald E. Johnson and Eldon W. Kienholz. Animal
Science Department, Colorado State University,
Fort Collins, Colorado

ABSTRACT

This research was initiated to examine effects of sewage sludge disposal on agricultural land and possible transmission of sludge contaminants into the food chain. Soils, forages, and cattle grazing a sewage sludge disposal site were examined for trace metals and refractory organics content.

Soils at the sludge disposal site had higher concentrations of metals than non-sludged soils. Forages grown at the sludge disposal site had higher concentrations of Zn, Cu, Ni, and Cd, and lower Pb concentrations than forages from non-sludged areas. Nine organic residues were found in these soils, but were at trace levels. No organics were found in any of the forage tissues examined.

A study of trace metal concentrations (Zn, Cu, Cd, Pb, and Fe) determined in kidney and liver tissues from cattle of known age and no exposure to sewage sludge was conducted so that information concerning "normal" tissue metal levels could be developed. Kidney Cd was found to increase with increasing age of the animals, while liver Cu concentrations appeared to be a function of diet. Tissues from cattle grazing the sludge disposal site showed no significant increases in trace

*This project was funded in part by the Municipal Environmental Research Laboratory, U.S. Environmental Protection Agency, Cincinnati, OH, and the U.S. Food and Drug Administration, Bureau of Foods, Division of Chemical Technology under EPA Contract No. 68-03-2210.

metals or organics over control animals.

Cattle fed diets containing 0, 4, or 12% digested Denver or Ft. Collins sludge for a 3 month period, showed metal increases of Cd, Cu, or Pb in either kidney, liver, or bone tissues. There were no measurable increases in muscle tissues. Fat tissues showed small increases in organics. However, all levels were below tolerance levels that have been established by FDA for fish or other food products.

Cows and calves were fed Ft. Collins sewage sludge for 9 months, followed by a 4 month withdrawal period. Concentrations of Cd, Cu, and Pb increased in kidney, liver, or bone tissues. Cadmium did not change during the withdrawal period, while liver Cu and Pb decreased. Muscle Cd concentrations increased slightly due to sludge ingestion, while other metals were unaffected.

The 3 sludge feeding studies showed that sludge had no positive or negative effects on animal health or performance other than to act as a diet diluent. These studies showed that metal or organic uptake into cattle tissues is in proportion to the contaminant level of the sludge. Different sludges with different metal concentrations may affect the availability of sludge metals for absorption and retention into cattle tissues. If sewage sludge is utilized as an agricultural fertilizer and incorporated into the soil, there would appear to be little hazard that metals or organics would be elevated above levels that would be considered normal.

INTRODUCTION

The idea of recycling waste materials to the land is certainly not new and has been an accepted agricultural practice since the inception of agriculture and civilization millennia ago. Because sewage sludge is mostly organic matter, it contains a number of valuable nutrients and humus materials which can be utilized to increase soil fertility. The addition of organic matter to soil improves soil tilth, raises the soil CEC, water holding capacity, and consequently may increase crop yields over what would normally be achieved through the use of conventional inorganic fertilizers.

However, there are problems associated with the recycling of sewage sludge. Much of our sewage is not derived from garbage or fecal material, but is the product of an industrial society. Our society tends to centralize and concentrate industry, which is reflected in the residues of wastewater treatment plants; thus, heavy metals and persistent organics such as PCB's and pesticides, many of which are products or by-products of industry, turn up in sewage sludge. In addition to these elements and compounds there is a disease potential. The problems associated with many of these elements, compounds, and biological agents become particularly acute when they are potentially harmful to crops, animals, or the human food chain.

The Metropolitan Denver Sewage Disposal District No. 1 has been recycling sewage sludge to a dedicated waste disposal site called the Lowry Bombing Range (LBR) since 1969. The sludge has been applied to the soil surface, plowed under and planted to forages. These forages have been subsequently grazed by a herd of beef cattle which were on the site when sludge disposal operations began. These animals were exposed to sewage sludge for six years and thus had an opportunity to ingest sludge contaminants either from forages or by direct ingestion of soil and sludge. This represented a unique opportunity to examine some long-term effects of sludge exposure to cattle from a somewhat typical ongoing practice.

The primary objectives of this study were to examine effects of sewage sludge disposal on agricultural land, and possible transmission of sludge contaminants into the food chain. To accomplish this objective, cattle from the LBR were examined, slaughtered, and tissues analyzed for heavy metals and persistent organics content. Cattle from the LBR were then compared with a large group of control cattle that had never been exposed to sewage sludge or elevated levels of heavy metals. Concurrent with the examination of the LBR cattle, a survey of soils and plants at the LBR was conducted in order to document the variation and magnitude of sludge contaminants.

Several additional studies were designed around the premise that the application of sludge to soil or pasture land will result in the direct ingestion of sludge constituents. Three separate studies were undertaken to examine the effects of direct ingestion of measured amounts of sludge by cattle and assess the magnitude of uptake of heavy metals and persistent organics into their tissues. In the first study, young steers were fed a feedlot ration consisting of 0, 4, and 12% anaerobically digested Metro Denver sludge for 3 months. The second feeding study was conducted with a similar group of cattle, however, anaerobically digested sludge with a higher Cd and Cu content was fed. The third feeding study, which utilized the high Cd-Cu sludge, examined uptake rates in young and older animals over a 9 month feeding period, followed by a 3 month withdrawal period. Specific elements examined in soil, plant, and animal tissues were As, Cu, Cd, Mo, F, Pb, Hg, Ni, Se, and Zn; however, only Zn, Cu, Cd, and Pb will be reported here. These 4 heavy metals are of concern because of their known health effects, and because other elements showed little or no increase in either forages grown on the sludge disposal site or cattle tissues after direct ingestion of sewage sludge.

METHODS OF ANALYSIS

Samples of the various animal tissues were collected during slaughter. Each sample was collected using stainless steel knives and extreme caution to avoid contamination with gut or foreign material. Whole kidneys were taken for analysis while slices of other tissues were taken transversely through each organ to obtain a composite sample of the whole organ. Bone tissues were from the proximal half of the tarsal bone, and muscle (*ulnaris lateralis*) samples were taken from the same forearm. No attempt was made to remove bone marrow. Tissue samples were rinsed in deionized water and stored in acid washed plastic containers for heavy metals analysis. Identical samples were stored in acid washed glass containers and covered with aluminum foil for refractory organics analysis. All samples were stored frozen until analysis.

Animal tissues from the LBR, 3 month Metro. Denver and Fort Collins sludge feeding trials were analyzed by the Colorado State University Analytical Facility. When samples were received in the lab they were chilled in liquid nitrogen and then pounded (wrapped in plastic bags) until the sample had shattered into many small pieces. These pieces were then coned and quartered. Representative aliquots were taken for moisture determination which permitted wet weight results to be converted to a dry weight basis. Tissue samples were digested in nitric perchloric acids with potassium dichromate as a catalyst (Feldman, 1974). Copper and Zn were determined directly by flame AAS when kidney and liver tissues were involved. Other tissues were concentrated after digestion in triple distilled perchloric plus nitric acid, using 1% ammonium-pyrollidine dithiocarbamate (APDC)(Boyle, et al, 1975). The APDC chelate was isolated on a 0.22 µm Millipore filter and stored for later analysis. At the time of analysis, the filter with APDC was dissolved in simmering nitric acid and diluted to a 10-ml volume. Metals were then determined directly by flameless AAS.

Kidney, liver, and muscle tissues from the 9 month sludge feeding trial and control study examining older cattle of known age were analyzed utilizing a slightly different method developed by Metro. Denver Sewage. Whole kidneys and 0.5 to 1-kg samples of liver were homogenized in a blender. Aliquots of the homogenized kidney or liver tissue were transferred to tared 100-ml volumetric flasks to give a dry weight of 1- to 2-g. The aliquot was oven dried at 105°C. Concentrated HNO_3 (10-ml) was added and allowed to digest for 2 hours at room temperature. The flasks were then placed on a hot plate (163-191°C) and digested for 45 minutes. Samples were cooled and 5-ml of concentrated H_2SO_4 (suitable for Hg determinations) added. The samples were then placed on the hot plate (260-288°C) for 45 minutes. Samples were cooled and 10-ml of concentrated HNO_3 was added, then returned to the hot plate for 60 minutes. Samples were cooled again and 10-ml of concentrated HNO_3 acid added. Samples were then returned to the hot plate until the solution turned clear yellow or orange and for 5 minutes thereafter. The digestate was transferred to a 25-ml volumetric flask and diluted to 25-ml. Samples were then determined directly using flame AAS.

Plant samples were washed in a dilute solution of lauryl sulfate, dried, ground to 20 mesh in a stainless steel Wiley mill, coned and quartered, and then stored in plastic bags until analysis. Soil samples were air dried, crushed, sieved through a 32 mesh stainless steel sieve, coned and quartered, and stored in plastic bags until analysis.

Extraction of metals (Zn, Cu, Ni, Cd, Pb) in plant and soil samples were accomplished using the same technique. Samples were oven-dried overnight at $105^{o}C$, 1- to 5-g weighed into 250-ml beakers, and ashed at $480^{o}C$ for 14 hours. Ashed samples were then refluxed with 10-ml of concentrated HNO_3 and taken to near dryness, followed by addition of 10-ml 3N HCl. Ashed samples were refluxed again and then filtered through #43 acid washed Whatman filter paper. The filtrate was then volumetrically diluted to 50-ml with deionized water, after which the metals were determined directly with flame AAS. All metals were analyzed utilizing a background correction deuterium lamp.

The Colorado Epidemiologic Pesticides Studies Center performed all the refractory organics analyses. Methods employed were essentially those described in an E.P.A. manual (E.P.A., 1974).

SOIL AND PLANT SURVEY OF LBR SLUDGE DISPOSAL SITE

The LBR sludge disposal site is located in Arapahoe County, a few miles east of the city of Denver. The site was originally natural pasture land typical of the high plains of eastern Colorado. Soils at the site were formed on loessol deposits, have an average pH of 7.0, an average CEC of 28-meq/100-g, and range from clay loam to loam in texture.

Sewage sludge has been applied to the soil surface at the rate of 67-mT/ha (dry weight) and plowed under. The average heavy metal composition of the sludge cake disposed of at the LBR is shown in Table 1. In order to assess the impact of sludge on soils at the LBR, 16 soil and forage samples were collected from the perimeter of the site which had never received any sludge applications. Samples were also collected from 20 fields which had received a range of sludge applications. Six equally spaced soil samples, to a depth of 0.9-m, were taken from a transect through each field.

Table 1. Heavy Metal Composition of Metro. Denver Sludge*

	μg/g (Dry Weight)	
	Anaerobic**	Waste Activated[†]
Zn	1443	1028
Cd	23	18
Cu	961	703
Ni	222	190
Pb	1016	352

* Values from the analysis of 90 composite samples.
** Anaerobically digested primary sludge.
[†] Mixture of raw primary plus waste activated sludge treated with $FeCl_3$ and lime.

Table 2 shows the mean concentration of heavy metals in surface soil samples from the perimeter of the LBR where no sludge had been applied and the median and range of surface soil samples taken from the 20 sludged fields. The median is employed to describe the sludged areas because it is a more realistic description of central tendencies of the heavy metal status of the soils. Table 2 shows that the addition of

Table 2. Mean, Median and Range of Heavy Metals in Surface Soil Samples from the LBR

	μg/g (Dry Weight)			
	No Sludge Applications		Sludge Applied	
Element	Mean	Range	Median	Range
Zn	50.8	39.5 - 62.5	93.0	40.0 - 252
Cd	0.43	0.18 - 0.72	1.45	0.43 - 4.40
Cu	10.6	6.75 - 18.8	11.0	8.6 - 50.0
Ni	11.8	6.0 - 43	21.8	6.00 - 43.0
Pb	18.2	14.0 - 22.0	36.2	14.2 - 101.0

sludge to the LBR has significantly raised the soil concentration of these heavy metals. Analysis of subsurface samples showed that no heavy metals leached below the depth of sludge incorporation.

The effect of the sludge applications on forages at the LBR are shown in Table 3. The mean heavy metal concentrations of forage tissues (mainly buffalo grass) sampled on the perimeter of the LBR which never received sludge applications are compared with the median and range of forage tissues (winter wheat seedlings) from the 20 fields that had received

Table 3. Heavy Metal Concentrations of Forages from the LBR

	µg/g (Dry Weight)		
	No Sludge Applications	Sludge Applied	
Element	Mean	Median	Range
Zn	11.7	75.8	37.2 - 153
Cd	0.11	1.08	0.25 - 3.70
Cu	2.28	12.5	3.80 - 22.0
Ni	0.56	2.80	0.25 - 15.5
Pb	2.3	0.75	0.65 - 4.0

sludge applications. Concentrations of Zn, Cd, Cu, and Ni increased significantly with the application of sludge, however, the lead content of forages decreased with sludge applications. The decrease in Pb content of forage tissues, despite increasing soil Pb, may have been the result of several factors. The increased soil P or organic matter content from sludge applications may have rendered the Pb less available for plant uptake. Forages taken from areas of sludge application were green and actively growing, while forages taken from the control areas were growing poorly in a dry nutrient deficient soil. Because of the slow growth habits of the native pasture, the decrease in Pb may have been a dilution effect, rather than a decrease in plant availability of soil lead.

Soils, sludges, and forages were analyzed for 22 refractory organics. The concentration of organic residues detected in sludge samples is shown in Table 4. All of these residues were found in one or more soil samples, but at trace levels below practical detection limits. Plant tissues contained no detectable levels of any of the 22 refractory organics analyzed for.

Table 4. Concentration of Refractory Organic Residues Found in Sewage Sludges Disposed of at the LBR

	ppb (Wet Weight)		
	DNS(2)*	DNS(1)**	F & L Cake[†]
p,p' - DDT	<40.0	NF[§]	202
p,p' - DDE	300	51	94
p,p' - DDD	497	11	65
αBHC	16	NF	20
Dieldrin	86	101	35
HCB	NF	5	5
Oxychlordane	NF	94	NF
Chlorine	1535	1345	636
PCB's (AR-1254)	4252	2570	751

* Air dried anaerobically digested primary sludge (∼5% moisture content).
** Anaerobically digested primary sludge cake (∼80% moisture content).
[†] Waste activated sludge cake (∼80% moisture content).
[§] Not Found.

HEAVY METAL CONTENT OF KIDNEY AND LIVER TISSUES OF CONTROL ANIMALS

A study of heavy metal concentrations found in kidney and liver tissues of cattle that had never been exposed to sewage sludge was established so that information concerning "normal" levels of metals found in these tissues could be developed. This control study consisted of 25 range cattle from eastern Colorado, 4 range cows from eastern Wyoming, 15 dairy cows from Colorado State University, and 12 steers that were used as controls in sludge feeding trials. Animals were selected principally on the basis that their ages were known. The 29 range cows varied in age from 5 to 15 years, the steers were approximately 18 months old at slaughter, and the dairy cows ranged in age from 2 to 8 years. Table 5 shows the mean concentration of Zn, Cu, Cd, Pb, and Fe in kidney and liver tissues of the control cattle.

Table 5. Mean and Standard Deviation of Trace Metal Concentrations in Liver Tissues of Control Cattle

Tissue	μg/g (Dry Weight)				
	Zn	Cu	Cd	Pb	Fe
Kidney	87±20	17.6± 4.2	10.0 ±8.1	2.9±2.1	246±75
Liver	131±32	65.0±84	0.74±0.56	1.5±1.4	153±91

The variation in tissue trace metal content between animals was large in some cases, particularly with kidney-Cd and liver-Cu concentrations. The concentration of Cd in kidneys was found to increase with the age of the animal. Data shown in Figure 1 represents all animals used in the control study. The range of kidney Cd concentrations was from 0.83 ppm in a two year old dairy cow to 31.3 ppm in a thirteen year old range cow.

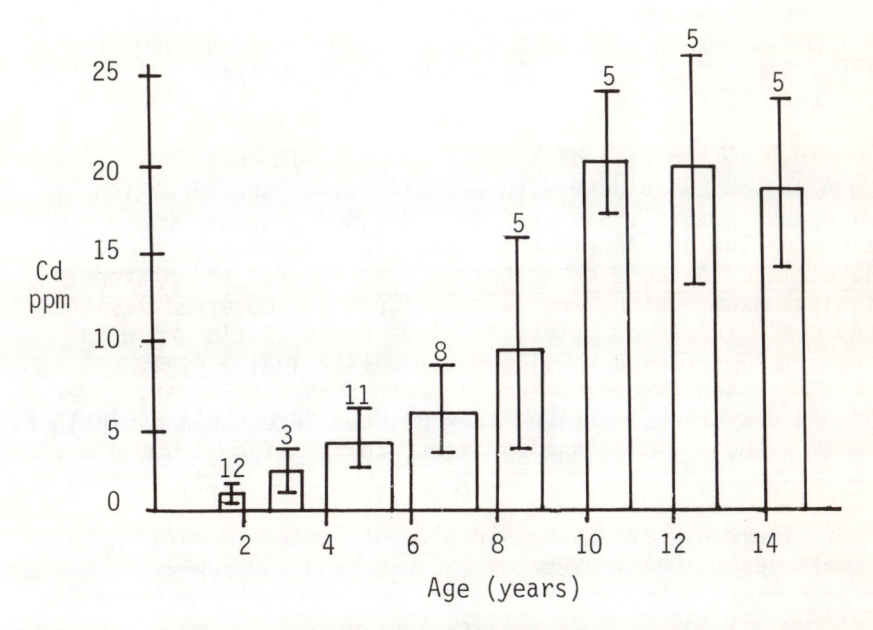

Figure 1. Mean concentration of Cd in kidney tissues (dry weight) as a function of cattle age. Numbers above bars represent number of animals for each age group. Tick marks represent standard deviation for each group.

Liver Cu concentrations ranged from 5.9- to 398-ppm (dry weight). However, this large range could not be explained as being due to age. When the dairy cattle and Colorado range cattle were examined separately there was a significant difference in mean liver Cu concentrations between the two groups (9.5-ppm Cu in range cows and 216-ppm Cu in dairy cows). The Cu status of the range cows was below the 100- to 400-ppm Cu (dry weight) that is considered normal for cattle (Underwood, 1977). These low levels would tend to indicate that these animals were suffering from Cu deficiency; however, they were not suffering any overt symptoms of Cu deficiency, nor has there been any indication of Cu deficiency in this region. Information regarding normal Cu liver concentrations is limited; thus, the levels reported here may not be as low as thought at first glance. The livers of normal adults of many other species contain 10- to 50-ppm Cu. The differences in liver Cu concentrations between these two groups is most likely diet related. The dairy cattle may have received trace element salt supplements in their feed, while it is unlikely that the range cattle would have received similar supplements.

TRACE ELEMENT LEVELS IN TISSUES OF CATTLE GRAZING LBR SLUDGE DISPOSAL SITE

During the fall of 1975, a group of 12 cows that had been on the LBR sludge disposal site since the inception of sludge disposal (1969) were examined and slaughtered for tissue anlysis. Antemortem inspection revealed no overt symptoms of disease or pathology. Postmortem and histopathological examinations were also conducted. Observed lesions and abscesses were typical of older range cattle and were not symptoms caused by sludge components in the diet.

The mean trace element concentrations in kidney, liver, bone and muscle tissues of the LBR cows, are shown in Table 6. Cadmium and Pb, metals of greatest concern from a human health

Table 6. Mean Concentration of Trace Elements in Tissues of Cattle Grazing the LBR Sludge Disposal Site

Tissue	µg/g (Dry Weight)			
	Zn	Cu	Cd	Pb
Kidney	93±9	16.1±2.4	16.0 ±6	0.8 ±0.5
Liver	129±22	4.6±1.8	1.4 ±0.7	0.26±0.08
Bone	68±5	1.3±0.8	<0.01	3.1 ±1.4
Muscle	247±43	2.5±0.1	<0.04	<0.2

aspect, were not elevated over levels that would be expected (see table 5). The only tissue metal that was found to be significantly different from the control group was liver Cu. Liver Cu concentrations of the LBR cattle were significantly less than the control group of cattle. Reasons as to why the LBR sludge exposed animals had lower liver Cu concentrations are uncertain. This may be a sludge effect, however, this seems unlikely. If the LBR cattle were consuming significant quantities of sludge, liver Cu concentrations would most likely have risen from the increased dietary Cu. The most probable explanation is that the low liver Cu concentrations are a reflection of lower dietary intake of Cu.

Of the tissues examined (fat, liver, kidney, muscle, bone, brain, lung, spleen, and blood) only fat, liver, and muscle tissues showed any detectable levels of persistent organics, and except for alpha-BHC, all were below practical detection limits. Table 7 shows the concentrations of organics found in fat tissues of the LBR herd which was compared to a nearby control herd that had never been exposed to sludge. The only difference between the two herds was that the control herd showed a significantly higher level of alpha-BHC in fat tissues.

Table 7. Concentration of Refractory Organic Residues Determined in Fat Tissues of Range Cattle*

Residue	ppb (Wet Weight)	
	LBR Cattle**	Control Cattle†
HCB	<10	<10
α-BHC	10§	30
p,p'-DDE	<10	<10
Dieldrin	<10	<10
AR-1254	<500	<500

* Mean % moisture of fat tissue = 16%.
** Mean of 12 animals.
† Mean of 6 animals.
§ Significantly different at the 1% level using Student's t-test.

Evidence from this study would suggest that the disposal of sewage sludge at the LBR has had little or no impact on cattle health, or increased cattle tissue levels of the heavy metals and persistent organics examined.

HEAVY METAL AND PERSISTENT ORGANIC CONTENT OF BEEF CATTLE FED METRO. DENVER SEWAGE SLUDGE

Metro. Denver anaerobically digested primary sewage sludge was fed to 15-month old steers. Three diet mixtures were pelleted, which consisted of a basal concentrate mix (BCM) for control steers (Table 8), and two sludge mixtures for the sludge fed steers. The BCM plus sludge mixtures, were made up to approximate a final sludge diet of 0, 4, and 12% sludge. Steers (6 per treatment) were started on 0.45-kg of pellets per head per day, plus silage and hay ad lib. During a one month period, the diet was changed to 1.81-kg of

Table 8. Ingredients of Pelleted Basal Concentrate Mix

Ingredient	%
Corn	87.0
Cottonseed Meal	7.2
Molasses	5.0
Limestone	0.7
Salt	0.1
Vitamin A	*

* Vitamin polmitate to supply 40,000 IU/animal/day.

silage per head per day and pellets were fed ad lib. The amount of diet components consumed by the cattle during the 94-day trial is shown in Table 9, and the trace element composition of the sludge and consumed diet is shown in Table 10.

Table 9. Amount of Diet Components Consumed by Cattle During Metro. Denver Sludge Feeding Trial

	kg (Dry Weight)		
	0% Sludge	4% Sludge	12% Sludge
BCM	4056	4045	3497
Sewage Sludge	0	198	587
Silage	1042	1059	1031
Hay	99	102	102

Table 10. Trace Element Composition of Sludge
and Diets Amended with Metro. Denver Sewage Sludge

	µg/g (Dry Weight)			
Element	Sludge	0% Sludge	4% Sludge	12% Sludge
Zn	1,500	25.0	79.0	190.
Cd	21	0.03	0.79	2.4
Cu	710	3.2	29.0	89.0
Pb	780	0.5	29.0	80.0
Fe	12,800	138.	600.0	1,560.

Cattle ingesting sludge gained less weight than control animals. The feed efficiency (gain:feed ratio) of the sludge fed steers averaged 2.2 and 2.0% for the 4 and 12% sludge diets, which was lower than the 2.6% observed with the controls. However, the feed efficiencies were equal when calculated using non-sludge components. All animals remained healthy throughout the experiment and no signs of pathology related to the sludge component of the diet was evident. The feeding trial indicated that the anaerobically digested sludge had no positive or negative effects on general performance of the cattle other than to act as a diet diluent.

As expected, the tissues that retained the highest concentrations of all metals, except Zn, were kidney and liver. Table 11 shows the mean concentration of Zn, Cd, Cu, and Pb in kidney and liver tissues; Cu decreased slightly in kidney and increased in liver; and Pb increased in kidney, liver, and bone tissues. Zinc concentrations of all the tissues examined were for the most part unaffected by the dietary sludge. While liver Zn appeared to increase in this study, it did not increase in any of our other sludge feeding studies. Piscator (1972) and Elinder, et al, (1976) have shown that as animals age they accumulate increasing concentrations of kidney Zn and Cd. They hypothesized that both elements accumulate due to the sequestering effect the kidney has on metallothionein, which tends to bind both metals in equimolar amounts. However, increasing kidney Zn was not seen in this study, nor was it seen in either of the Fort Collins sludge feeding trials, or the control study utilizing animals of known age.

Table 11. Mean Concentration of Heavy Metals in Tissues of Cattle Fed Metro. Denver Sewage Sludge

Tissue and % Sludge in Diet	Zn	Cd	Cu	Pb
Kidney - 0	84a*	1.1a	17a	0.9a
Kidney - 4	ND†	2.5b	18a	12.2b
Kidney - 12	82a	2.4b	15b	15.8c
Liver - 0	87a	0.2a	124a	0.2a
Liver - 4	99ab	0.5b	260b	3.3b
Liver - 12	101b	0.4b	240b	4.6c
Bone - 0	64a	<0.004	1.8a	0.75a
Bone - 4	ND	ND	ND	3.7b
Bone - 12	58a	<0.005	1.4a	11.0c
Muscle - 0	253a	<0.006	2.4a	<0.2a
Muscle - 12	273a	<0.01	2.4a	<0.2a

µg/g (Dry Weight)

* Means followed by different letters are significantly different at the 5% level using Student's t-test.
† ND - Not Determined.

Fat tissues were the only tissues that showed a significant increase in persistent organics due to the sludge ingestion. Trace levels of HCB, alpha-BHC, beta-BHC, lindane, oxychlordane, and TNC were found in fat tissues from all treatments, however, significant increases of p,p'-DDE, dieldrin, and AR-1254 resulted from sludge ingestion (Table 12).

Table 12. Concentration of Refractory Organics in Metro. Denver Sludge, Feed, and Concentration in Fat Tissues of Cattle Fed Metro. Denver Sludge

ppb (Wet Weight)

Compound	Sludge* (range)	12% Sludge** Diet	Fat Tissues[†] 0% Sludge	12% Sludge
p,p'-DDE	30 - 70	<20	<20	56
Dieldrin	80 - 140	ND[§]	<10	40
AR-1254	2150 - 2850	<500	<500	1000

* % moisture of sludge ~30%.
** % mositure of feed ~10%.
[†] % moisture of fat ~11%, mean % lipid of all fat tissues was 70%.
[§] ND - Not Detected.

INGESTION OF FORT COLLINS SEWAGE SLUDGE BY BEEF CATTLE

A second feeding trial was conducted using sewage sludge from the city of Fort Collins, Colorado. The sludge was anaerobically digested primary plus waste activated which had dried in earthen drying beds for approximately one year. Two different pelleted mixtures were made, consisting of a BCM (Table 8) for the control steers, and a sludge mixture to provide a total diet of approximately 12% sludge (dry weight). The pelleted diet mixtures were fed _ad lib_, with a constant amount of roughage (1.81-kg of silage and hay per head per day). The amounts of individual diet components that were consumed by the cattle during the experiment are shown in Table 13.

Table 13. Amount of Diet Components Consumed by Cattle During the Fort Collins Sludge Feeding Trial

	kg (Dry Weight)	
	0% Sludge Diet	12% Sludge Diet
Basal Concentrate Mix	4,240	3,879
Sewage Sludge	0	652
Silage	1,084	1,095
Hay	68	68
Sludge Consumption %	0	11.5

The cattle used in this study were similar to cattle used in the first Metro. Denver sludge feeding trial. They were approximately 15 months old and were selected for uniformity of breed (Herefords), weight, and divided into two treatment groups of six animals each. The feeding trials began in November, 1976, and were terminated 106 days later in February, 1978, after the cattle had reached a finishing weight of about 465-kg.

The principal differences between the Metro. Denver sludge and the Fort Collins sludge were that the Fort Collins sludge contained 98 ppm Cd and 1700 ppm Cu (dry weight) (Table 14), whereas the Metro. Denver sludge contained 21 ppm Cd and 710 ppm Cu. Concentrations of other metals were not significantly different between the two sludges.

Table 14. Concentration of Heavy Metals Contained in Fort Collins Sludge and Sludge Amended Cattle Diets

	μg/g (Dry Weight)		
Element	Ft. Collins Sludge*	0% Sludge Diet	12% Sludge Diet
Zn	1,700	26.3	236.
Cd	98	0.14	11.3
Cu	1,700	7.7	213.
Pb	470	0.86	56.6
Fe	10,000	130.0	1,300.

* Mean of six composite samples.

As in the Metro. Denver sludge feeding trial, the sludge fed cattle gained less weight than the controls. The sludge fed animals had a lower feed efficiency ratio at 2.1% than the controls at 2.4%. However, the expected gains agreed with the observed gains when the sludge component of the diet was assumed to have no energy value; thus the sludge, as in the Metro. Denver sludge trial, appeared to act as a diet diluent. All animals remained healthy throughout the experiment and no signs of pathology related to the sludge component of the diet was detected.

The tissues that retained the highest concentrations of Cd, Cu, and Pb were the kidney and liver, while muscle tissues had the highest concentration of Zn. Table 15 shows the concentration of Zn, Cd, Cu, and Pb in kidney, liver, bone and muscle tissues. Cadmium increased significantly in both kidney and liver tissues. Lead increased in kidney, liver, and bone tissues, and Cu showed a statistically significant, although very small reduction in kidney tissues. None of the metals examined showed any increase in muscle tissues due to the sludge ingestion.

Table 15. Mean Heavy Metal Concentration in Tissues of Beef Cattle Fed Fort Collins Sewage Sludge

Tissue Type and %Sludge in Diet	μg/g (Dry Weight)			
	Zn	Cd	Cu	Pb
Kidney - 0	93a	1.2a	23a	0.95a
Kidney - 12	96a	14.0b	21b	11.0b
Liver - 0	142a	0.19a	127a	0.31a
Liver - 12	132a	4.9b	113a	4.3b
Bone - 0	86a	<0.02a	0.37a	5.0a
Bone - 12	71a	<0.01a	1.5a	7.2b
Muscle - 0	340a	<0.01a	3.6a	<0.01a
Muscle - 12	267a	<0.03a	3.2a	<0.01a

1. Means followed by a different letter are significantly different at the 5% level using Student's t-test.

Cadmium concentrations increased significantly in both kidney and liver tissues after sludge ingestion. If the mean kidney Cd concentration of 14 ppm is compared with the kidney Cd concentration of 2.4 ppm from the ingestion of a similar amount of Metro. Denver sludge, a dose response is evident; thus, a five-fold increase in dietary Cd has led to an equivalent increase in kidney Cd.

Retention of dietary Cd in the form of salts has been estimated to be somewhere between 2 and 9% (Friberg, et al, 1974; McLellan, et al, 1978; Flanagan, et al, 1978). Total body retention of Cd in this study was impossible to determine, because tissues which make up as much as 50% of the body (muscle, bone, fat) often contained levels of metals that were below our detection limits. Other tissues such as the mucosa of the alimentary canal were not examined and may have contained significant levels of metals; therefore, estimates of metal retention detailed here are not for the whole body and should not be construed as such, but rather indicate trends and differences between the sludge diets.

Estimates of Cd retention were made by estimating the average amount of increased retention in kidney and liver tissues of cattle. Estimates of Pb retention were calculated by estimating the increased retention in kidney, liver, and bone tissues. Estimates of specific tissue weights were obtained from Moulton, et al, (1922).

Table 16 shows the estimated amounts of Cd and Pb contained in kidney, liver, and bone tissues of cattle fed the sludge diets along with the consequent increases due to the ingestion of the two sludges. If the increased metal retention is compared with the amount of Cd and Pb actually consumed during the treatment periods, a measure of retention can be calculated. The steers fed Metro. Denver sludge consumed about 2-g of Cd and 70-g of Pb from the sludge. The steers fed Fort Collins sludge consumed about 10-g of Cd and about 54-g of Pb from the sludge. The estimated body retention of sludge-born Cd and Pb was 0.02% and 0.85% for the

Table 16. Estimated Amount of Cd and Pb Contained in Kidney, Liver, and Bone Tissues of Cattle Fed Diets of 0 or 12% Sewage Sludge

	\multicolumn{3}{c}{mg}			
	Metro Denver Sludge			Ft. Collins Sludge		
Element	0%	12%	Increase Due to Sludge	0%	12%	Increase Due to Sludge
Cd	0.43	0.82	0.39	0.44	8.4	7.9
Pb	44.0	633.0	589.0	285.0	407.0	122.0

Metro. Denver sludge and 0.07% and 0.23% for the Fort Collins sludge. Thus, it appeared that the Cd contained in the Fort Collins sludge was somewhat more available for absorption and retention into cattle tissue than the Metro. Denver sludge. While the retention values shown here are admittedly less than "true" retention values, it is doubtful that they are off by much more than one order of magnitude; thus, it would appear that Cd contained in both sewage sludges is less available for absorption and retention than that which has been shown in previous studies utilizing Cd salts (McLellan, et al, 1978; Flanagan, et al, 1978; Doyle, et al, 1974).

Abdominal fat tissues showed significant increases in a few residues due to the ingestion of Fort Collins sludge. Table 17 shows the mean concentration of refractory organics detected in the sludge and the increase in fat tissues due to the sludge ingestion. Apparent increases in p,p'-DDE, α-BHC, dieldrin, oxychlordane, HCB, and AR-1254 resulted from sludge additions to the diet. All of these organic compounds, as in the Metro Denver sludge feeding trial, are below tolerance levels that have been established by FDA for fish or other

Table 17. Mean Concentration of Refractory Organics Determined in Fort Collins Sludge and Fat Tissues of Cattle Fed Fort Collins Sludge*

	ppb (Wet Weight)			
			Fat Tissues	
Compound	Sludge**	12% Sludge Diet	0% Sludge	12% Sludge
p,p'-DDE	121	<10.0	<10	74
α-BHC	<4.0	<10.0	ND[†]	<10
β-BHC	<19	ND	ND	ND
Dieldrin	39	<10	<10	30
Heptachlor Epoxide	ND	ND	<10	<10
Oxychlordane	28	ND	<10	18
TNC	ND	ND	<10	<10
HCB	<10	ND	ND	<10
Chlordane	1047	ND	ND	ND
AR-1254	4434	<500	ND	1672

* % moisture of sludge ~5%, % moisture of fat ~11% with mean fat content ~84%.
** Mean of 6 samples.
† ND - Not Detected.

food products. Tolerance levels, however, have not been established for TNC, HCB, or PCB's in red meat.

TRACE ELEMENT CONTENT OF BEEF CATTLE TISSUES AFTER A NINE MONTH PERIOD OF SLUDGE INGESTION

It is known that young animals absorb greater percentages of dietary heavy metals (particularly Pb) than older animals, but the extent to which this might occur in cattle has been unknown. In the previous studies described, feeding trials were conducted utilizing young animals that presumably would have higher trace metal uptake rates than older animals. In a typical cow-calf operation where sludge would be added to pasture land, the cows would be exposed to sewage sludge for a number of years while the younger cattle would be sold off. Thus, the greatest exposure to sewage sludge would be to those breeding animals that are kept for a number of years.

Another aspect which was not examined in the previous studies was the possibility that the cattle could eliminate some portion of the sludge-born heavy metals or persistant organics after being removed from the sludge diet. The U.S.D.A. had reported severe joint deterioration in cattle from grazing pastures after surface applications of liquid sludge (Feasibility of Using Sewage Sludge for Plant and Animal Production, 1976-1977). Subsequent work showed that the joint deterioration they observed was probably related to the high iron sludge that was being used in their grazing studies (Feasibility of Using Sewage Sludge for Plant and Animal Production, 1978-1979). But, because of their original findings, we were interested in the health effects to cattle from long term exposure to sewage sludge.

Two groups of cattle (16 each) were selected for this study; one group of older Hereford cows ranging in age from 4 to 7 years, and a younger group of 6 month old Hereford steers. The cows were lotted into two uniform groups of 8, on the basis of weight and age, so that each group had an average age of 5. Steers were lotted into two groups of 8 according to weight. One group of cows and steers were placed on a control diet and one group of cows and steers were fed a diet containing approximately 12% (dry weight) Fort Collins anaerobically digested sludge for a period of 9 months. The cow diets were fed at approximately maintenance levels, while the steer diet was fed and adjusted to maintain a 0.77-kg per head per day weight gain over the entire feeding period. The amounts of diet components consumed by the cattle during the 9 month sludge feeding period are shown in Table 18.

Table 18. Amount of Diet Components Consumed by Cattle During Nine Month Sludge Feeding Trial

	\multicolumn{4}{c}{kg (Dry Weight)}
	Cows		Steers	
	0%	12%	0%	12%
Silage	9,689	9,392	11,772	10,205
Alfalfa	1,142	1,144	1,528	1,170
Protein Supplement	0	0	973	886
Sludge	0	1,571	0	1,690

Both the control cows and calves gained slightly more weight than the sludge fed animals. Control cows had an average daily gain of 0.3-kg versus 0.2-kg for the sludge fed cows. Control steers had an average daily gain of 0.8-kg versus 0.7-kg for the sludge fed steers. These results were similar to the other sludge feeding trials in that the sludge showed no positive or negative effects on cattle performance, other than to act as a diet diluent.

At the conclusion of the 9 month sludge feeding period, 4 animals from each group were slaughtered, and samples of kidney, liver, and muscle tissues taken for heavy metals analysis. The remaining animals were kept on test for another 4 months, but sludge was removed from the diet. These 16 remaining animals were then slaughtered and sampled at the end of this 4 month "withdrawal" period.

All animals remained healthy throughout the trial with no overt signs of pathology. Inspection of bone joints at the time of slaughter revealed no joint deterioration.

Table 19 shows the concentration of heavy metals determined in kidney tissues of the cows and steers after 9 months of sludge ingestion, and 4 month "withdrawal" period.

Table 19. Heavy Metals Concentrations in Kidney Tissues after Nine Months of Sludge Ingestion and Four Month Withdrawal Period

Treatment	μg/g (Dry Weight)				
	Cd	Pb	Zn	Fe	Cu
Cows:					
Control*	7.4a[§]	1.4a	77a	320a	17a
Sludge Fed**	54.0b	4.3b	88a	281a	15a
Sludge Withdrawn[†]	69.0b	3.4b	131b	343a	17a
Steers:					
Control	3.5a	1.1a	82a	255a	19a
Sludge Fed	57.0b	5.2b	98a	290a	16a
Sludge Withdrawn	64.0b	3.4b	116a	247a	16a

* Average of all control animals, 9 month and 4 month withdrawal period.
** Cattle fed diet containing 12% sludge and slaughtered at 9 months.
[†] Cattle fed diet containing 12% sludge and slaughtered after 4 month withdrawal period.
[§] Values within a group followed by different letters are significantly different at the 5% level using Student's t-test.

There were significant increases in kidney Cd and kidney Pb, while Zn, Fe, and Cu did not appear to be affected by the sludge ingestion. Kidney Zn showed a significant increase in the cows removed from sludge for 4 months, but it is doubtful that this was a true treatment effect. There appeared to be no differences in uptake rates between the cows and young steers, and there was no evidence that Cd or Pb concentrations decreased after sludge was removed from the diet.

Table 20 shows the concentrations of metals found in liver tissues from these animals. Cadmium, Pb, and Cu increased significantly due to the sludge ingestion. Copper and Pb significantly decreased during the sludge withdrawal period, while Cd remained unchanged. As in the kidney, the cows showed a significant increase in liver Zn after being

Table 20. Heavy Metals Concentrations in Liver Tissues after Nine Months of Sludge Ingestion and Four Month Withdrawal Period

	μg/g (Dry Weight)				
Treatment	Cd	Pb	Zn	Fe	Cu
Cows:					
Control*	1.22a[§]	.65a	108a	434a	63a
Sludge Fed**	19.0b	4.9b	118a	525a	212b
Sludge Withdrawn[†]	22.0b	1.2c	141b	596a	118c
Steers:					
Control	0.91a	<0.5a	116a	228a	32a
Sludge Fed	20.0b	4.1b	138a	330a	393b
Sludge Withdrawn	18.0b	1.5c	130a	234a	126c

* Average of all control animals, 9 month and 4 month withdrawal period.
** Cattle fed diet containing 12% sludge and slaughtered at 9 months.
[†] Cattle fed diet containing 12% sludge and slaughtered after 4 month withdrawal period.
[§] Values within a group followed by different letters are significantly different at the 5% level using Student's t-test.

withdrawn from the sludge, however, it doubtful that this was a sludge effect. The uptake rates and depletion rates of Pb and Cu did not appear to be different between the young steers and older cows.

Table 21 shows the heavy metal concentrations found in muscle tissues. From the data, it appears that Cd significantly increased in muscle tissues due to the sludge ingestion, with no significant decrease in the cows during the "withdrawal" period, however, there was an apparent decrease back to control levels in the steers. There were no other differences in metal concentrations between the treatments.

Table 21. Heavy Metals Concentrations in Muscle Tissues after Nine Months of Sludge Ingestion and Four Month Withdrawal Period

	μg/g (Dry Weight)				
Treatment	Cd	Pb	Zn	Fe	Cu
Cows:					
Control*	0.12a[§]	<0.4a	280a	87a	3.7a
Sludge Fed**	0.27b	<0.2a	265a	98a	3.9a
Sludge Withdrawn[†]	0.21b	<0.4a	286a	92a	2.9a
Steers:					
Control	0.14a	<0.4a	293a	86a	4.4a
Sludge Fed	0.43b	<0.5a	280a	89a	5.5a
Sludge Withdrawn	0.15a	<0.3a	297a	81a	3.2a

* Average of all control animals, 9 month and 4 month withdrawal period.
** Cattle fed diet containing 12% sludge and slaughtered at 9 months.
[†] Cattle fed diet containing 12% sludge and slaughtered after 4 month withdrawal period.
[§] Values within a group followed by different letters are significantly different at the 5% level using Student's t-test.

CONCLUSIONS

All of these studies have shown that the direct ingestion of sewage sludge by cattle will increase the levels of sludge contaminants of tissues to some extent. The amount of increased metal or persistent organic uptake is in proportion to the metal or contaminant level of the sludge. Sludges from different sources or different metal concentrations may effect to some extent the availability of the metals for absorption and retention in cattle tissues.

However, if sewage sludge is incorporated into the soil and utilized as an agricultural fertilizer, there would appear to be little hazard that heavy metals or persistant organics would be elevated in cattle tissues above levels that would be considered normal. These studies demonstrated that there were no detrimental health effects to the cattle from exposure or ingestion of rather large amounts of sludge, even for extended periods of time.

REFERENCES

1. Boyle, E.A. and J.M. Edmond. "Determination of Trace Metals in Aqueous Solution by APDC Chelate Co-precipitation," Anal. Meth. in Oceanography, (1975), pp. 44-45.

2. Doyle, J.J., W.F. Pfander, S.E. Grebing, and J.O. Pierce. "Effects of Dietary Cadmium on Growth, Cadmium Absorption, and Cadmium Tissue Levels in Growing Lambs," J. Nutr. 104:160-166 (1974).

3. Elinder, C.G. and M. Piscator. "Cadmium and Zinc in Horses," to be published in Skall publiceras i Proc. 3rd Int. Symp., Trace Element Metabolism in Animals and Man (1978).

4. "Analysis of Pesticide Residues in Human and Environmental Samples," Environmental Protection Agency, Pesticides and Toxic Substances Effects Laboratory, National Environmental Research Center, U.S. E.P.A. (Research Triangle Park, N.C., 1974).

5. "Feasibility of Using Sewage Sludge for Plant and Animal Production," Cooperative Research Project of University of MD and U.S.D.A., Final Report 1976-1977. Wash. Suburban San. Comm.

6. Ibid., Final Report 1978-1979. Wash. Suburban San. Comm.

7. Feldman, C. "Perchloric Acid Procedure for Wet-Ashing Organics for the Determination of Mercury (and other metals)," *Anal. Chem.* 46:1606-1609 (1974).

8. Flanagan, P.R. et al, "Increased Dietary Cadmium Absorption in Mice and Human Subjects with Iron Deficiency," *Gastroenterology*. 74:841-846 (1978).

9. Friberg, L., M. Piscator, G.F. Nordberg and T. Kjellstrom. *Cadmium in the Environment*, 2nd ed. (CRC Press, Inc., 1974).

10. McLellan, J.S., P.R. Flanagan, M.J. Chamberlain, and L.S. Valberg. "Measurement of Dietary Cadmium Absorption in Humans," *J. Toxic. and Environ. Health* 4:131-138 (1978).

11. Moulton, R., P.F. Trowbridge and L.D. Haigh. "Studies in Animal Nutrition. II. Changes in Proportions of Carcass and Offal on Different Planes of Nutrition," Agri. Exp. Sta. Res. Bull. #54. University of Missouri (1922).

12. Piscator, M. and B. Lind. "Cadmium, Zinc, Copper, and Lead in Renal Cortex," *Arch. Environ. Health*. 24:426 (1972).

13. Underwood, E.J. *Trace Elements in Human and Animal Nutrition* (New York: Acad. Press, 1977), p. 58.

CHAPTER 14

Health Effects of Sewage Sludge for Plant Production or Direct Feeding to Cattle, Swine, Poultry or Animal Tissue to Mice

G.T. Edds, D.V.M., Ph.D.*, O. Osuna, D.V.M., Ph.D.*, C.F. Simpson, D.V.M., Ph.D.*, Plus Cooperating Project Leaders**

INTRODUCTION

The research program planned, and partially supported by EPA, included determination of the fate of various toxic elements, bacterial and viral pathogens with municipal sludge application to soil-water-plant systems. This involved major physical, chemical transformations and transport of such elements in selected soil types. Another major area examined was the bacterial, viral, pathologic, growth and reproductive responses of cattle, swine and poultry fed forages or grains from soils receiving sludges or fed dried sewage sludges or recycled cattle manure. Liver, kidney and muscle tissues were examined for elements translocated through the plants or grain or retained from the sludges when fed directly or when pure chemicals were substituted for the levels in the treated diets. The sludges and field-grown feeds used were examined for presence of bacteria, drugs, mycotoxins, parasites, metals and viruses.

Finally, liver and kidney tissues were collected from control and treated groups of both cattle and swine at termination of the feeding trials, lyophilized, powdered and incorporated into the diets of mice for growth and reproductive performance studies. The cadmium and lead levels present in the collected tissues as well as in the liver and kidneys of the 3 generation mouse trials were determined.

CATTLE TRIALS

Early trials with beef cattle consuming feeds grown on soils supplemented with liquid Pensacola sludge, LPS, or

---

*Faculty, Department of Preventive Medicine, University of Florida, Gainesville, FL 32610.
**J. E. Bertrand, D. L. Hammell, C. E. White, B. L. Damron, R. L. Shirley and K. C. Kelley.

rations to which 100, 250 or 500 g dried Pensacola sludge, DPS, per head per day were added for feeding periods of about 5 months indicated there were neither differences in growth rates, feed efficiency nor carcass quality at slaughter. However, when Dried Chicago Sludge, DCS, was added to the cattle rations at 500 g. per head per day or when the corn in the ration was replaced with corn from soil previously amended with more than 3 acre inches equivalent DCS prior to planting delayed adverse health effects were observed.

Analysis of Variance showed that gammaglutamyl transferase γGT, and glutamicoxalacetic transaminase, GOT, were significantly increased ($p<0.5$ and 0.1 respectively) for the steers receiving the 500 g. DCS per head per day or when fed corn grown on the DCS amended soil. Clinically, this suggests that the presence of higher levels of the toxic metals, cadmium or lead, or presence of increased damage to the liver from the fluke infection or lowered resistance resulting in bacterial infection with increased white cell count may have contributed to the rise in serum enzyme levels.

Sludge, feed and serum collected during these trials were shown to be free of 8 sulfonamides, sulfanilamide, sulfathiazole, sulfisoxazole and acetyl-sulfisoxasole and 3 tetracyclines, chlortetracycline, tetracycline hydrochloride and oxytetracycline except where intentionally included.

The 1979 steer trial allowed the control animals to graze untreated pasture while one group of steers grazed on forage from soils pretreated with 1½ acre inches LPS sprayed on during the trial. The other group of steers consumed forage from soils pretreated with 3 acre inches before planting and treated with 3 acre inches during the trial.

Among cattle grazing on the untreated forage and at slaughter only 6 of 17 or 35% contained sarcosporidia in the cardiac muscle. The cardiac muscles of those grazing forage sprayed with 1½ acre inches PLS or 3 acre inches PLS contained sarcocysts in 19 of 32 or 60% of the steers. This preliminary report of the association of increased incidence of sarcosporidia in animal tissues grazing forage sprayed with liquid sludge may partially account for high incidences reported in the literature in feedlot cattle.

This _may be_ of _public health significance_.

Incorporation of dried Pensacola sludge, DPS, in the diets of steers and the consumption of forages produced on land treated with DPS consistently decreased the liver concentrations of copper and zinc. The iron absorption and/or storage was also affected. Increased levels of lead were observed in both liver and kidney tissues of steers fed DPS in their ration. Animals consuming rations supplemented with DCS contained higher than normal levels of both cadmium and lead in the liver and kidneys.

SWINE

An early trial performed at Florida, 1976, evaluated the effect of incorporation into the rations of 4 gilts/group containing either 0, 10, or 20% dried University of Florida sludge, DFS, in a metabolism trial. Later, thirty-three York-Hampshire crossbred gilts were divided into 3 groups, 11 each and fed one of these levels of DFS for 12 months. Milk from lactating sows and tissues from their randomly selected weanling pigs, market weight offspring and dams from dietary groups were analyzed for lead, cadmium, nickel, copper, chromium, zinc, manganese, aluminum and iron.

Seventy-two weanling pigs of each first and second litters were randomly selected, divided into 3 groups and fed either 0, 10 or 20% DFS-containing growing-finishing diets. At market weight, animals from each group were slaughtered and blood, liver, kidney, spleen and muscle tissues collected for assay.

The mean values for total digestible nutrients were 79.4, 73.7 and 55.0%; those for metabolizable energy were 3.36, 2.25 and 1.15 Mcal per kg. diet; and those for nitrogen retained were 42.8, 44.0 and 25.3%, respectively. Feeding of the sewage sludge diets to sows resulted in more live pigs being farrowed and weaned per litter from sows fed the 20% sludge diets. However, the 21 day weaning weights were lower in pigs from sows consuming the sludge-containing diets. Growth performance in weanling pigs to market weight was less in those on the sludge-treated rations. There were no increases in the 9 mineral elements in sow blood or milk samples. However, the kidney cadmium levels of the sows receiving 10 or 20% levels of DFS were significantly increased, i.e., 4 ppm vs 17 and 24 ppm; both lead and cadmium were increased in the livers and kidneys of weanling pigs but not in the blood nor muscle. Again, the levels of cadmium were significantly increased in the liver and kidney tissues of the finishing swine at slaughter; liver levels were 0.33, 0.84 and 3.18 ppm. dry basis; and kidney levels, 1.40, 8.60 and 12.5 ppm., respectively.

A further trial was performed in 1977 at Live Oak, Florida which paralleled the above trial but the 33 gilts were fed in outside dirt lots with no access to pasture. Gilts in the 3 sludge-ration groups, 0, 10 and 20% DFS were allowed 5 pounds of feed daily but during lactation the sows were fed free-choice. Offspring were continued on the same diets as their dams and fed ad libitum. At slaughter, 6 sows or 6 pigs were selected and liver, kidney and muscle samples taken for metal analyses. In other trials, dried Chicago sludge, DCS, was substituted into the sludge amended rations. (See Figure 1)

SUMMARY OF REPRODUCTIVE PERFORMANCE OF GILTS FED SLUDGE

Sewage sludge, % DGS	0	10	20
Gilts mated	11	11	11
Gilts farrowed	11	9	9
Breeding wt., lbs.	265.4	262.6	253.9
Farrowing wt., lbs.	372.8	352.6	332.1
Total pigs farrowed/litter	10.54	9.44	9.67
Live pigs farrowed/litter	9.18	8.11	9.00
Birth weight/pig (live), lbs.	3.10	3.27	2.94
Weaning weight/pig (21-day), lbs.	11.86	11.52	10.39
Pigs weaned/litter	7.91	6.89	7.22

Gilts selected from 2 successive generations were fed diets containing 0, 10 or 20% DFS in a basal corn-soybean meal formation. Second generation females were fed the same dietary regimen as their dams.

Such continued feeding over this extended period adversely affected many criteria used in evaluating reproductive performance. Breeding, farrowing and rebreeding weights were reduced. Lactation and gestation weight changes were lower with fewer pigs surviving at 21 days as compared to those from sows fed the basal diet. In each parity, pigs farrowed by sows fed DFS in diets displayed depressed average daily gains. Reproductive performance was more suppressed in the second generation sows than in the first.

Finally, in the last breeding and performance trial at Live Oak, the 3 groups of sows fed rations supplemented with DFS at 0, 10 or 20% levels were slaughtered after farrowing. As before, liver, kidney and muscle tissues were collected for metal analyses. In this instance, special attention was given to examination for the possible translocation of sarcosporidial infection by way of the DFS. No sarcocyts were observed in cardiac tissues from the sows on the control ration while of 7 sows fed 10% DFS rations, 4 had sarcosporidia (57%) and 2 of 4 sow's hearts fed the 20% DFS or 50% had sarcosporidia

SUMMARY OF REPRODUCTIVE PERFORMANCE OF SECOND LITTER
SOWS FED SLUDGE

Sewage Sludge, %, DFS	0	10	20
Sows mated	11	11	11
Sows farrowed	10	11	10
Breeding wt., lbs.	317.0	324.2	289.8
Farrowing wt., lbs.	422.4	418.0	357.1***
Total pigs farrowed/litter	10.6	10.0	9.1
Live pigs farrowed/litter	9.90	9.63	8.30
Birth weight/pig (liver), lbs.	3.32	3.48	3.24
Weaning weight/pig (21-day), lbs.	11.25	20.70	11.54
Pigs weaned/litter	8.00	6.27*	6.30*

*** 357.1 significantly less (P<0.1) than 418.0 and 422.4
* 6.30 and 6.27 significantly less (P<.10) than 8.00

Sludge and Ration Mineral Analysis

		Grower or Finisher Diet	
Mineral ppm	DCS	Control	10% Sludge	20% Sludge
Hg	1.79	0.02	1.85	5.05
Cu	1330.00	18.90	107.10	200.94
Fe	49862.00	236.60	2616.00	5244.40
Zn	3230.00	236.70	349.10	557.66
Cr	3350.00	2.43	179.85	324.12
Ni	380.00	2.72	31.77	57.12
Cd	216.00	0.10	9.26	17.15
Co	17.91	2.00	2.74	4.31
Pb	715.00	1.00	38.76	86.10

Detection Limits	PPM Dry Matter
Aluminum	0.50
Cadmium	0.25
Cobalt	0.10
Chromium	0.10
Copper	0.05
Iron	0.05
Lead	0.10
Magnesium	0.01
Manganese	0.05
Mercury	0.002
Nickel	0.10
Selenium	0.10
Zinc	0.01

Sludge and Ration Mineral Analysis

		Gestation Ration		
Mineral ppm. DM.	DFS	Control	10% Sludge	20% Sludge
Cu	555.9	8.65	46.43	102.10
Fe	9366.6	269.00	3750.00	984.00
Zn	1216.7	123.22	568.33	303.83
Cr	217.6	1.46	18.96	37.52
Ni	24.9	1.04	3.00	7.61
Cd	9.1	.365	3.81	2.21
Co	7.9	1.67	4.17	4.17
Pb	416.8	0.83	19.58	76.25

First Generation
N=11 (0% DSS)
N=11 (10% DSS)
N=LL (20% DSS)

Parity 1   Parity 2   Parity 3

Second Generation
N=11 (0% DSS)
N=11 (10% DSS)
N=11 (20% DSS)

Parity 1   Parity 2

Figure 1. Diagram of experimental design used to evaluate effect of feeding digested sewage sludge on long-term sow reproductive performance

POULTRY

An experiment on growth was performed in broiler chicks utilizing corn grain harvested from soils at Jay which had been previously amended by adding 22.5 cm. of PLS, prior to planting. This corn was used to replace either 50% or 100% of the corn in the broiler diet as compared to the effects of corn from plots fertilized with commercial fertilizer. Feed and water were provided ad libitum. At the end of a 21-day feeding period, the 30 chicks from each group were weighed and feed consumed determined. With one exception, there were no significant differences between the growth performance of the chicks in the 3 groups.

A parallel trial was performed in which Leghorn hens, at 6 months of production were provided a laying diet containing 0, 50 or 100% replacement of regular corn in their diets with the "PLS" corn. Two trials were run, one at 84 days, the other at 112 days. Eight replicate groups of 5 individually caged hens were assigned to each of the 3 dietary treatments. The performance data indicated that the partial or total substitution of sludge fertilized corn for that produced with commercial fertilizer had no significant effects on laying or egg hatchability.

Further growth trials with Cobb broiler chicks compared the effects of poultry rations with 0, 3 and 6% Dried Chicago Sludge, DCS. The 6% sludge level was chosen as a maximum amount which would not result in altered nutritional values. In addition, 4 other treatments were included containing the amounts of cadmium, chromium, copper and iron equivalent to that present in the 6% DCS. The diet containing the iron equivalent, 2922 mg/kg in the feed, produced significant depression of body weight as compared to the other treatment groups and the controls. Addition of an equivalent level of chromium also significantly depressed bodyweight.

Tissue data indicated no accumulation of minerals occurred in muscle tissue by feeding of sludge nor any of the supplemental elements. There were increased levels of cadmium in the liver and kidneys in those chicks receiving the increased DCS levels.

In a similar trial, Leghorn hens were provided a layer ration containing 0, 3.5 and 7% DCS in the basal diet. None of the production criteria, i.e., production, daily feed intake, feed efficiency, egg weights nor body weights were significantly affected.

The potential hazards in urban sewage sludge may be minimized by its use for soil amendment with the grain or forage utilized in animal feeds. It was believed important to establish the toxic effects of some of the sludge ingredients in 2 or more animal model systems to evaluate the potential hazard to human health. As indicated earlier, direct feeding trials in cattle, swine and poultry indicated some of the metals present, especially lead and cadmium, accumulated in liver and kidney tissues at above normal levels. Therefore, a series of trials were performed in swine using the Dried Chicago Sludge, DCS, or equivalent levels of cadmium for determining the direct effects on health.

COMPARISON OF SLUDGES VS EQUIVALENT CADMIUM LEVELS IN SWINE

In a preliminary experiment, 1977, it was demonstrated that young male pigs consuming swine starter rations containing 50% DCS for an 8 week period produced signs of toxicity including decreased weight gains, leucocytosis, lowered erythrocyte counts and packed cell volume along with increased serum transaminase levels. Liver and kidney assays for lead and cadmium showed high levels with lowered levels of copper and zinc as compared to normal values in the control groups. In part, the progressive liver damage may have resulted from a deficiency of available protein, other essential nutrients and vitamins. The starter ration contained a protein level of 17.2% while the sludge ration contained only 9.0%. Such lesions, regardless of cause, may have predisposed to immunosuppression in the pigs exposed to the high levels of cadmium and lead.

Cadmium interferes with the availability and usefulness of necessary elements in body enzyme pathways. A lethal dose of cadmium may inhibit mitochondrial oxidative phosphorylation and lead to death. Other toxic effects include anemia, enlarged joints, scaly skin, kidney damage and testicular degeneration, reduced growth rate and increased mortality. Neither blood nor urine are useful for assaying for cadmium hazard; highest concentrations are found in the kidneys.

One lot of DCS was assayed for cadmium; the level present was 165 ug/g. Forty eight, 4 week-old hybrid pigs were allotted to 2 replicates of 2 treatments each, 12 pigs per group. The pigs were fed a basal starter ration, 18% crude protein, or a 50% DCS ration or a 83 ug/g cadmium ration. Feed and water were provided <u>ad libitum</u> for 9 weeks. Body weights, blood and feed samples were taken weekly for analysis. Liver, kidney and muscle samples were collected for metal analysis and pathologic studies.

Depressed growth and feed consumption were evident in groups receiving 83 ug/g cadmium ration as compared to controls. An extreme microcytic, hypochromic anemia was observed by day 42 of the trial ($P < 0.0001$) in the cadmium supplemented group. In the second replicate, cadmium was again significantly higher in the tissues of the pigs consuming the DCS ration than in the controls.

Cadmium was at significantly higher levels in the liver and kidney tissues of pigs on the $CdCl_2$ supplemented ration; they were also higher than the control levels in those consuming the DCS mixed ration. Also, in the DCS supplemented rations, the iron levels were 8846.9 ug/g in the control; not in the $CdCl_2$ supplemented rations. Thus, the DCS sludge could provide a supply of iron to prevent development of the iron-deficiency anemia.

The purpose of using 0 or 83 ug/g cadmium diets in pigs was to determine whether animals might be predisposed to higher health risk when exposed to such feeds containing a naturally occurring feed contaminant aflatoxin B_1, or warfarin.

CADMIUM EFFECTS ON AFLATOXIN-WARFARIN TOXICITY IN SWINE

Thirty-six weaned barrows, averaging 9 kg BW, were assigned at random to 6 treatment groups, 6 per group. Cadmium (Cd) was provided daily through the diets in 3 groups during the 40 day experiment. Aflatoxin B_1 or Warfarin was given at 0.2 mg/kg per os for 5 days to treated and nontreated Cd groups during the 5th week and effects followed for 10 days.

Aflatoxin B_1 in the 0 cadmium group significantly increased values of AP, SDH, aspartate aminotransferase (SGOT), PT, activated partial thromboplastin time (APTT) and significantly decreased values in serum total protein, α-globulin, β-globulin, γ-globulin and fibrinogen. The hepatotoxic effects of aflatoxin B_1 correlated with lobular fatty infiltration in 3 out of 4 pigs in the group. Aflatoxin B_1 was very toxic as was Warfarin.

Warfarin in the 0 Cd group was more effective in producing earlier and significantly higher values in PT and APTT than those receiving AFB_1 by the second day. Values in the pigs receiving Warfarin returned to normal after dosing had been suspended, while those in the aflatoxin B_1 group increased significantly by the 10th day. No significant liver or kidney damage was induced by Warfarin.

Cadmium "treatment" prevented the fatty hepatocytic infiltration but not hydropic degeneration induced by aflatoxin B_1. Cadmium treated diets resulted in differences in the intenstiy and duration of response on PT and APTT when pigs were dosed with aflatoxin B_1 or Warfarin.

The Cd containing diet resulted in 35.13 ug/g and 6.80

ug/g Cd tissue accumulation in kidney and liver respectively at the 4th week of the experiment as compared to nondetectable levels in the controls.

This was the first demonstration of the Cd blocking effect on the microsomal enzyme system in pigs. There were increased concentrations of Cd in tissues of the treated groups associated with decreased iron utilization resulting in microcytic hypochromic anemia. Therefore, there is an inhibitory metabolic effect on aflatoxin B_1, and enhanced synergistic toxic effect with Warfarin when Cd is present in the diets of young pigs at 83 ug/g.

HEALTH EFFECTS OF METAL RESIDUES FROM CATTLE AND SWINE ON MICE

Animal tissue intended for human consumption, including muscle, liver and kidney may, on occasion, contain higher than normal levels of cadmium, lead, chromium, nickel, or iron with reduced levels of copper and zinc. Such imbalances could predispose to faster accumulation of cadmium or lead in human body tissues. Having demonstrated that increased levels of cadmium were, in fact, present in cattle and swine consuming feeds from sludge amended soils, the following experiment was performed in mice.

Kidney and livers were collected from cattle and swine at termination of the 1978 and 1979 trials and frozen. These frozen tissues were sliced on a hand-saw, ground in a Hobart grinder, freeze-dried, powdered and assayed for protein contents. The final ration contained 5% kidney tissue and adequate casein to yield a diet with a protein level of 15%. The control ration also contained 15% protein equivalent.

Swiss mice, 45 days of age, were fed freeze-dried liver and kidney from cattle or swine fed a control diet, sewage sludge, corn grain or sorghum forage, or from land treated with liquid sewage sludge. Kidney and liver of F_0 mice had increased levels of Cd when they were fed diets that contained liver or kidney from cattle fed 500 g/head/day of DCS in their diets. Copper in muscle of mice was depressed in sludge treatments. Copper, Fe, Co, Zn, Pb, Hg, Se, Cr and Ni showed no differences due to prior sludge treatments in mice kidneys, livers nor muscles. Male mice had a higher concentration of Cu, Fe, Co and Pb in their kidney than mice fed 10% cattle liver in diets. Nickel was higher in kidney, liver and muscle; and Cr higher in kidney of mice fed diets that contained 5% kidney from cattle. F_1 females fed kidney had a decrease in number weaned due to DCS; F_1 females fed liver had a decrease in number born in both sludge treatments.

F_0 mice fed kidney and liver from cattle fed sorghum

that received 15.2 or 22.8 cm/ha PLS showed no changes in minerals in kidney, liver or muscle due to treatment or diet. F_0 males had an increase in liver Cu but no changes in minerals occurred in kidney and muscle. Number born per litter was greater in F_0 mice fed kidney in the 22.8 cm/ha PLS treatment but there was no effect when liver was fed. The number weaned was not affected.

Liver from swine F_0 fed 0, 10 or 20% University of Florida digested sewage sludge (DFS) in diets of mice showed an increase in the Pb content of mice livers and muscles at the 20% level while Cr was lower in liver tissue; in kidney from swine, female mice had higher levels of Fe than males.

Mice fed liver from swine F_{1-2} in diets had higher levels of Cd in liver, kidney and muscle tissues in the 10% DFS group. Pb in mice liver was increased in the 10% DFS and in muscle in both the 10 and 20% groups. Female mice had a higher level of Ni in kidneys than males. Mice fed kidneys from swine had high levels of Cd in liver, but no changes due to diet were observed in kidney and muscle.

Abstracts of Research by Others on Project
Please Refer to Specific Papers

Bitton, Farrah et al, "Enteroviruses were readily recovered from sludges, including lagooned sludge at Jay. No virus could be detected following topsoil monitoring for 8 months. Studies with soil columns under saturated flow conditions have shown that sludge associated viruses are well retained within the soil matrix. The virus sludge complexes are effectively retained in the soil surface which may explain why viruses were not detected from soil cores or lysimeters Virus survival in sludge-amended soils is controlled primarily by desiccation and soil temperature. Moreover, enteroviruses were able to survive only up to 9 days following sludge application at the Jay, Florida site. Methods used during this trial were efficient for detecting enteroviruses. Rotaviruses, associated with infant diarrheas, should also be carefully monitored."

Farrah, Bitton et al, "While enteroviruses were readily detected in grab samples of sludge from the lagoon, they were not detected in water from deep wells located on the sludge disposal site or near the lagoon."

Hortenstine - "Sludge applications increased soil Zn, Cd, Cu, Pb and Ni along with an increase in Ca with some movements of other elements below the 15 cm. depth. Sludge application at 24 ton/ha rate compared favorably with mineral fertilizer as a source of plant nutrients."

Richter - "Inclusion of dried cattle manure in feeder cattle rations at 20% for approximately 200 days produced improvement in growth performance with no evidences of toxic effects on carcasses or their quality."

Thompson - "Little if any chlorinated hydrocarbon residues were present in sludge used for research in this project."

SUMMARY

1. Digested urban sewage sludges from Pensacola and Chicago, when applied to soils, were equivalent to commercial fertilizers in production of corn, soybeans or sorghum as animal feeds.
2. Incorporation of such forage or grain into cattle rations produced no significant growth differences nor carcass quality from those fed such products from commercially fertilized soils.
3. However, after 4 to 5 months, some evidence of liver or kidney damage occurred as evidenced by increased serum levels of released cellular enzymes.
4. Cattle consuming pasture or forage sprayed with Pensacola Liquid Sludge had higher incidences of the protozoa sarcosporidia in cardiac muscle than in the control animals.
5. There were no indications of increased incidence of residue hazards due to bacteria, viruses, parasites, pesticides, drugs nor mycotoxins in the "treated" groups.
6. Addition of dried sewage sludges to the rations of swine resulted in suppression of growth, feed efficiency, and reproductive performance.
7. While the animals on control rations were normal, those on DFS amended rations developed significant numbers of sarcospiridia in the cardiac muscle tissue. This may be of public health significance.
8. Incorporation of equivalent levels of cadmium as cadmium chloride in swine rations resulted in hypochromic, microcytic anemia due to iron-deficiency. A similar level of cadmium provided in sludge did not produce anemia; in fact, the "chelated" iron present corrected anemias.
9. Broilers or layers fed grain from soils pretreated with urban sewage sludge were not affected as to growth, egg laying or hatchability.
10. When liver or kidney tissues were collected from steers or swine at slaughter that had been previously fed DCS or DFS, higher levels and/or cadmium were present.
11. Incorporation of liver or kidney tissues into mouse diets at 5% levels, 15% protein levels in both controls and modified diets, resulted in reduced reproductive performance, along with increased levels of cadmium and lead in their liver and kidney tissues. No pathologic lesions were observed.
12. These results suggest further research on translocation of hazardous metal residues as well as sarcosporidia from urban sludges should be performed.

SECTION V

ABSTRACTS

A. University of Florida Project

Bertrand, J.E. et al., Agric. Res. Ctr., Jay, FL

"HEALTH EFFECTS OF SEWAGE SLUDGE AND FEEDS FROM SLUDGE-TREATED SOILS WITH BEEF CATTLE"

Beef steers were fed digested municipal sludges incorporated into feedlot diets and feeds (corn grain, forage sorghum silages, and bahiagrass pastures) produced on land treated with sludge to determine the effects on animal performance, carcass quality, and concentrations of selected metals in liver, muscle, and kidney tissues. The performance and carcass data of treated steers in all of the studies were not different from the data obtained with the control steers.

The feeding of dried Pensacola liquid digested sludge (DPS) in the diets of steers and the consumption of forages produced on land treated with DPS consistently decreased the liver tissue concentrations of copper (Cu). This suggested that the DPS had a detrimental effect at the absorption site or on the liver storage mechanism for Cu. The liver data indicated that iron (Fe) absorption and/or storage was also affected. There were some accumulations of lead (Pb) in both kidney and liver tissues of steers fed DPS in their diets.

Accumulations of cadmium (Cd) and Pb in liver and kidney tissues were observed from feeding the diet containing dried Chicago digested sludge (CDS), which was considered to be a high Cd sludge. Since Cd exposure can cause kidney damage, the Cd content of a sewage sludge could determine the amount that may be safely applied to agricultural land.

There were no differences among treatments in the concentrations of selected metals in muscle tissues. The concentrations were all within presently acceptable tolerance or guideline limits.

Bitton, G., Farrah, S.R., Overman, A.R., Gifford, G.E., Pancorbo, O.C., and Charles, J.M., Environ. Eng. Sci., Gainesville, FL

"FATE OF VIRUSES FOLLOWING APPLICATION OF MUNICIPAL SLUDGE TO LAND"

Land disposal of sewage effluents and residuals appears to be an acceptable method for the utilization and disposal of sludge. Advantages include addition of plant nutrients, water conservation, and improvement of soil physical properties. The survival and transport of microbial pathogens and viruses remains a concern in sludge application to land.

Enteroviruses were readily recovered from sludges, including lagooned sludge at Jay, Florida. However, no virus could be detected following topsoil monitoring for 8 months. Allowing the liquid sludge to dry before being mixed with the soil results in inactivation of all or most of the viruses present. Sludge has been applied for several years at the Jay site and this study demonstrated that enteroviruses represent a minimal hazard, either through translocation through grain or forage or with regard to groundwater contamination.

Damron, B.L., Osuna, O., Suber, R.L. and Edds, G.T., Poultry Science, Gainesville, FL

"HEALTH EFFECTS OF SEWAGE SLUDGE AND GRAIN FROM SLUDGE TREATED SOILS IN POULTRY"

Duplicate experiments of 21 days duration were conducted with day-old broiler-type chicks to study the influence of replacing one-half or all of the normal dietary corn complement with corn grown on soil fertilized with municipal sludge.

Neither level of sludge corn had any adverse effect upon final body weights or daily feed intake. The feed conversion values of experiment 1 were not significantly influenced by treatment; however, a statistically significant decrease of efficiency was noted for the all-sludge corn treatment of experiment 2.

In two studies with laying hens, the partial or total substitution (50 or 100%) of sludge fertilized corn for that produced with commercial fertilizer had no statistically significant effects upon any of the production parameters measured in experiment 1. In experiment 2, the 100% sludge corn treatment was associated with a significantly increased daily feed intake and final body weight. Hatchability parameters and taste panel results for eggs indicated no significant relationship to dietary treatment.

Mineral assays of blood samples and liver, kidney and muscle tissues from hens and broilers were not influenced by dietary treatment.

Levels of 0, 3 or 6% Chicago sludge were substituted into the diet of broiler chicks while equivalent nutrient levels were maintained. In addition, four other treatments in experiment 1 and five in experiment 2 contained the amounts of cadmium, chromium, copper and iron from reagent sources equivalent to the levels of these elements coming from 6% sludge.

In experiment 1, the feeding of iron or chromium resulted in significant body weight depressions. Only the feed intake of the birds receiving the iron treatment was significantly below that of the control group. Both iron levels in experiment 2 (2993 mg/kg and 2196 mg/kg) significantly depressed body weights and daily feed intake. The cadmium and iron treatments of both studies resulted in elevated liver and kidney levels of these minerals. There was also a trend of increasing cadmium residues in the liver and kidney resulting from increasing sludge levels, however the utilization rate from sludge appeared to be only approximately 20%.

Levels of 3.5 and 7% Chicago sludge were fed to hens in two experiments. In addition, amounts of cadmium, chromium or copper equivalent to those found in the 7% sludge diet were fed from reagent sources.

In the first experiment, none of the production criteria were significantly influenced. In experiment 2, the addition of iron resulted in a numerical depression of egg production. Daily feed intake was significantly reduced by the iron level fed. Hatchability data was not found to be consistently influenced by dietary treatment in either experiment.

Edds, G.T., Ferslew, K.E. and Bellis, R.A., Preventive Medicine, Gainesville, FL

"FEEDING OF URBAN SEWAGE SLUDGE TO SWINE (PRELIMINARY REPORT)"

The inclusion of 50% dried urban sewage sludge in a normal swine starter ration for an 8 week period produced signs of toxicity including decreased weight gains, leucocytosis, lowered erythrocyte counts and packed cell volume, along with increased serum transaminase levels. There was an increase in lead and cadmium levels in the liver and kidneys with a decrease in the copper and zinc levels.

Farrah, S.R., Bitton, G., Lanni, O., Lutrick, M.C., Bertrand, J.E. and Pancorbo, O.C., Microbiology, Gainesville, FL

"SURVIVAL OF ENTEROVIRUSES ASSOCIATED WITH LAGOONED SLUDGE"

Enteroviruses associated with aerobically and anaerobically digested sludge were determined before addition of the sludge to a sludge lagoon. The fate of sludge-associated viruses was followed during detention of sludge in the lagoon and after application of sludge to land for disposal. While digested sludge was being added to the lagoon, enteroviruses were readily detected in grab samples of sludge from the lagoon. The level of sludge-associated viruses dropped to low or undetectable levels following disposal of sludge on land and during periods when addition of digested sludge to the lagoon was suspended. Enteroviruses were not detected in water from deep wells located on the sludge disposal site or near the lagoon.

Hoffmann, E.M., and Bertrand, J., Microbiology, Gainesville, FL

"FEED-LOT CATTLE: BACTERIAL ANALYSIS OF FEED, SLUDGE, FECES, AND ANIMAL TISSUES"

Samples of sludge, feed, feces, and animal tissues (kidney, liver, spleen, and blood) were analyzed for pathogenic bacteria. Kidney, liver, spleen, and blood samples were tested for the presence of bacteria in general, (including Mycobacteria), while analysis of sludge, feces, and feed was restricted to pathogenic enteric bacteria. Since large numbers of bacteria were isolated from tissue samples, identification of most isolates was superficial. However, gram negative rods were subjected to more detailed analysis using selective screening media, and the "Enterotube II" system (Roche Diagnostics, Nutley, N.J.). Non-lactose fermenting, gram negative rods from feces, feed, and sludge were also identified using the "Enterotube II" system.

Contamination was a major problem with tissue samples taken under slaughterhouse conditions. The same was true with blood samples taken at the farm site. No enteric pathogens or Mycobacteria were isolated from these kinds of samples. There was one isolation of Staphylococcus aureus from 612 blood samples, one isolate of S. aureus from 96 tissue necropsy samples, and two isolations of Streptococcus pyogenes from the latter tissue samples. The S. aureus and St. pyogenes were isolated from different animals.

Two group B Salmonella enteritidis isolates were obtained from feces of animals fed on a sludge amended diet, and three group C S. enteritidis isolations were made from the same group, but at a later date. There were 208 samples in the group.

Hortenstine, C.C., Soils, Gainesville, FL

"SLUDGE EFFECTS ON YIELD AND CADMIUM UPTAKE OF COASTCROSS I BERMUDAGRASS"

Cadmium measurements in and adsorption by a soil treated with sewage sludge of known content, measurement of its uptake by Coastcross I Bermudagrass (Cynodon dactylon) and comparison of sludge as a source of plant nutrients as a mineral fertilizer were performed. Additional potassium was applied to the sludge-treated plots.

Sludge applications increased soil Zn, Cd, Cu, Pb and Ni along with an increase in Ca with some movements of other elements below the 15 cm. depth. The cadmium content of the bermudagrass increased greatly at a second harvest period after prior one-half sludge application in May. Second application on August 1st was associated with a further cadmium increase at the harvest on August 21st.

These trials indicated that sludge application at the 24 ton/ha rate compared favorably with mineral fertilizer as a source of plant nutrients.

Lutrick, M.C. and Cornell, J.A., Agric. Res. Ctr., Jay, FL

"THE UPTAKE OF CERTAIN METALS BY CORN GROWN ON SOIL TREATED WITH CHICAGO OR PENSACOLA SEWAGE SLUDGE"

Land spreading of sewage sludge is probably the most practical means of disposal for municipalities and cities. However, uptake of certain metals by forage and grain crops from land treated with sludge creates health risks. A 3-year study was conducted to determine the uptake of copper (Cu), zinc (Zn), manganese (Mn), lead (Pb), and cadmium (Cd) by corn leaves and grain from soil treated with Chicago and Pensacola sewage sludge. The Chicago sludge contained large quantities of Cu, Zn, Pb, and Cd. The Pensacola sludge was high only in Zn.

The quantities of metals extracted from sludge-treated soil were proportional to the quantities added from the sludges. The metal uptake by the corn plant was directly associated with soil pH. The higher the soil pH the smaller the quantity of metal uptake. The quantity of metals extracted from the sludge-treated soil after one year was somewhat less than from the soil where sludge had been recently applied. The quantity taken up by the corn leaves was much less from the residual sludge treatment than from the same treatment where sludge had been recently applied.

The concentration of metals in the grain was always much less than the concentration found in the corn leaves from the same sludge treatment. Corn plants from the Chicago sludge treatment probably would have contained too much Cd to be utilized for forage. The Pb and Cd concentration in the grain was below detectable limits from all treatments.

Osuna, O. and Edds, G.T., Preventive Med., Gainesville, FL

"FEEDING TRIALS OF DRIED URBAN SLUDGE AND THE EQUIVALENT CADMIUM LEVEL IN SWINE"

In swine production with modern feeding and management conditions, cadmium toxicity is relatively rare. However, borderline toxicities are possible where animals ingest recycled waste materials, such as urban sewage sludge, in which cadmium may be concentrated. Two percent sludge has been found to provide a satisfactory source of vitamin B_{12} for the pig. However, at higher levels, cadmium is of extreme importance because of its economic effect upon food-producing animals and from a public health standpoint as a toxic residue for human consumption.

Toxicity from feeding dried sewage sludge included in a normal swine starter ration, may occur from a deficiency of available protein or other essential nutrients, or from the accumulation of hazardous chemical residues. One lot of the sewage sludge from Chicago was found to contain high levels of cadmium (165 ppm).

This trial compared the effects of feeding weanling pigs a starter ration containing 50% dried, activated, Chicago sewage sludge with a standard 18% crude protein basal diet, 83 ppm cadmium, for 9 weeks.

Forty-eight 4 week-old, hybrid pigs were allotted to 2 replicate experiments of 2 treatment groups each. Body weight, feed consumption and blood samples were determined weekly including PCV, RBC, WBC, MCV, Hb and serum levels of 4 enzymes, AP, γGT, GOT and GPT. Feed and fecal samples were collected weekly and analyzed for cadmium content. Tissue samples were provided at the slaughter time on days 38, 42 and 56 for metal analysis and pathologic evaluations.

Depressed growth and feed consumption were evident in pigs consuming 50% Chicago sludge and 83 ppm of cadmium.

Cadmium exposure induced microcytic and hypochromic anemia. At necropsy, pale, white muscles and kidneys were observed.

Osuna, O., Edds, G.T., and Simpson, C.F., Preventive Med., Gainesville, FL

"TOXICOLOGY OF AFLATOXIN B_1, WARFARIN AND CADMIUM IN YOUNG PIGS"

The purpose of using 0 or 83 μg/g cadmium diets in pigs was to determine whether animals may be predisposed to higher health risk when exposed to such feeds containing aflatoxin B_1, dihydrofuranocoumarin or warfarin, 3-(α-acetonyl-benzyl)-4 hydroxycoumarin.

Thirty-six weaned barrows, averaging 9 kg BW, were assigned at random to 6 treatment groups, 6 per group. Cadmium (Cd) was provided daily through the diets in 3 groups during the 40 day experiment. Aflatoxin B_1 and warfarin were given at 0.2 mg/kg per os for 5 days to treated and nontreated Cd groups during the 5th week and effects followed for 10 days.

Aflatoxin B_1 in the 0 cadmium group significantly increased values of AP, SDH, aspartate aminotransferase (SGOT), PT, activated partial thromboplastin time (APTT) and significantly decreased values in serum total protein, αglobulin, βglobulin, γglobulin and fibrinogen. The hepatotoxic effects of aflatoxin B_1 correlated with lobular fatty infiltration in 3 out of 4 pigs in the group. Aflatoxin B_1 was very toxic as was warfarin.

Warfarin in the 0 Cd group was more effective in producing earlier and significantly higher values in PT and APTT than those receiving AFB_1 by the second day. Values in the pigs receiving warfarin returned to normal after dosing had been suspended, while those in the aflatoxin B_1

group increased significantly by the 10th day. No significant liver or kidney damage was induced by warfarin.

Cadmium treatment prevented the fatty hepatocytic infiltration but not hydropic degeneration induced by aflatoxin B_1. Cadmium treated diets resulted in differences in the intensity and duration of response in PT and APTT when pigs were dosed with aflatoxin B_1 or warfarin. Cadmium also induced differences in the activity of AP, SDH, SGOT, in pigs exposed to aflatoxin B_1 or warfarin.

The Cd diet resulted in 35.13 µg/g and 6.80 µg/g Cd tissue accumulation in kidney and liver respectively at the 4th week of the experiment as compared to nondetectable levels in the controls.

This is the first demonstration of the Cd blocking effect on the microsomal enzyme system in pigs. There were increased concentrations of Cd in tissues of the treated groups associated with decreased iron utilization resulting in microcytic hypochromic anemia. Therefore, there is an inhibitory metabolic effect on aflatoxin B_1, and enhanced synergistic toxic effect with warfarin when Cd is present in the diets of young pigs at 83 µg/g.

Popp, J.A., Osuna, O., Edds, G.T., Preventive Med., Gainesville, FL

"PATHOLOGIC LESIONS IN CATTLE, SWINE AND POULTRY RECEIVING DCS, DFS OR CORN FROM DCS AMENDED SOILS"

Beef cattle receiving DCS or corn grown on DCS supplemented soils demonstrated delayed increased serum transaminase levels. These were associated with parenchymal liver disease, degeneration, cellular infiltration with multiple foci of lymphocytes and eosinophils and necrosis in kidneys also.

Swine receiving 10 and 20% DFS in their rations also demonstrated increased serum transaminase levels plus parenchymal liver disease with cellular and organelle damage at the 20% level. The cellular damage in the 10% group was primarily cytoplasmic. The chronic effects from cellular degeneration and necrosis resulted in decreased transaminase levels; the cells could no longer synthesize the enzymes.

Broilers and hens receiving diets containing 50% or 100% of the corn ingredient from Chicago sludge supplemented soils demonstrated lipid infiltration of hepatocytes, vacuolation or hepatic lipidosis.

Richter, M.F., Agric. Res. Ctr., Ona, FL

"HEALTH EFFECTS OF RECYCLED CATTLE MANURE"

Two of three one-year feeding trials have been completed. In each trial 20 animals were divided into 5 groups of 4 animals: group 1, initial slaughter group; group 2, control ration; group 3, manure silage, no withdrawal; group 4, manure silage, 10 days withdrawal; group 5, 20 days withdrawal. All animals were adapted to the feed and feeding facilities before the start of the trial. At the beginning of the trial the initial slaughter group was sacrificed for baseline data on the body compostion of the animals. The remaining 16 animals were placed on treatment for approximately 200 days. Manure was withdrawn from the rations of two of the groups 10 and 20 days before slaughter. The control ration contained corn grain, citrus pulp, cottenseed meal, pelleted bagasse, molasses and minerals. The manure ration contained ensiled cattle manure at a level such that 20% of the dry matter of the ration was from manure. The silage was made by mixing raw cattle manure with pelleted bagasse to give a dry matter of 50%. At slaughter, liver, muscle, and kidney samples were taken and analyses are in progress for drugs, pesticides, and heavy metals.

Animals on the control ration consumed an average of 7.1 kg of feed per day and gained an average of .85 kg per day; manure fed animals consumed an average of 10.7 kg of feed per day and gained .76 kg per day.

The control animals consumed 6.1 kg of dry matter; the manure fed animals, 6.4 kg, of which 4.4. kg was concentrate feed identical to the control ration. The feed conversion (kg feed dry matter/kg gain) was 7.2 for the control animals and 8.4 for the manure animals. However, if only concentrate feed dry matter is considered, the concentrate feed conversion for the manure fed animals is 5.8. Therefore, the manure silage significantly contributed to the performance of the animals consuming it ($P<.05$). The manure fed steers gained 89% as much as the control animals while consuming only 72% as much concentrate feed. More detailed evaluations of the utilization of the rations will be available when the data from metabolism trials are available.

Preliminary results on the carcass analysis show no

unusual pathologies or differences in blood parameters. This would indicate that the manure silage did not affect the health of the animals. Data on the mineral composition of selected tissues show no accumulation of Cd, Co, Cr, Cu, Fe, Ni, Pb, or Zn. Mercury and Se were present only at the detection limits of the instrument. Analysis of this data is not complete at this time.

Simpson, C.F., Osuna, O. and Edds, G.T., Preventive Med., Gainesville, FL

"EFFECT OF FEEDING DRIED SLUDGE TO ANIMALS"

Swine, cattle and poultry were fed dried sewage sludge from Chicago, Pensacola and Gainesville, or forage grain produced on land supplemented with such sludges. Histologic lesions were not found in the tissues of these animals at necropsy except in the case of hearts of swine and cattle. Of 7 pigs fed 10% Gainesville sludge (DFS), 4 had Sarcosporidia in the myocardium, and the hearts of 2 of 4 pigs fed 20% DFS contained the parasite. None of the swine fed control feed had Sarcosporidia in the myocardium. Among cattle fed Pensacola sludge, 19 of 32 contained Sarcosporidia in the cardiac muscle, but the cardiac muscle of 6 of 17 controls was parasitized. It was concluded that the presence of Sarcosporidia in hearts of swine fed sludge may be of public health significance. The presence of the parasite in cattle myocardium probably was not of significance because of the high incidence of the agent in hearts of control animals.

The several tissues of mice fed liver and kidney of cattle and swine fed sludge did not contain microscopic lesions.

Suber, R.L. and Edds, G.T., Preventive Med., Gainesville, FL

"HIGH PERFORMANCE LIQUID CHROMATOGRAPHIC DETERMINATIONS OF SULFONAMIDES BY IONIC SUPPRESSION"

A high pressure liquid chromatography procedure is reported for extraction and quantitation of 8 sulfonamides in stock solutions and in vitro samples. This assay consists of a single, one-step extraction of sulfonamides from plasma and is sensitive to 10.0 ng/ml at 254 nm without additonal concentration of the sample. Four sulfonamides (sulfamerazine, sulfamethazine, sulfapyridine and sulfathiazole) were separated from the plasma matrix by either mobile phase regardless of pH. The sulfonamides with the highest pKa, sulfanilamide (10.5) and sulfaguanidine (11.3), were only separable from plasma in a 50% water/50% methanol mobile phase at pH 7.45. The sulfonamide with the lowest pKa, sulfisoxazole (4.9), and its metabolite, acetylsulfisoxazole (N^4), were separated from plasma by either mobile phase, 50/50 or 60/40 water/methanol, when acetate buffer reduced the pH to 4.00. Standard concentration curves of peak height were the most sensitive at 254 nm when a 60% water/ 40% methanol mobile phase at pH 4.00 was used. Sulfanilamide and sulfaguanidine were the most responsive to ultraviolet quantitation at 254 nm regardless of ionic suppression or polarity of the mobile phase.

Thompson, N.P., Osuna, O. and Edds, G.T., Pesticides, FS, Gainesville, FL

"PESTICIDE RESIDUES IN FEED, SLUDGE, SOIL/SLUDGE, AND ANIMAL TISSUES"

Samples of sludge, soil/sludge mixture, feed and animal tissues (kidney, liver, fat, muscle) were analyzed for chlorinated hydrocarbon pesticide residues and also polychlorinated biphenyls. The analytical method is typical of that used for determination of persistent pesticide residues in environmental samples and includes Soxhlet extraction with petroleum ether; clean-up by gel permeation chromatography, florisil and silicic acid columns; and detection by electron capture gas chromatography. The sensitivity of the method is 0.01 ppm.

Results indicate that little if any chlorinated hydrocarbon residues were present in sludge used for research in this project. The available tables list the residues found in ppm in sludge, feed and various animal tissues. It can be concluded that sludge as used in experiments associated with this project presents no hazard from the aspect of pesticide residues.

White, C.E., Hammell, D.L. and Osuna, O., Agric. Res. Ctr., Live Oak, FL

"EFFECT OF FEEDING DIGESTED SEWAGE SLUDGE ON LONG-TERM SOW REPRODUCTIVE PERFORMANCE"

Feeding sewage sludge on reproductive performance in female swine during successive gestation-lactation periods was evaluated. Gilts selected from two generations in succession were fed diets containing 0, 10, or 20% Dried Florida Sludge (DFS) in a basal corn-soybean meal formulation. In order to assess long-term effects, continuity in the experiment was maintained by feeding second generation females the same dietary regimen fed to their dams.

Data collected indicated that feeding DFS to female swine over an extended period adversely affected many criteria used in evaluating reproductive performance. Breeding, farrowing and re-breeding weights were reduced. Lactation and gestation weight changes were lower and fewer pigs were farrowed in sow groups receiving 10 and 20% sewage sludge in diets. First and second generation sows fed diets containing 10 and 20% DFS weaned lighter pigs at 21 days when compared with sows fed the basal diet. In each parity, pigs farrowed by sows fed DFS in diets displayed depressed average daily gain. A comparison of the data from both generations receiving diets containing sludge indicated that reproductive performance was more diminished in second generation sows than in the first.

B. Abstracts of Papers Presented at the Symposium

Bastian, E. and Brams, E., Soil Science, Prairie View A & M U., Prairie View, Texas

"CADMIUM AND LEAD IN BROILER CHICKENS FED SORGHUM GRAIN PRODUCED ON ADULTERATED SOIL"

Sandy loam soil treated with soluble salts of Cd and Pb applied together to a soil depth of 16 cm not exceeding 8 and 63 ppm, respectively, induced a direct and significant accumulation of Cd in sorghum grain from 0.47 to 12.1 ug/g, but soil Pb did not accumulate in the grain which averaged 2.0 ppm over all treatments. Cadmium in grain rations averaged from 0.5 to 32.0 ppm. The dose response to ingestion of grain at the highest level of Cd by broilers for 2 weeks resulted in concentrations at 7.0, 18.0 and 1.5 ppm Cd in liver, kidney, and flesh respectively and indicated a significant increase in the liver and kidney over the control which averaged 0.15 and 0.14 ppm. No dose response to ingestion of Pb from grain was detected in liver, kidney or flesh which averaged 1.10., 2.31, and 1.0 ppm Pb, respectively. Feeding treatments continued for 6 weeks induced levels of Cd in liver, kidney and flesh comparable to the 2 weeks feeding period. No treatment effect was noted for Pb in any tissue, although continued ingestion of grain for 6 weeks induced significantly higher levels of Pb in all tissues averaging 4.1, 4.4, and 3.9 ppm in liver, kidney, and flesh. Concentrations of Cd in liver of broilers fed grain containing Cd above 12 ppm should be viewed with caution as a possible health hazard in human food.

Beaudouin, J., Shirley, R.L. and Hammell, D.L., Animal Nutrition, Gainesville, FL

"EFFECT OF SEWAGE SLUDGE DIETS FED SWINE ON NUTRIENT DIGESTIBILITY, REPRODUCTION, GROWTH AND MINERALS IN SWINE"

Twelve female swine were fed in a 3 x 4 cross-over design metabolism trial corn-soybean grower diets that contained 0, 10 or 20 percent sewage sludge over three 19-day periods. The mean values for total digestibile nutrients were 79.4, 73.7 and 55.0%; those for metabolizable energy were 3.36, 2.25 and 1.15 Mcal per kg diet; and those for nitrogen retained were 42.8, 44.0 and 25.3%, respectively. Sewage sludge (0, 10, 20%) diets were fed to 31 sows approximately equally divided in the dietary groups during their first two pregnancies and to their offspring from weaning to market weight. More live pigs were farrowed and weaned per litter from sows fed 20% sludge diets than from the control group. However, 21-day weaning weights of pigs were lower with sows fed the sludge-containing diets. Offspring of both first and second litters fed growing finishing diets containing sludge from weaning to market weight had decreased daily weight gains and feed efficiency. There were no increases in the nine mineral elements (Pb, Cd, Ni, Zn, Cr, Cu, Mn, Fe and Al) in sow milk or blood. Offspring of sows fed sludge diets showed increases of several elements in selected tissues at weaning and after consuming sludge diets to market weight.

Brams, E., Soil Science, Prairie View A & M U., Prairie View, Texas

"CADMIUM AND LEAD IN POULTRY FED WHEAT FROM ADULTERATED SOIL: AN ASSESSMENT OF THE FOOD CHAIN"

Wheat grain as a biological source of Cd and Pb produced on soil adulterated with Cd and Pb at 3 levels: 1, 2, 3 and 5, 10, 30 ppm Cd and Pb respectively, was fed mixed with supplement to poultry and the concentrations of the metals in the soil-root zone, whole grain, egg yolk and albumin, flesh and liver determined during a feeding experiment of 64 days to quantitatively assess the movement of Cd and Pb in the food chain along select pathways. Soil-Cd in excess of 2.0 ppm induced significant increases in wheat-Cd to values between 3.0 and 5.0 ug/g whole grain. Where soil Cd ranged from near background levels (0.64 ppm) to 2.0 ppm, wheat-Cd averaged 1.6 ug/g grain. No effect of soil Pb for any treatment was detected in grain which averaged 2.9 ug/g whole grain. Wheat fed to poultry during a 64 day period did not result in any significant accumulation of Cd or Pb in egg components, liver and flesh as a function of treatments. However, continuous ingestion of experimental wheat rations in excess of 10,000 ug Cd and 14,000 ug Pb resulted in significant increases of Cd in albumin, and Pb in albumin and yolk. No increase in these metals was evident in liver or flesh after ingestion of 14,000 ug Cd and 24,000 ug Pb. Assessment of the food chain with respect to Cd and Pb in soil and respective tissues show that wheat grain accumulates only 0.24 percent of the Cd and 0.06 percent of the Pb in the soil available to the wheat. Poultry accumulates 0.003 and 0.006 percent of soil Cd and Pb respectively in liver and flesh. The major point of entry for toxic metals into the food chain is the soil and the source is the soil pool of available metals. Short of curtailing environmental pollution by reducing the production and/or release of pollutants into the enviroment, agronomic practices to curtail Cd and Pb in the food chain should be effective in reducing the amounts of available metals in the soil that could move into the plant. The results here have shown that the quantity of metals moving from the soil pool into grains and finally into poultry tissues is extremely minute relative to the soil pool. However, even at these low concentrations in tissues, dose-response relationships relative to human health from ingestion of these tissues must be elucidated for Cd and Pb before any definitive guidelines can establish tolerable limits in food plants. Soil adulterated with Cd and Pb can remain a source of these metals for long periods and care must be exercised as to management practices using amendments that contain these toxic metals.

Davidson, J.P., Chaney, R.L., Decker, A.M., Hammond, R.C., Doty, K.T. and Machis, A., Dept. Vet. Sci., U. of Maryland, College Park, MD

"GROSS AND MICROSCOPIC LESIONS IN ANGUS CATTLE MAINTAINED ON LIQUID AND COMPOSTED SEWAGE SLUDGE-TREATED PASTURES"

On 3 successive years, Angus cows, calves and steers were necropsied following an approximate 6 month grazing trial on tall fescue grass. A 4-paddock rotational system was utilized. Pasture treatments were: 1) NH_4NO_3, applied 21 days before grazing; 2) liquid sludge applied 21 days before grazing; 3) liquid sludge, applied 1 day before grazing; and 4) composted limed-raw sludge applied 2-3 times/season. Gross lesions were quantitatively and qualitatively noted at the time of death, and microscopic lesions were similarly evaluated from tissues collected at necropsy.

Lesions were essentially limited to the digestive system, and were directly related to either the iron content of the applied sludge, or the interval between sludge application and grazing. Stainable ferritin was observed in the forestomachs, duodenum, anterior jejunum, mesenteric lymph nodes, and on occasion in the liver, and kidney. As an example, liver iron rose from 200 ppm in controls (pasture treatment 1) to 17,700 ppm in animals maintained on pastures treated with high iron, liquid sludge (pasture treatment 3).

A feed-lot trial, designed to study the effects of composted limed-raw sludge in a pelleted diet was subsequently conducted. Gross and microscopic lesions were similar to observations and conclusions made on the initial 3-year study.

Decker, A.M., Chaney, R.L., Davidson, J.P., Rumsey, T.S., Mohanty, S.B., Hammond, R.C., Doty, K.T. and Machis, A., Dept. Agron., U. Maryland, College Park, MD

"FORAGE PRODUCTION AND ANIMAL PERFORMANCE AS AFFECTED BY LIQUID AND COMPOSTED SEWAGE SLUDGE"

Cows, calves, and steers grazed tall fescue on a 4-paddock rotational system (7 days grazing-21 days rest). Animal performance and health were evaluated and pathologic lesions and metal residues measured at necropsy. Pasture treatments were: 1) NH_4NO_3 21 days before grazing; 2) liquid sludge (L.S.) 21 days before grazing; 3) L.S. 1 day before grazing; and composted limed-raw sludge applied 2-3 times/season.

Excellent plant growth resulted from both liquid and compost applications; forage quality was not reduced by sludge fertilization. Forage acceptability was excellent for all except day-1 liquid. Spray-applied sludge adhered strongly to forage; major source of metal consumption was from sludge adherence and direct consumption from soil surface. Parasitic ova in feces were within normal range and virus isolations appeared to be associated with animal stress rather than pasture treatment per se. Animal performance was similar on NH_4NO_3-, compost -, and liquid (4.4% FE)-treated pastures but was markedly reduced on day-1 and day-21 "high" FE (11%) liquid sludge. While metal accumulations were observed, statistically significant sludge-related increases were found only with Fe. Although dietary Cd was considerably increased by sludge and compost (5-10 ppm Cd), kidney Cd was not increased. Ferritin accumulated in duodenum, lymph nodes, liver, and other tissues.

Subsequently, a feedlot trial was conducted to evaluate effects of composted limed-raw sludge in a pelleted diet. Similar results and conclusions were observed.

Dorn, R.C., Lamphere, D., Crowl, T., Hamparian, V., Ottolenghi, A., Gaeuman, J., Lanese, R. and Powers, J., Dept. Preven. Med., Ohio State U., Columbus, Ohio

"ANIMAL AND PUBLICH HEALTH STUDIES ON PRIVATE OHIO FARMS RECEIVING LAND APPLICATION OF URBAN SEWAGE SLUDGE: A PRELIMINARY REPORT"

Health studies are being conducted in three areas of Ohio to evaluate land application of sludge. Fifty farms have been identified to receive sludge and fifty randomly chosen farms serve as controls. The health status of farm residents and animals on these farms is monitored each month by interviewers who record all illnesses using an epidemiologic questionnaire. Other objectives of the investigation are to compare tuberculin skin test response of residents on sludge receiving and control farms, <u>Salmonella</u> infection, parasite infection, and translocation of heavy metals in the food chain. The design of the research project and its progress will be reported.

Doyle, E.A. and O'Connor, G.A., Dept. Agron., N. Mex. State U., Las Cruces, NM

"THE EFFECT OF SEWAGE SLUDGE AMENDMENT ON THE ADSORPTION AND DEGRADATION OF 2,4-D IN SOIL"

The proposed application of sewage sludge to agricultural soils as a means of disposal has resulted in interest in the interaction of agricultural chemicals with sludge. In this study, effects of sludge additions were determined on the adsorption and degradation of 2,4-D in soils. Thermoirradiated sludge was added to three soils at rates of 0, 10, and 20 T/A. Batch adsorption studies were conducted at initial herbicide concentrations of 0.1, 1.0, and 10.0 ppm. Degradation studies utilized carboxyl-labeled 2,4-D and trapped $^{14}CO_2$ as a function of time.

Sludge additions increased overall 2,4-D degradation, with the greatest effects observed in sludge-soil mixtures preincubated for 2 months. Adsorption of 2,4-D was not significantly affected by sludge additions made immediately before the study. Preincubating sludge with soil for 2 months, however, increased herbicide adsorption.

R.H. Dowdy, D.E. Pamp, and R.D. Goodrich, USDA, SEA, AR, Soil Sci., U. Minn., St. Paul, Minnesota

"AN EVALUATION OF THE ACCUMULATION OF SEWAGE SLUDGE-BORNE METALS IN THE FOOD CHAIN"

This study assessed the transfer of biologically active trace metals into the food chain by way of goat milk and market lamb tissues. A sewage sludge containing ~ 150 µg Cd/g sludge was applied annually to a silt loam soil at 0, 15, 30, and 45 tonnes dry matter/ha. To insure high levels of Cd in the corn silage ration, elemental S was applied to all sludge treatments at rates commensurate with the $CaCO_3$ equivalent of added sludge.

Cadmium concentrations in corn leaf samples were < 0.1,

0.9, 1.0, and 1.2 µg/g for the control, low, medium, and high treatments, respectively, the first year and increased to < 0.2, 3.3, 4.8, and 7.1 µg/g for comparable treatments after a second sludge application. This resulted in corn silages containing < 0.02, 0.71, 1.27, and 1.73 µg Cd/g dry matter the first year and < 0.04, 1.84, 2.68, and 4.2 µg Cd/g dry matter the second year. Similar increases in silage Zn concentrations were noted, reaching 115 µg Zn/g dry matter the second year.

The Cd concentration in milk from dairy goats was not enhanced as a result of feeding a Cd-enriched corn silage ration for 2 years. Data for blood and market lamb tissues are presented. These data are significant because: i) we fed high levels of biologically active Cd on a year-round basis, and ii) to achieve these high Cd concentrations, sludge-borne Cd addition to the soil far exceeded current USEPA regulations for Cd "... application to land used for the production of food-chain crops."

Fairbanks, B.C. and O'Connor, G.A., Dept. Agron. N. Mex. State U., Las Cruces, NM

"ADSORPTION OF POLYCHLORINATED BIPHENYLS (PCBs) BY SEWAGE SLUDGE AMENDED SOILS"

Polychlorinated biphenyls (PCBs), a class of toxic compounds bio-accumulated by animals, may be introduced to the terrestrial environment by the application of contaminated sewage sludge to soils. To assess the PCB pollution hazard of sewage sludge applications to soils, the extent of PCB attentuation by soil-sludge systems was examined.

Aliquots of a saturated solution of ^{14}C-labeled PCBs were equilibrated with various soils amended with 0, 10, or 20 T/A or thermo-irradiated sewage sludge. Sludge was either preincubated with the soils for 2 months or applied immediately prior to PCB additions. PCBs were very strongly adsorbed in non-amended soils, and in sludge amended soils regardless of whether the sludge was preincubated with, or added immediately prior to, soils.

Farrah, S.R., Bitton, G., Lanni, O., Lutrick, M.C., Bertrand, J.E. and Pancorbo, O.C., Microbiology, Gainesville, FL

"SURVIVAL OF ENTEROVIRUSES ASSOCIATED WITH LAGOONED SLUDGE"

Enteroviruses associated with aerobically and anaerobically digested sludge were determined before addition of the sludge to a sludge lagoon. The fate of sludge-associated viruses was followed during detention of sludge in the lagoon and after application of sludge to land for disposal. While digested sludge was being added to the lagoon, enteroviruses were readily detected in grab samples of sludge from the lagoon. The level of sludge-associated viruses dropped to low or undetectable levels following disposal of sludge on land and during periods when addition of digested sludge to the lagoon was suspended. Enteroviruses were not detected in water from deep wells located on the sludge disposal site or near the lagoon.

Feder, W.A., Chaney, R.L., Hirsch, C.E. and Munns, J.B., Dept. Agron., U. of Mass., Warthon

"DIFFERENCES IN Cd AND Pb ACCUMULATION AMONG LETTUCE CULTIVARS AND METAL POLLUTION PROBLEMS IN URBAN GARDENS"

Nine lettuce cultivars were grown on NPK fertilized control soil and calcareous incinerator bottom ash 60 days in raised concrete benches in a greenhouse. The control and ash soils contained 0.5 and 53 ppm Cd, 46 and 5580 ppm Pb and had pH levels of 5.8 and 7.8 respectively. Lettuce shoot Cd ranged from 3.8 to 8.1 ppm dry weight on the ash soil; Valmaine Romaine lettuce was lowest and Summer Bib lettuce highest in Cd. Lettuce shoot Pb ranged from 10.6 to 27.4 ppm dry weight on the ash treatment; Valmaine was lowest and Tania (a Butterhead cv) lettuce highest in Pb. Shoot zinc and shoot Cd were similarily ordered among cultivars.

In a separate study, 6 lettuce cultivars were grown in field plots and pots in a growth chamber. Soil treatments at both pH 5.5 and 6.5 included varied Cd, and varied Cd:Zn to learn whether soil Cd, Cd:Zn, and/or pH alter the relative Cd accumulation among lettuce cultivars. Plants are presently being analysed.

Recent research has shown widespread Pb (and other heavy metal) enrichment of urban soils. Much of this soil Pb results from exterior and interior house paint; garden soil Pb concentrations are as high as 17,000 ppm. These levels appear to pose a substantial Pb risk to children because children play in urban soil and ingest soil through hand to mouth contact.

Results on metals in soils and crops from Boston and Baltimore will be discussed.

Foster, David R., Robbins, Louis K., Greenman-Pederson, Assoc., Annapolis, MD

"UNDIGESTED SEWAGE SLUDGE ENTRENCHMENT PRACTICES IN MONTGOMERY AND PRINCE GEORGE'S COUNTIES, MARYLAND

Undigested, lime-stabilized, sewage sludge from the Blue Plains Wastewater Treatment Plant, Washington, D.C. is currently being disposed of in narrow trenches, on prepared entrenchment sites, in Montgomery and Prince George's Counties.

The pratices employed by the engineer in inventorying, evaluating, and designing sewage sludge entrenchment sites are reviewed in this paper.

Environmental, geohydrologic, and traffic impact evaluations are conducted to assess the suitability and accessibility of a potential site. The results of these evaluations are then used in selecting and designing the land containment sites. Additionally, current practices employed in designing a site's surface drainage, holding ponds, haul roads, and field office facilities are presented.

Fries, George F., USDA, Beltsville, MD

"AN ASSESSMENT OF POTENTIAL RESIDUES IN ANIMAL PRODUCTS FROM APPLICATION OF SEWAGE SLUDGE CONTAINING POLYCHLORINATED BIPHENYLS TO AGRICULTURAL LAND"

Laboratory studies and field observations involving polychlorinated biphenyls (PCB) and related compounds were used in making this assessment. Animals can become exposed to PCB applied to land by three routes; ingestion of plants directly contaminated during sludge application, ingestion of plants contaminated by uptake or volatilization from the soil, and ingestion of contaminated soil when grazing. The last route has been greatest potential for producing residues in animal products from a given rate of application of PCB contaminated sludge to land. Feeding experiments with PCB indicate that the steady state milk fat concentrations are about 5 times the diet concentrations (dry basis) in dairy cattle and that the steady state body fat concentration will be substantially similar in nonlactating animals. Soil consumption by grazing dairy cows can be as high as 14% of dry matter intake when the amount of available forage is low and no supplemental feed is used. This could cause milk fat residues of 0.7 ppm for each ppm of PCB in surface soil. There is little, if any, plant uptake and translocation of PCB. However, plants can become contaminated by volatilization if PCB occurs in the surface soil. Field experience with other halogenated hydrocarbons suggest that this will not be the limiting factor in determining rates of PCB - containing sludge that can be applied to land.

Hammond, L.C., Hortenstine, C.C. and Street, J.J., Soils, Gainesville, FL

"TRANSPORT OF SEWAGE SLUDGE CONSTITUENTS IN A SANDY SOIL AND UPTAKE OF METALS BY BERMUDAGRASS"

Commercial Chicago sewage sludge (dried) was applied to bermudagrass growing on a well-drained sandy soil (coated, hyperthermic Typic Quartzipsamment) at Gainesivlle, Florida. Annual rates for 3 years on four respective treatments were 0, 12, 24, and 48 tons/ha. The no-sludge treatment received 6 tons/ha annually of a chemical fertilizer (10-4. 4-8.3, N-P-K). Forage was harvested for yield and chemical analysis at about 6-week intervals during the growing season. Frost-killed sod remained undisturbed during the winter. Rainfall, irrigation, and estimated evapotranspiration data were used in a computer model to estimate the leaching potential for the soil-plant-climate system over the 3-year period.

Forage yields increased each year and in 1979 they were 20, 14, 20 and 26 thousand kg/ha for the 0, 12, 24, and 48 tons/ha sludge rates. For these respective treatments in 1979 uptake of Cd was 3 , 22 , 46 , and 101 mg/ha and uptake of Zn was 413 , 501 , 788 , and 1530 mg/ha.

Metals have accumulated in the 0 to 15-cm soil profile; Cd= 6.3 µg/g and Zn= 83 µg/g at the highest sludge rate. Transport into the 15 to 30-cm zone has occurred; Cd= 0.8 µg/g and Zn= 12 µg/g. These results are in line with the low leaching potential calculated from the rainfall amounts and distribution over the 3-year period.

Kelley, K.C., Osuna, O. and Shirley, R.L., Animal Nutrition, Gainesville, FL

"PERFORMANCE AND TISSUE MINERAL ANALYSES DATA OF MICE"

<u>Swiss</u> mice 45 days of age were fed freeze-dried liver and <u>kidney</u> from cattle or swine fed a control diet, sewage sludge, corn grain or sorghum forage from land covered with liquid sewage sludge. Kidney and liver of F_0 mice had increased levels of Cd when they were fed diets that contained liver of kidney from cattle fed 500 g/head/day of

Chicago digested sludge (C.D.S.) in their diets. Cu in muscle of mice was depressed in sludge treatments. Copper, Fe, Co, Zn, Pb, Hg, Se, Cr and Ni showed no differences due to sludge treatments in mice kidney, liver and muscle. Male mice had a higher concentration of Cu in liver and higher levels of Cu and Fe in muscle than females. Mice fed diets containing 5% cattle kidney had higher levels of Cu, Fe, Co and Pb in their kidney than mice fed 10% cattle liver in diets. Nickel was higher in kidney, liver and muscle; Fe, Co and Pb higher in liver; Cu, Co and Pb higher and Zn lower in muscle; and Cr higher in kidney of mice fed diets that contained 5% kidney from cattle. F_1 females had a six-fold increase in liver Cd in the C.D.S. treatment, while F_2 females showed no differences in tissue minerals due to C.D.S. treatments. F_0 females fed kidney had no differences in number weaned. F_1 females fed liver had a decrease in number born in both sludge treatments.

F_0 mice fed kidney and liver from cattle fed sorghum that received 15.2 or 22.8 cm/ha Pensacola liquid sludge (P.L.S.) showed no changes in minerals in kidney, liver and muscle due to treatment or diet. F_0 males had an increase in liver Cu but no changes in minerals occurred in kidney and muscle. Number born per litter was greater in F_0 mice fed kidney in the 22.8 cm/ha P.L.S. treatment but there was no effect when liver was fed. The number weaned was not affected by tissue fed or sludge treatment.

Liver from Swine F_0 fed, 0, 10 or 20% University of Florida digested sewage sludge (D.S.S.) in diets of mice showed an increase in the Pb content of mice liver and muscle at the 20% D.S.S. level while Cr was lower in liver tissue; in kidney, female mice had higher levels of Fe than males.

Mice fed liver from swine F_{1-2} in diets had higher levels of Cd in liver, kidney and muscle tissue in the 10% D.S.S. group. Pb in mice liver was increased in the 10% D.S.S. and in muscle in both the 10 and 20% D.S.S. groups. Female mice had a higher level of Ni in kidney than males. Mice fed kidney from swine F_{1-3} had high levels of Cd in liver, but no changes due to diet were observed in kidney and muscle. Sex and sludge treatment of mice had no effect.

Kothary, M.H., MacMillan, J.D. and Chase, T., Jr., Dept. Biochem & Microbiol., N.J. Agric. Exp. Sta., Rutgers U., New Brunswick, NJ

"SALMONELLAE IN COMPOSTED SEWAGE SLUDGE AND THE EFFICIENCY OF PLATING MEDIA IN DETECTION"

As a part of a study to assess whether composting of raw sewage sludge was effective in killing pathogens, compost was analyzed regularly for salmonellae. Hektoen enteric agar (HE), bismuth sulfite agar (BS), brilliant green agar (BG) and xylose lysine desoxycholate agar (XLD) were evaluated for their abilities to detect salmonellae in actively composting, curing and finished compost. Samples were pre-enriched in lactose broth, selectively enriched in tetrathionate broth, plated out on the four media and identified using biochemical and serological methods. All the samples containing confirmed salmonellae were from the surface of curing piles, and composting piles constructed with cured compost as a bulking and insulating agent. No salmonellae were detectable in composting piles bulked with wood chips, and in finished compost. A 3 tube MPN procedure (using 1, 0.1 and 0.01 g samples) failed to detect salmonellae in curing piles indicating that the numbers were less than 0.3/g compost. BS was the single most effective medium detecting 68.6% of all the samples confirmed as positive for *Salmonella* (using a combination of all four media). BG, HE, and XLD detected 65.7%, 60% and 17.1%, respectively, of the total *Salmonella*-positive samples. Our studies indicate that compost on the surface of curing piles is susceptible to reinfection with salmonellae and consequently its use as an insulating agent may be undesirable.

Kothary, M.H., MacMillan, J.D. and Chase, T.,Jr., Dept. Biochem. and Microbiol., NJ Agric. Exp. Sta., Rutgers U., New Brunswick, NJ

"INCREASED CONCENTRATIONS OF ASPERGILLUS FUMIGATUS IN AIR DUE TO COMPOSTING OF SEWAGE SLUDGE"

Aspergillus fumigatus is an opportunistic fungal pathogen causing pulmonary aspergillosis, otomycosis, cellulitis, mycetoma and sinusitis in man, pulmonary diseases in birds and mycotic abortion in cattle and sheep. Abilities to metabolize a wide variety of complex substrates and grow at high (50°C) temperatures favors selective enrichment of *A. fumigatus* during composting. High numbers are detected in compost and in the air surrounding a composting facility. As a part of our studies to assess potential public health hazards of composting sewage sludge, we have enumerated *A. fumigatus* in air samples taken at the composting facility in Camden, New Jersey. Oxgall-antibiotic agar plates were exposed in a New Brudnswick Scientific Co., Inc. air sampler to air at various locations. Plates were incubated at 50°C for 3 days. Colonies were then counted and identified. On calm days with a minimum of mechanical activity, counts were 300 to 4160 colony forming units/m^3, 1m from composting piles. Highest counts (e.g., 4160) were downwind. Lower counts (0.5-30) were obtained after a rainfall. The numbers decreased with distance, typically 4160 (1 m), 3720 (10 m),

1650 (40 m) and 450 (50 m). The levels at the facility were compared to background levels (0.5-7.5) in air at various other locations.

Lutrick, M.C. and Cornell, J.A.*, Agric. Res. Ctr., Jay, FL

"THE UPTAKE OF CERTAIN METALS BY CORN GROWN ON SOIL TREATED WITH CHICAGO OR PENSACOLA SEWAGE SLUDGE"

Land spreading of sewage sludge is probably the most practical means of disposal for municipalities and cities. However, uptake of certain metals by forage and grain crops from land treated with sludge creates health risks. A 3-year study was conducted to determine the uptake of copper (Cu), zinc (Zn), manganese (Mn), lead (Pb), and cadmium (Cd) by corn leaves and grain from soil treated with Chicago and Pensacola sewage sludge. The Chicago sludge contained large quantities of Cu, Zn, Pb, and Cd. The Pensacola sludge was high only in Zn.

The quantities of metals extracted from sludge-treated soil were proportional to the quantities added from the sludges. The metal uptake by the corn plant was directly associated with soil pH. The higher the soil pH the smaller the quantity of metal uptake. The quantity of metals extracted from the sludge-treated soil after one year was somewhat less than from the soil where sludge had been recently applied. The quantity taken up by the corn leaves was much less from the residual sludge treatment than from the same treatment where sludge had been recently applied.

The concentration of metals in the grain was always much less than the concentration found in the corn leaves from the same sludge treatment. Corn plants from the Chicago sludge treatment probably would have contained too much Cd to be utilized for forage. The Pb and Cd concentration in the grain was below detectable limits from all treatments.

---

*Professor (Soil Chemist), University of Florida, IFAS, Agricultural Research Center, Jay, Florida 32565 and Professor, Department of Statistics, University of Florida, IFAS, Gainesville, FL 32611

Makeig, K.S., Dames & Moore, Washington, DC

"THE SUITABILITY OF CONTAINMENT SOILS FOR ATTENUATING SLUDGE LEACHATE"

The huge quantities of solid wastes generated daily severely tax existing municipal waste treatment plants. An economic alternative to waste treatment consists of disposing of these wastes in the form of an undigested slurry, or sludge, in on-land containment sites. These sites provide a setting for natural attenuation mechanisms in the unsaturated soil horizons to act upon the sludge leachate and render it harmless to the environment. A thorough knowledge of the mechanisms of attenuation allows for the protection of the ground and surface water supplies, thereby reducing the health risks associated with land containment of sludge.

Previous studies have concentrated on different components of the leachate, such as pathogens, heavy metals, and exchangeable bases, and how certain soil properties serve to inhibit or enhance their migration. But these studies have failed to integrate their findings into a useful, predictive tool that can be used for design purposes when assessing the suitability of one containment site over another. This paper demonstrates a methodology for determining the effectiveness of soil properties such as pH, cation exchange capacity, temperature, moisture content, permeability, organic content, and texture (both independently and in combination) in immobilizing the primary components of sludge. Based on these criteria, soils are characterized according to their suitability for sludge containment.

Data for this paper are based on site evaluation studies in Montgomery and Prince Georges' Counties, Maryland, by the Washington Suburban Sanitary Commission and Dames & Moore.

Manning, William J. and Spitko, Roberta A., Plant Path., U. of Mass., Amherst, Mass.

"LAND APPLICATION OF IRRADIATED DIGESTED SEWAGE SLUDGE: EFFECTS ON HEAVY METAL UPTAKE BY FIELD-GROWN VEGETABLES"

Liquid, irradiated (electron beam), digested sludge, from Boston's Deer Island Treatment Plant, was applied to the surface of limed field plots (Merrimack fine sandy loam) in May of 1977 at the Suburban Experiment Station, in Waltham, Mass. Sludge was applied at 4.7 metric tonnes/ha and at 9.4 metric tonnes/ha. Other plots received either 10-10-10 inorganic fertilizer, at 1.1 metric tonnes/ha, or no fertilizer or sludge at all. Plots were then planted to field and sweet corn, bush bean, beet, lettuce, onion and tomato. The same procedure and plantings were also done in May of 1978. Elemental analyses of sludges and edible plant parts were performed in both years. With the exception of lead, which exceeded normal concentrations only in lettuce (cv. Salad Bowl), concentrations of all other metals during both years of study were within normal ranges observed for plants grown without sludge, as reported by EPA. Lead concentrations were high in all plants, including controls. Part of the lead is probably aerial in origin and the rest is probably taken up from the sludges. Up to 10 metric tonnes/ha/yr. of irradiated digested sludge could be applied annually for two years without health risks from accumulated heavy metals in vegetables grown on the sludge-treated land.

Naylor, L.M., Loehr, R.C., Tyler, L.D., McBride, M.B., Lisk, D., and McDuffie, B.*, Dept. Agric. Eng., Cornell U., Ithaca, NY

"UPTAKE OF CADMIUM AND ZINC BY CORN GROWN ON ACID SOILS FERTILIZED WITH A HIGH CADMIUM SLUDGE"

From June 1975 to December 1979 about 7,900 dry tons of an urban-industrial sludge from the Binghamton/Johnson City Joint Sewage Treatment Plant were applied to 85 acres of strongly acid silt loam agricultural land near Vestal, New York. Data indicate that during this period the cadmium content of the sludge varied from 24 ppm to 330 ppm, with an aveargeof 115 ppm.

During the lifetime of the site about 50 lb Cd/acre is estimated to have been added to the soil. As a result, the cadmium content of the soil has increased by a factor of 50 to 80 over a control site. Based on the CEC of the soils at the site (6 to 8 meq/100 g) the maximum lifetime cadmium loading recommended by USEPA should be limited to about 10 lbs/acre.

The environmental effects of sludge application at this site had been the focus of a two year study (1976 to 1978) led by Bruce McDuffie. McDuffie showed that uptake of cadmium into corn grain grown on the sludge amended soil was only slightly greater than that of corn grown on a control site. There was, however, an increase in the cadmium content of the corn leaves over the control plants. No contamination of surface water or groundwater was detected.

Because of the intense interests of the New York State Departments of Environmental Conservation, and Agriculture and Markets, a monitoring program was continued in 1979. Soil tests indicated that several metals including cadmium had moved through the soil column to a depth of 18 to 24 in.

However, metals such as copper appeared to be much less mobile and were retained in the top 12 in. of soil.

The crop harvested for corn grain was initially impounded, tested, and then released for use as an animal feed. Corn grain contained 0.14 to 0.53 mg/kg Cd. In all samples, the Cd content of the grain was less than 1% of the Zn content. The fodder contained 6.0 to 12.5 mg/kg Cd. Grain and fodder harvested from a control site contained no detectable Cd. These data indicate that even when Cd application rates exceed EPA recommendations, Cd content of corn grain may be less than the maximum recommended by FDA.

---

*Departments of Agricultural Engineering (L.M.N., R.C.L.), Agronomy (L.D.T., M.B.M.), and Food Science, Pesticide Residue Laboratory (D.J.L.) Cornell University, Ithaca, N.Y. Department of Chemistry (B.M.), State University of New York, Binghamton, N.Y.

Pal, Dhiraj, Dept. Biol. & Ag. Eng., N.C. State U., Raleigh, NC

"CHEMICAL CHARACTERIZATION OF SEWAGE SLUDGES FOR PREDICTION OF POTENTIAL HEALTH RISKS IN LAND TREATMENT SYSTEMS"

Complete characterization of municipal sludges for toxic metal forms, individual organic species or isomers and pathogenic agents is required to assess the health risks associated with continuous use in land farming systems. At present, analytical data on municipal sludges are limited and bioassays of soil-waste systems are in developmental stage. There exists strong need for (i) establishment of a regional data base on toxic organics, metals and pathogens found in municipal sludges and wastewaters and (ii) invention of a classification scheme for various municipal wastes based on the characteristics, suitability for land treatment, and potential health risks.

All municipal wastes carry similar constituents in various proportions. These constituents are characterized as: water, nitrogen, phosphorus, sulfur, salts, acids, bases, anions, metals, oils and greases, specific organic groups, biomass, and pathogens. Each category of constituents may include a galaxy of individual chemical species and isomers in various forms and states of transformation. A partial list of parameters, for which every municipal waste must be characterized to assess the potential health risks and the success of land treatment system, is prepared. Complexity of variability among different municipal sludges and wastewaters is simplified by development of a classification scheme based on sound judgement of practical limits on the (i) proportions of various constituents composing wastes, (ii) suitabiltiy of municipal wastes for land treatment, and (iii) potential health risks associated with long or short term use of waste in land farming. Developmental phases of municipal waste classification scheme will be discussed from utilization and safety points of view.

Pal, D. and Overcash, M.R., Dept. Biol. & Ag. Eng., N.C. State U., Raleigh, NC

"LAND APPLICATION AND ASSIMILATION OF MUNICIPAL WASTE ORGANIC PRIORITY POLLUTANTS"

Municipal wastes contain many organic contaminants that may be considered broadly as toxic, hazardous, infective or pathogenic and are on the recommended list of consent decree priority pollutants. Dispersive assimilation of these organic substances by land application has been examined from environmental safety, agricultural productivity and economic points of view. Transformations and fate of the few representative toxic organics have been evaluated with respect to physical, chemical and microbiological properties of the total soil-waste-plant system.

Development of land application design criteria for soil assimilation of these waste organics has been proposed. The need for establishing the critical levels of each organic pollutant isomer in soil-plant system has been recognized with reference to effects on soil population, plants, animals and human subjects utilizing the produce of the system. Tolerance of the plants to the organic pollutants has been discussed in relation to the phytotoxicity, biomagnification, and entrance into food chain.

Greater research efforts must be directed in achieving higher rates of biodegradation of toxic organics in soil-plant system by proper understanding of enhancement factors (nutrients, moisture, areation, temperature, etc.). The results of studies underway will be discussed with respect to design of successful land treatment systems for municipal wastes. In order to help improve the quality of water, life, and environment and to minimize the potential health risks, municipal wastes must be characterized for organic priority pollutants that may otherwise overload the system.

Reimers, R.S., Bowman, D.D., Englande, A.J., Jr., Little, R.D., Wilkinson, R.F. and Leftwich, D.B., Env. Hlth. Sci.,

Tulane U., New Orleans, LA

"FATE OF PARASITES IN DRYING BED STUDIES"

Results of the investigation of parasites in southern domestic waste sludges indicate that, in general, conventional domestic waste treatment processes do not effectively destroy parasite eggs in sewage. Both aerobic and anaerobic digestion (stabilization processes) appear to be ineffective in destroying parasites and dewatering processes tend to only concentrate the parasite eggs. The only sludge process which appears to effectively destroy parasites is the drying bed.

Previous reports have indicated that when the moisture content of a drying bed was less than 5 percent, all _Ascaris_ eggs present were destroyed. In our studies complete destruction of _Ascaris_ and _Toxocara_ eggs was noted when moisture content in drying beds was as high as 20 percent. It appears that moisture is one of several variables (solar radiation, temperature, time, etc.) that can influence the destruction of parasites in drying beds. This paper will report field data indicating the above observations.

Smith, G.S., Kiesling, H.E., Ray, E.E., Hallford, D.M. and Herbel, C.H., N.M. State University, Las Cruces, NM

"SEWAGE SOLIDS AS SUPPLEMENTAL FEED FOR RUMINANTS: BIO-ASSAYS OF BENEFITS AND RISKS"

New Mexico State University has evaluated the nutritive benefits and toxicological risks from usage of dried, "raw" (undigested) sewage solids as a supplemental feed for ruminants subsisting on poor-quality roughage or dormant rangeland forage. Sewages were dried to about 90 percent DM and subjected to about 1 mega rad _gamma_-irradiation. Analyses for nutrients and the toxicants, heavy metals, aflatoxins, and halogenated hydrocarbons preceded bioassays and feeding trials. Rat growth and reproduction bioassays were used to "screen" for evidence of acute or sub-acute toxicity prior to nutritive assays with ruminants.

After demonstrations of substantial nutritive benefits in short-term trials with sheep and steers, sewage solids were fed as ten to fifteen percent of diet to growing-finishing cattle in feedlot, over periods of 68 or 84 days. Representative animals were slaughtered immediately and others were fed for 56 or 50 additional days to evaluate "withdrawal" following ingestion of sewage solids. Blood, livers and kidneys were analyzed for halogenated hydrocarcarbons. Neither heavy metals nor refractory organic residues in tissues from cattle fed sewage solids and slaughtered immediately exceeded levels reported in recent literature; beef from cattle fed sewage solids was compared with beef from cattle fed conventionally when fed as 0, 10, 20 or 30 percent of diets to albino rats from weaning through adulthood and one cycle of reproduction. Rats fed such beef grew faster and reproduced more efficiently than rats fed a commercial lab chow.

During 1978 and 1979 seventy-four beef cows were managed as a single herd on semi-desert rangeland near Las Cruces, NM. One-third received no supplemental feed, one-third received supplemental cottonseed meal (CSM) and one-third received experimental supplement, RS#2, which was comprised of dried, irradiated sewage solids (62%), milo grain (22%), molasses (12%), urea (2.9%) and Bentonite plus flavors (1.4%). Other supplements were fed at about 2 kg per head per week when indicated. "Calf Crops" for both years (total calves/total cows) were 66% for unsupplemented controls; 84% for cows fed CSM and 82% for cows fed RS#2. Cows re-bred in November 1979 were 61% for unsupplemented controls; 88% for cows fed CSM and 88% for cows fed RS#2. Samples of blood, milk and livers from these cows had no detectible increase in contents of Cd, Hg or Pb; although some apparently beneficial increases in Fe, Cu, Mn and Zn were observed in cows fed RS#2.

In sheep fed RS#2 as total diet for about 3 months, excessive urination, glucosuria and slightly enlarged livers were regarded as evidence of chronic, early toxicity; although body weight gains were not adversely affected during the 3-month period.

In Research (Spring, 1980) breeding performance of crossbred ewes fed a protein-deficient basal mixture was equally improved by 7 percent dietary sewage solids or 3.5 percent cottonseed meal. Sewage sludge as 7% of diet from mid-gestation through lactation until weaning exerted no detectible detrimental influence on the productive characteristics.

Nutritive benefits from usage of sewage solids as supplemental feed for ruminants could be substantial while demonstrating that risks to animal health and the human food chain appear manageable.

Sopper, William E. and Kerr, Sonja N., Forest Hydrology, Pa. State U., Univ. Park, PA

"HEALTH RISKS ASSOCIATED WITH THE USE OF MUNICIPAL SLUDGE FOR STRIP MINE LAND RECLAMATION"

Treated municipal sludges were used to revegetate barren mined land in both the bituminous and anthracite coal regions of Pennsylvania. Types of sludges used in the projects were (1) liquid digested sludge, (2) dewatered by centrifuge, vacuum filter, and sand bed drying, (3) heat dried, and, (4) composted with wood chips. Application rates varied with the type and quality of sludge. Extensive monitoring and sampling were conducted at each site to determine the effects of the sludge applications on the soil, vegetation, and groundwater. Special attention was placed on chemical and bacteriological analyses in terms of the potential health risks to animals and humans. These projects are part of a cooperative program in the Commonwealth to introduce the concept of using municipal sludges for the reclamation of mined land. This paper will discuss the results of these demonstration projects.

Stoewsand, G.S., Reid, J.T., Haschek, W.M., Telford, J.N. and Lisk, D.J., Dept. Food Science, NY State Agric. Exp. Sta., Geneva, NY

"HEPATIC CHANGES IN LAMBS FED CORN SILAGE GROWN ON MUNICIPAL SLUDGE AMENDED SOIL"

Dorset wether lambs were fed corn silage grown on a field plot amended with 280 metric tons per hectare of municipal sludge from the Ley Creek Sewage Treatment Plant, Syracuse, NY. The length of the feeding period was 274 days. Analysis of 43 elements and polychlorinated biphenyls showed that only silver, cadmium, rubidium and zinc were elevated in the sludge grown corn silage compared to control corn. Cadmium and zinc were found higher in most tissues of the animals fed sludge-grown corn silage, especially in kidneys and liver as compared to control lambs. Hepatic mixed function oxidase was slighlty elevated in the lambs fed the sludge-grown corn, but electron microscopy of these livers revealed greatly enlarged and swollen mitochondria and hepatic necrosis. The cause of this histopathological condition is unknown, but similar changes have been observed in livers of laboratory animals exposed to cadmium salts.

Ward, N. Robert, Kowinski, J. and Litsky, W., Depts. Envir. Sci. & Health, U. of Mass., Amherst, MA

"THE EFFECT OF pH ADJUSTMENT BY LIME ON THE SURVIVAL OF SALMONELLA IN SLUDGE PRIOR TO LAND APPLICATION"

A field study was conducted to determine the survival of salmonella and various indicator organisms in agricultural soil following the application of liquid (approximately 3% solids) and sand bed-dried (approximately 40% solids) aerobically-digested, municipal sludge. Salmonella could not be isolated after 7 days in plots receiving liquid sludge, but survived at least 266 days in those receiving dried sludge. The prolonged survival of salmonella in sludge has onerous public health implications, but these risks may be minimized if a reduction or removal of pathogens is effected prior to land application. Two systems were studied for their potential in decreasing the number of pathogens in sludge. The first determined the die-off of salmonella and indicator organisms during the drying of sludge in sand beds. The results demonstrated that salmonella could not be isolated after 19 days when the solids went above 40%. In this study and in the field study, fecal coliforms and fecal streptococci were not useful indicators of the presence of salmonella. The second involved the die-off of 5 serotypes of salmonella and *E. coli* ATCC 25922 as determined by D values in liquid sludge, pH-adjusted with lime. Initial testing using steam-sterilized sludge adjusted above pH 10 revealed that the D values for Salmonella were less than the D values for *E. coli*. Assays using unsterilized, pH-adjusted sludge showed that the D values for fecal coliforms corresponded well with the D values obtained for *E. coli* and that fecal coliforms can predict the die-off of salmonella. D values for salmonella were usually less than 15 minutes for pH adjustments above 11.

INDEX

actinomycetes, thermo-
   philous   246,248,251,
   252,258,259
acute toxicity, chlorinated
   insecticides   93
aerated pile composting   216
Aeromonas sp.   28
aerosols   228,249,250,252,
   253,254,258,259
   animal health, and   268,
   269
aflatoxin   321-322,335
airborne microorganisms   219,
   220,246,259
allergens   247,250,251,252,
   258
   response to   246,247,251,
   252,258
andersen sampler   220,227
animal health   267-284
anthrax   182
antibiotic resistance   28
antibody titers   250
aquatic animals
   effect of sludge feeding
   on   257
arsenic
   phytotoxicity of   69
   sludge, in   69
Ascaris   185,204
   drying beds, in   357
   egg survival in soils   52
   ova   246,255
Ascaris lumbricoides   20,23,
   49,50,51,52,144,278
Ascaris suum   49,50,51,52
aspergillosis   220,221
Aspergillus flavus   217,223,
   225

Aspergillus fumigatus   217,
   219,223,225,226,229,231,
   232,235-241,246,247,248,
   250-252,254-256,258,259,
   351
Aspergillus niger   223,225,
   239,240

Bacillus anthracis   182
Baylisascaris, skunks   204
birds
   effect of sludge feeding
   on   157
Brucella sp.   180,278

Cadmium
   aflatoxin toxicity, effect
   on   321-322
   animal tissues, in   10,359
   cattle, in   208
   Chicago sludge, in   334
   crops, in   159
   dose-effect relation-
   ships   71
   dose-response model   72
   dose-response relationship   71-73
   food chain, in   345-346
   food chain assessment   342
   goats, in   208
   health effects in
   humans   71-73
   kidney and liver, in   273-
   275,277,278,280
   lettuce, in   347
   mice, effect on   159
   phytotoxicity of   70,71
   pigs, effect on   334,335,336
   poultry, in   341,342

362 INDEX

sheep, effect on 159,208
sludge, in 9,11,154
sludge-amended soil 354
sorghum, in 341
swine 320-322
transport through soil 349
uptake by Coastcross I Bermudagrass 333
uptake by corn 352,354, 359
warfarin toxicity, and 321-322

Carcinogens
  biassay program 94
  cancer risks 95
cattle performance 297,300, 304
Chicago sludge 267-284
  effect on animals 156
chlorinated compounds
  in sludge 88,89
Citrobacter sp. 26
Clean Water Act 2
Clostridium sp. 23,24
Clostridium botulinum 23
Clostridium perfringens 23
Clostridium tetani 23
coliform, fecal 22,23,24
composition of sludge 105-108
compost 216,217,221,224, 225,229,231,239,240, 241,245-247,249-259, 344,350-351
consumer product safety commission 10
copper
  bioavailability of 64
  phytotoxicity of 66
  risks to humans 67
  toxicity in animals 66
corn 108-114,117,121,122, 127
crops
  animal health, and 158-159
  fiber crops 147
  parasites 147

root crops 147
  Salmonella 147
cysticercosis 54,56
Cysticercus bovis 203

DDT
  sheep, in 154
dioxin 88

Echinococcus sp. 203
Echinococcus granulosus 51, 53,54
Echinococcus multilocularis 54
endotoxin 217,218,225,240, 252,258
Entamoeba histolytica 143
enteric pathogens
  bacteria 257
  cysts 257
  ova 257
  viruses 257
Enterobacter 26
enteroviruses, survival in sludge lagoon 346-347
enzyme-linked immunosorbent assay 225,226,235,239, 240
eosinophils 238
Escherichia coli 142,181,206
excreta
  composting 18
  fertilizer 18
  nitrogen conservation 17-18

fecal coliforms 228,231,233, 235,240
fecal streptococci 228,231, 233,235
fluorine
  sludge, in 69
  uptake by plants 69
Food and Drug Administration 7,10
franconi syndrome 71
fungi 217,220,221,223,225, 227,228,236

INDEX   363

gamma-irradiated sewage solids
   supplemental feed, as   357-358
Giardia sp.   202
Giardia lamblia   143
guidelines, land disposal
   of sludge   129-130

heavy metals
   animal tissues, in   273-278
   bone tissues, in   294,298,301,302
   forages, in   291
   kidney, in   293,296,298,301,305
   liver tissues, in   293,294,298,301,302,306
   muscle tissues, in   294,298,301,307
   sludge, in   290,297,300
   soil, in   290
helminth ova   48,50,246,258
Hymanolepsis nana   23
hypersensitivity   225,238
hypersensitivity pneumonitis   248

iron
   animal tolerance to   68
   interaction with Cu and Zn   68,69
   phytotoxicity of   68
iron sludge
   effect on animals   156
irradiated sludge
   land application of   353-354
insecticides, half-lives of   91

Klebsiella sp.   24,26,32

land application
   irradiated sludge, of   353-354
land disposal
   field studies   22-24
   history   16-19
   pathogen load   21

   soil conditioning   20
   soil deterioration   19
   soil percolation   19
   water conservation   18-19
lead
   bones, in   276,277
   broiler chickens, in   341
   food chain assessment   342
   ingestion by children   61
   kidney, in   277
   lettuce, in   347
   levels in sludge   154
   poultry, in   342
   phytotoxicity of   67
   sludge, in   9
   sorghum, in   341
   toxicity to animals   67
   uptake by corn   352
   vegetables, in   354
   see also metals, metal concentration in crops, metal uptake by crops
Leptospira sp.   183,278
leptospirosis   183
lipopolysaccharides (LPS)   217,218,222,223,237,238,239,240,241
Loeffler's syndrome   53

manganese
   uptake by corn   352
manure, recycled cattle
   health effect of   337
mercury
   see metals, metal concentration in crops, metal uptake by crops
metals
   adsorption to soils   323
   animal tissues, in   329
   bioavailability of   64
   concentration in sludge   106-107
   co-precipitation   106
   forms in sludge   106
   health concerns   59-83,108
   interactions between metals   64-71
   interactions with dietary constituents   64-71

    interactions with sludge
        constituents 64-71
    mice tissues, in 349-350
    phytotoxicity of 63
    transfer to the food chain
        60-64
metal concentrations in crops
    corn 108-114,117,121,124,
        127
    small grains 111-112,117,
        121
    soybean 111-113,117,121,
        125
    vegetables 111-112,117,
        123,128
metal uptake by crops 62
    bermudagrass, by 349
    corn 333-334,345-346,352
    cultivar 113-114
    environmental factors 113
    plant factors 108-115
    rate 122-128
    soil cation exchange
        capacity 116,129-130
    soil factors 115-118
    soil pH 115-117,119,129,
        130
    species 110-112
    vegetables, by 353-354
mice
    health effects of metals
        on 322-323
    tissue metal analyses
        349-350
mice feeding study 350
molybdenum
    animal tolerance to 70
    sludge, in 70
    uptake by plants 70
Mycobacteria 332
Mycobacterium sp. 20,21,24,
    178
Mycobacterium fortuitum 32,33
Mycobacterium gordonae 32,33
Mycobacterium kansasii 32,33

nickel
    see metals
    see metal concentration
        in crops

    see metal uptake by crops
    phytotoxicity of 67
    toxicity to animals 67
nitrogen applied to soil
    129-130
nosocomial infections 28

organics 4
    fat tissues, in 295,298,
        302,303
    forages, in 291
    health hazards of 90
    pollutant pathways 92
    soils, in 291
    sludge, in 86,292,299,
        303
organochlorine insecticides
    acute toxicity of 93
    banning by EPA 95
    cancer risk of 95
    carcinogenicity of 94
    half lives of 91
    nomenclature of 100
    sludge, in 87

parasites 47-58,143-144,185
    definition 47
    drying beds, in 357
    health risks 55-56
    sludge, in 48-54
pathogens 3
    animal exposure 175
    animal health risk 173
    control of 256
    land disposal 20
    primary 245,246,249,250,
        253-258
    secondary 245,246-248,
        249,250-252,254-256,
        258
    sludge, in 141,144,174
pathology, effect of sludge
    on animals 280-282
pesticides
    feed, soil, sludge and
        animal tissues, in 339
phytoferritin 68
phytotoxicity 118-121
P.L. 92-500 153
pollutant pathways 92

polybrominated biphenyls 160
polychlorinated biphenyls 9,
    11,105,160,207
    animal tissues, in 339
    dairy cattle, in 348
    feed, in 339
    soil, in 339
    sludge, in 87,154,339
    uptake from soil 92
polyhalogenated compounds,
    in sludge 87
polynuclear aromatic hydro-
    carbons 90
priority pollutants, organic
    356
protozoan cysts 48,246,258
Pseudomonas sp. 24
pulmonary extrinsic allergic
    alveolitis (PEAA) 247,
    252

rats
    effect of sludge feeding
    on 157
recovery from sludge
    enteric bacteria 24
    enteric viruses 24
refractory organics 160-161
regulations, 40 CFR part 257
    175
relay feeding studies 11
renal tubular dysfunction 71
reovirus 184
Resources Conservation and
    Recovery Act 10
risk 5
risk assessment 148-149,187
    parasites 148
    viruses 148
risk measurement 34
rotavirus 184
ruminants
    grazing by ruminants
    272-273

safety judgment 34
Salmonella 24,142,176,205
    compost detection in 350-
    351
    lime, effect of 360

Ohio farmers, in 345
    pH, effect of 360
Salmonella dublin 30,32
Salmonella enteritidis
    isolation from animals
    333
Salmonella heidelberg 32
Salmonella typhimurium 30,32
Salmonella typhosa 20,21
Sarcosporidia
    animal tissues, in 338
selenium
    sludge, in 70
    uptake by plants 70
serology 223
Serratia 26
sewage solids
    supplemental feed, as 357
Shigella 24,142
skin testing 223
sludge
    adherence to forage crops
    60,61
    aerobic digestion of 145,
    146
    anaerobic digestion of
    145,146
    animal health and 344-345
    application of 146-147
    application to strip mine
    land 359
    chemical characterization
    355-356
    chlorinated compounds in
    88,90
    composting 146
    economic value of 2
    history 1
    irradiation of 146
    lagooning 146
    Ohio farms, in 344-345
    organochlorine insecticides
    in 87
    pasteurization 146
    polyhalogenated compounds
    in 87
    polynuclear aromatics in
    90
    poultry, effects on 319-
    320,330-331

366 INDEX

production 144
regulatory aspects of 7-12
sources of pathogens in 144
swine, health effects on 313-318
toxicity 11
treatment 144-145
treatment costs 2
sludge amendment
effect on 2,4-D adsorption and degradation 345
sludge-borne pathogens, persistence of 28
sludge disposal site 285, 289-292
sludge entrenchment practices 348
sludge feeding 277
animals, effect on 155-158, 311-325, 338
cattle 329, 343
swine, effect on 331, 332, 334, 339-340
sludge ingestion 285, 296, 299, 303
small grains 111-112, 117, 121
soils
adsorption of 2,4-D to 345
bacterial movement in 22
containment soils 352-353
degradation of 2,4-D in 345
ingestion of 61-62, 154, 172
parasites, as intermediate host for 52
sludge application to 268
survival in 21-22
transport of sludge constituents 349
virus movement in 22
"soil-plant barrier" 63, 69, 70, 71
solid waste, municipal
effect on animal health 160
soybeans 111-113, 117, 125
Staphylococcus aureus 24
isolation from blood and tissues 333

streptococci, fecal 23
Streptococcus faecalis 23, 24
Streptococcus pyogenes
isolation from animal tissues 333
strip mine land 359
Strongyloides stercoralis 23
sulfonamides, detection of 338
swine, effect of sludge feeding on 156-157, 341, 342

Taenia sp. 186, 203
Taenia saginata 51, 144
infection 54
life cycle 53, 54
Taenia solium 51, 144
cysticercosis 56
infection 54
life cycle 53, 54
TCDD (dioxin) 88
temperature-by-time criteria 256-257
thermophilic microorganisms 217, 229, 231, 245, 250, 251
toxic chemicals, effect on animal health 153-171
toxic pollutants in water 86
toxicological tests 93
Toxocara 49
dogs, cats 204
drying beds, in 357
Toxocara cati 49, 51, 53
Toxoplasma gondii 51, 55, 202
Trichuris
soils, survival in 52
Trichuris suis 51
Trichuris trichiura 49, 50, 51
Trichuris vulpis 51
tuberculosis 178
typhoid, etiology 16-17

vegetables 111-112, 117, 123, 128
metal uptake by 353-354
survival on 24, 30
Vibrio parahaemolyticus 30
viruses 22, 23, 24, 143, 184, 246, 255, 256, 257, 258

groundwater in   330-332
lagoons, survival in   332
sludges, in   330,332
soils, in   21-22,330
visceral larva migrans (VLM)
      53
vitamin A
    cadmium and   162
    cutin and   162
    effect of sludge ingestion
        on   157
    metabolism   161-162

warfarin   321-322,335
wastewater workers   215
Water Pollution Control Act
      10
water, toxic pollutants in
      86
wild animals, disease transmission in   201-208

Yersinia   142
Yersinia enterocolitica   30

zinc
    animal tissues, in   359
    induction of copper
        deficiency   65
    kidney and liver, in   273-275
    phytotoxicity of   65
    toxicity to animals   65
    transport through soils
        349
    uptake by corn   352,354,359
    wildlife, toxicity to 65